Lignin Biodegradation: Microbiology, Chemistry, and Potential Applications

Volume II

Editors

T. Kent Kirk

Research Scientist
Forest Products Laboratory
U.S. Forest Service
U.S.D.A.
Madison, Wisconsin

Takayoshi Higuchi

Director
Wood Research Institute
Kyoto University
Uji, Kyoto, Japan

Hou-min Chang

Professor
Department of Wood and Paper Science
North Carolina State University
Raleigh, North Carolina

CRC Press, Inc.
Boca Raton, Florida

Library of Congress Cataloging in Publication Data

Main entry under title:

Lignin biodegradation.

   Proceedings of an international seminar,
organized under the auspices of the United States
-Japan Cooperative Science Program, held May 9-
11, 1978 at the U.S. Forest Products Laboratory,
Madison, Wis.
   Bibliography: v.   p.
   Includes index.
   1, Lignin—Biodegradation—Congresses.
I. Kirk, T. Kent. II. Higuchi, Takayoshi.
III. Chang, Hou-min. IV. United States-Japan
Cooperative Science Program.
TS933.L5L5   676'.12   79-13667
ISBN 0-8493-5459-5 (v.1)
ISBN 0-8493-5460-9 (v.2)

Direct all inquiries to CRC Press, Inc., 2000 N.W. 24th Street, Boca Raton, Florida 33431.

© 1980 by CRC Press, Inc.

International Standard Book Number 0-8493-5459-5 (Volume I)
International Standard Book Number 0-8493-5460-9 (Volume II)

Library of Congress Card Number 79-13667
Printed in the United States

FOREWORD

Lignin is a generic name for the complex aromatic polymers that are major components of vascular plant tissues. Lignin is abundant; in terms of weight it is probably second only to cellulose among renewable organic materials, and in terms of energy content it might well be the single most abundant. The tremendous tonnages of lignin that annually accumulate through photosynthesis are balanced by the decomposition, by microorganisms, of approximately equal amounts. This unique biopolymer, therefore, occupies a central position in the earth's carbon cycle. In this light it seems surprising that in 1978 man does not possess a fairly good understanding of its biodegradation—as he does, for example, of the biodegradation of the co-occurring biopolymer cellulose. We estimate, however, that the present understanding of lignin biodegradation is about equal to that of cellulose biodegradation in the 1950s. Several reasons can be cited for this paucity of knowledge, but two are perhaps the major ones: (1) the lack of a good understanding of the chemistry of lignin until the 1960s, and (2) the lack of a strong practical incentive to generate research support.

Research on lignin biodegradation around the world has increased dramatically in the last 5 to 6 years, the number of research groups having at least trebled. Most of this increased research attention stems from hopes of eventual industrial application of ligninolytic systems in processing the earth's most abundant renewable materials, the lignocellulosics. Progress is now being made at an accelerating rate, and much has already been learned.

In August of 1976 a small group of researchers gathered in Seattle at the request of the Weyerhaeuser Company to discuss one possible application of ligninolytic microbial systems: biodelignification. (The company has published the proceedings: *Biological Delignification: Present Status—Future Directions,* Weyerhaeuser Company, Seattle, Washington). After that meeting it was agreed among the participants that a comprehensive conference on basic as well as applied aspects of lignin biodegradation should be arranged.

Consequently, an international seminar on lignin biodegradation was organized and was held May 9 to 11, 1978, at the U.S. Forest Products Laboratory in Madison, Wisconsin.

This book records the proceedings of that seminar and is meant to provide a state-of-the-art summary of research. Each speaker/author was asked to summarize his research, including his latest unpublished results, and to describe how his work fits into the overall picture. Following two orientation chapters, one a review of lignin structure and morphological distribution in plant cell walls, and the second a review of the microbial catabolism of relevant aromatics, the book is comprised of chapters in the three subject areas given by the book's title. It does, as intended, provide comprehensive coverage of research to date (ms. submitted to CRC Press in August 1978).

We are grateful to several organizations for making the seminar possible. It was organized under the auspices of the U.S./Japan Cooperative Science Program of the U.S. National Science Foundation and the Japan Society for the Promotion of Science. Additional funding, which allowed us to invite European researchers, was provided by six North American pulp and paper companies: Crown-Zellerbach, International Paper, McMillan Bloedel, St. Regis, Westvaco, and Weyerhaeuser. The U.S. Forest Service's Forest Products Laboratory provided excellent facilities and clerical and other services. (We wish to acknowledge in particular the help of Ms. Nancy Maves.)

Greatest credit for the book, of course, goes to the speaker/authors; we appreciate their prompt and courteous attention to manuscript preparation. Several scientists who also attended the seminar added substantially to the perspective of discussions; their names are included in the list of seminar attendees.

We intend for this book to clarify what man has managed to learn to date about the unusual processes whereby Nature's most recalcitrant major biopolymer is biodegraded; and, primarily for the book to stimulate enterprising scientists to complete the story.

T. Kent Kirk
Madison, Wisconsin

Takayosi Higuchi
Kyoto, Japan

Hou-min Chang
Raleigh, North Carolina

August 1978

THE EDITORS

T. Kent Kirk, Ph.D, received his Ph.D. in Plant Pathology and Biochemistry from North Carolina State University in 1968. After 1-1/2 years as a post doctoral researcher (organic chemistry) at Chalmers University in Sweden, he joined the staff at the U.S. Forest Service's Forest Products Laboratory in Madison, Wisconsin, where he is a Research Scientist. He holds two Adjunct Associate Professorships: Department of Bacteriology, University of Wisconsin, Madison; and Department of Wood and Paper Science, North Carolina State University, Raleigh. He serves on the Editorial Boards of *Enzyme and Microbial Technology* and *Biotechnology Letters.* He has over 50 scientific publications in the area of lignin biodegradation and the chemistry and biochemistry of wood decay and has presented numerous lectures at universities and research laboratories.

Takayoshi Higuchi, Dr. Agric., is Director of the Wood Research Institute, Kyoto University, Uji, Kyoto, Japan. He is also Professor and Head of the Division of Lignin Chemistry at the Institute

Dr. Higuchi received his B.S. in Plant Physiology in 1950 from Nagoya University and his Dr. Agric. from the University of Tokyo in 1959. From September 1960 to September 1962, and from September 1963 to October 1964 he was a Post Doctorate Fellow of the Prairie Regional Laboratory, NRC., Saskatoon, Canada, and a Professur Associé, Faculté des Sciences, Université de Grenoble, France, respectively.

He was an Associate Professor of Wood Chemistry from 1960 to 1967 and Professor of Tree Biochemistry from 1968 in Gifu University. He moved to the Wood Research Institute, Kyoto University as the first Professor of Lignin Chemistry in 1968 when the Division of Lignin Chemistry was founded. He has served as Director of the Institute from May 1978.

Dr. Higuchi is a Fellow of International Academy of Wood Science, a member of Advisory Board of Cellulose Chemistry and Technology in Romania, and a council member of the Japan Wood Research Society. He has published many papers in the field of the chemistry and biochemistry of lignin, and tree biochemistry.

Hou-min Chang, Ph.D., is a Professor of Wood Chemistry in the Department of Wood and Paper Science, North Carolina State University. He received his B.S. degree in Forestry in 1962 from National Taiwan University, Taipei, Taiwan, and his M.S. degree in chemistry and Ph.D. degree in wood chemistry in 1966 and 1968, respectively, from the University of Washington at Seattle.

Dr. Chang is a member of the American Chemical Society and the Technical Association of Pulp and Paper Industry. His research interests include the study of structure and reactions of lignin as it is applied to biological degradation, wood delignification, and pollution abatement in the pulp and paper industry. He has published many papers in these fields.

SEMINAR

Lignin Biodegradation: Microbiology, Chemistry, and Applications

Forest Products Laboratory
Madison, Wisconsin, U.S.A.

May 9-11, 1978

LIST OF ATTENDEES

Dr. Derek Abson
Weyerhaeuser Company
3400 13th Ave., S.W.
Seattle, Washington 98134

Dr. Palle Ander
Swedish Forest Products Laboratory
Box 5604
S-11486 Stockholm
Sweden

Dr. Mel Armold
Department of Chemistry
Montana State University
Bozeman, Montana 59715

Dr. Fernand Barnoud
Centre de Recherches sur les
 Macromolecules Végétables
Universite de Grenoble I
53X-38041 Grenoble
France

Dr. R. B. Cain
Biological Laboratory
University of Kent
Canterbury CT2 7NJ Kent
England

Dr. H. -m Chang
Department of Wood and Paper
 Science
Box 5516
North Carolina State University
Raleigh, N.C. 27607

Dr. C. -L. Chen
Department of Wood and Paper
 Science
North Carolina State University
Box 5516
Raleigh, North Carolina 27607

Dr. Ellis B. Cowling
Plant Pathology and Forest Resources
North Carolina State University
Box 5397
Raleigh, North Carolina 27607

Dr. D. L. Crawford
Department of Bacteriology and
 Biochemistry
University of Idaho
Moscow, Idaho 83843

Dr. Ronald Crawford
University of Minnesota
Freshwater Biological Institute
Box 100
Navarre, Minnesota 55392

Dr. Michel Desroches
PPRIC
570 St. John's Blvd.
Pointe Claire, Quebec
Canada H9R 3J9

Dr. Robert W. Detroy
Northern Regional Research
 Laboratory
1815 North University Street
Peoria, Illinois 61604

Dr. Karl-Erik Eriksson
Swedish Forest Products Research
  Laboratory
Box 5604
S-11486 Stockholm
Sweden

Dr. W. W. Eudy
Corporate R&D
International Paper Co.
Tuxedo Park, New York 10987

Dr. S. I. Falkehag
Westvaco Corp.
PO Box 5207
Charleston Research Center
North Charleston, South Carolina

Dr. Pat Fenn
Forest Products Laboratory
USDA-Forest Service
PO Box 5130
Madison, Wisconsin

Dr. Toshio Fukuzumi
Laboratory of Forest Chemistry
Faculty of Agriculture
University of Tokyo
Tokyo, Japan

Dr. Michael Gold
Oregon Graduate Center
Beaverton, Oregon 97005

Dr. Konrad Haider
Institut für Biochemie des Bodens
Forschungsanstalt für Landwirtschaft
Braunschweig-Wolkenrode
West Germany

Dr. Philip Hall
Department of Chemistry
Virginia Polytechnic Institute
Blacksburg, Virginia 24061

Dr. Takafusa Haraguchi
Laboratory of Wood Chemical
  Technology
Faculty of Agriculture and Technology
183 Saiwaicho
3-5-8 Fuchu, Japan

Dr. Miyato Higaki
Laboratory of Forest Chemistry
Tokyo Univ. of Agriculture
158 Setagaya-ku
Tokyo, Japan

Dr. Takayoshi Higuchi
Division of Lignin Chemistry
Wood Research Institute
Kyoto University
611 Uji, Kyoto, Japan

Dr. Tatsuo Ishihara
Forestry and Forest Products Research
P.O. Box 2, Ushiku
Ibaraki 300-12, Japan

Dr. Shojiro Iwahara
Laboratory of Fermentation Chemistry
Faculty of Agriculture
Kagawa University
Miki-cho, Kagawa-ken 761-07
Japan

Dr. Hidekuni Kawakami
Laboratory of Chemistry of Forest
  Products
Faculty of Agriculture
Nagoya University
464 Furocho, Chikusa-ku, Nagoya
Japan

Dr. Paul Keyser
Corporate R & D
International Paper Co.
Tuxedo Park, New York 10987

Dr. T. Kent Kirk
Forest Products Laboratory
USDA-Forest Service
PO Box 5130
Madison, Wisconsin 53705

Dr. Masaaki Kuwahara
Department of Food Science
Kagawa University, Miki-cho
Kagawa 761-07, Japan

Dr. James P. Martin
University of California
Citrus Research Center and Agriculture
  Experimental Station
Riverside, California 92521

Dr. Noriyuki Morohoshi
Laboratory of Wood and Chemical
  Technology
Faculty of Agriculture
Tokyo University of Agriculture and
  Technology
183 Saiwaicho
2-5-8 Fuchu, Japan

Dr. C. A. Reddy
Department of Microbiology and
  Public Health
Michigan State University
East Lansing, Michigan 48824

Dr. Ian Reid
Prairie Regional Laboratory
National Research Council- Canada
110 Gymnasium Road
University Campus
Saskatoon, Saskatchewan
Canada S7N 9W9

Dr. Steve Rosenberg
Lawrence Berkeley Laboratory
Energy and Environment Division
Bldg. 38, Rm. 306
Berkeley, California 94720

Dr. David Sands
Department of Microbiology
Montana State University
Bozeman, Montana 59715

Dr. William Scott
Department of Microbiology
University of Miami
Miami, Florida 33124

Dr. Mikio Shimada
Division of Lignin Chemistry
Wood Research Institute
Kyoto University
611 Uji, Kyoto, Japan

Dr. Veronica Sundman
Department of General Microbiology
University of Helsinki
Malminkatu 20
SF-00100 Helsinki 10
Finland

Dr. Sprague H. Watkins
Research Laboratories
Crown Zellerbach Corp.
Camas, Washington 98607

Dr. Paul Wollwage
St. Regis Paper Co.
Technical Center
West Nyack Road
West Nyack, New York 10994

Dr. H. H. Yang
Forest Products Laboratory
PO Box 5130
Madison, Wisconsin 53705

Dr. Oskar Zaborsky
Program Manager, NSF
Problem-Focused Research
  Applications
1800 G Street, N.W.
Washington, DC 20550

Dr. J. G. Zeikus
Department of Bacteriology
University of Wisconsin
Madison, Wisconsin 53706

CONTRIBUTORS

P. Ander, Ph.D.
Research Assistant
Swedish Forest Products Research
  Laboratory
Stockholm, Sweden

F. Barnoud, Ph.D.
Professor
Centre de Recherches sur les
  Macromolecules Végétales
University of Grenoble
Grenoble, France

R. B. Cain, Ph.D., D. Sc., F.L.S.
Reader in Biochemistry
University of Kent
Canterbury, Kent
United Kingdom

A. Cheh, Ph.D.
Scientist
Department of Microbiology
University of Minnesota
Freshwater Biological Institute
Navarre, Minnesota

C. -L. Chen, Ph.D.
Senior Research Associate
Department of Wood and Paper
  Science
North Carolina State University
Raleigh, North Carolina

T. Cheng, M.S.
Research Assistant
Department of Chemistry and
  Biochemical Sciences
Oregon Graduate Center
Beaverton, Oregon

D. L. Crawford, Ph.D.
Associate Professor
Department of Bacteriology and
  Biochemistry
University of Idaho
Moscow, Idaho

R. L. Crawford, Ph.D.
Associate Professor
Department of Microbiology
University of Minnesota
Freshwater Biological Institute
Navarre, Minnesota

S. Drew, Ph.D.
Associate Professor
Department of Chemical Engineering
Virginia Polytechnic Institute and State
  University
Blacksburg, Virginia

K. -E. Eriksson, Ph.D.
Professor and Head
Biochemical and Microbiological
  Research
Swedish Forest Products Research
  Laboratory
Stockholm, Sweden

W. W. Eudy, Ph.D.
Manager
Forest Services
International Paper Company
Tuxedo Park, New York

T. Fukuzumi, Ph.D.
Associate Professor
Department of Agriculture
University of Tokyo
Tokyo, Japan

W. Glasser, Ph.D.
Associate Professor
Department of Forest Products
Virginia Polytechnic Institute and State
  University
Blacksburg, Virginia

M. Gold, Ph.D.
Associate Professor
Department of Chemistry and
  Biochemical Sciences
Oregon Graduate Center
Beaverton, Oregon

K. Haider, Ph.D.
Scientific Director
Institute for Soil Biochemistry
Braunschweig, West Germany

P. L. Hall, Ph.D.
Associate Professor
Department of Chemistry
Virginia Polytechnic Institute
Blacksburg, Virginia

T. Haraguchi, Ph.D.
Professor
Department of Forest Products
  Technology
Tokyo University of Agriculture and
  Technology
Tokyo, Japan

H. Hatakeyama, Ph.D.
Director
Plant Materials Division
Industrial Products Research Institute
Ministry of International Trade and
  Industry
Tokyo, Japan

A. Hatakka, M.Sc.
Research Assistant
Department of Microbiology
University of Helsinki
Helsinki, Finland

T. Ishihara, Ph.D.
Chief
Microbiological Chemistry Laboratory
Department of Forest Products
  Chemistry
Forestry and Forest Products Research
  Institute
Ibaraki, Japan

S. Iwahara, Ph.D.
Associate Professor
Department of Agricultural Chemistry
Kagawa University
Kagawa-ken, Japan

H. Kawakami, Ph.D.
Research Associate
Department of Forest Products
  Chemistry
Nagoya University
Nagoya, Japan

K. Krisnangkura, Ph.D.
Research Associate
Department of Chemistry and
  Biochemical Sciences
Oregon Graduate Center
Beaverton, Oregon

M. Kuwahara, Ph.D.
Associate Professor
Department of Food Science
Kagawa University
Kagawa, Japan

J. P. Martin, Ph.D.
Professor
Department of Soil and Environmental
  Sciences
University of California
Riverside, California

M. Mayfield, B.A.
Research Assistant
Department of Chemistry and
  Biochemical Sciences
Oregon Graduate Center
Beaverton, Oregon

L. Robinson, Ph.D.
Research Associate
Department of Microbiology
University of Minnesota
Freshwater Biological Institute
Navarre, Minnesota

S. L. Rosenberg, Ph.D.
Research Microbiologist
Lawrence Berkeley Laboratory
Energy and Environment Divison
Berkeley, California

K. Ruel, Ph.D.
Research Associate
Centre de Recherches sur les
  Macromolecules Végétables
University of Grenoble
Grenoble, France

M. Salkinoja-Salonen, Ph.D.
Assistant Professor
Department of General Microbiology
University of Helsinki
Helsinki, Finland

E. Setliff, Ph.D.
Postdoctoral Associate
State University of New York
College of Environmental Sciences and
  Forestry
Syracuse, New York

M. Shimada, Ph.D.
Research Associate
Division of Lignin Chemistry
Kyoto University
Uji, Kyoto, Japan

L. Smith, Ph.D.
Research Associate
Department of Chemistry and
  Biochemical Sciences
Oregon Graduate Center
Beaverton, Oregon

V. Sundman, Ph.D.
Professor and Head
Department of General Microbiology
University of Helsinki
Helsinki, Finland

J. B. Sutherland, Ph.D.
Research Associate
Department of Bacteriology and
  Biochemistry
University of Idaho
Moscow, Idaho

J. Trojanowski, Ph.D.
Professor
Department of Biochemistry
Marie-Curie Sklodowska University
Lublin, Poland

L. Vallander, M. Ch.E.
Research Assistant
Swedish Forest Products Research
  Laboratory
Stockholm, Sweden

C. R. Wilke, Ph.D.
Professor
Lawrence Berkeley Laboratory
Energy and Environment Division
Berkeley, California

J. G. Zeikus, Ph.D.
Associate Professor
Department of Bacteriology
University of Wisconsin
Madison, Wisconsin

TABLE OF CONTENTS

VOLUME I

VOLUME II

Chapter 1

DEGRADATION OF LIGNIN AND LIGNIN-RELATED SUBSTANCES BY *SPOROTRICHUM PULVERULENTUM (PHANEROCHAETE CHRYSOSPORIUM)*

Paul Ander, Annele Hatakka, and Karl-Erik Eriksson

TABLE OF CONTENTS

I. INTRODUCTION

Next to cellulose, lignin is probably the most common organic compound cycled on earth. Much of this lignin from trees and other plants is converted to humus, thereby importing desirable properties to the soil. Humus influences the structure, aeration, and moisture-holding properties of the soil. It also functions as an ion exchanger and is able to store an release nutrients and carbon dioxide, which can be utilized by growing plants and trees.

Lignin, however, is also a major waste product in pulp manufacture, although large quantities are burned, rather inefficiently, for energy production and recovery of pulping reagents.

With a better understanding of the enzyme mechanisms involved in biological degradation of lignin and lignocellulosic materials, it will be possible to elucidate how lignin is decayed in nature. With this knowledge, it may be possible to use biological delignification processes for technical purposes, as well as to use waste lignins in the production of useful chemicals.[1]

II. DEGRADATION OF LIGNIN BY *S. pulverulentum* AND OTHER WHITE ROT FUNGI

A. The Influence of Carbohydrates and Cellobiose:Quinone Oxidoreductase on Lignin Degradation

For the production of biomechanical pulp (see Volume II, Chapter 14), as well as for delignification of straw and sugar cane bagasse to obtain a better ruminant feed, much research efforts are needed to understand how white rot fungi degrade lignin. One of the first goals in specific fungal degradation of lignin at the Swedish Forest Products Research Laboratory was to obtain mutants of white rot fungi which did not degrade cellulose or hemicellulose, but only degraded lignin. The methods to obtain such mutants were developed by Eriksson and Goodell[2] and are described in Volume II, Chapter 14.

Somewhat later a new enzyme called cellobiose:quinone oxidoreductase (CBQ) was found in culture solutions *Polyporus versicolor* and *S. pulverulentum.*[3,4] This enzyme reduces quinones produced from phenols by phenol oxidases and simultaneously oxidizes cellobiose to cellobiono-δ-lactone. It contains FAD as a prosthetic group and has a molecular weight of 58,000.[5] The enzyme is induced by cellulose or cellulose degradation products (not by glucose) and seems to be of importance both in cellulose and lignin degradation. This has been further investigated by Ander and Eriksson[6] with the wild-type and the cellulase- and CBQ-less mutant Cel 44 of *S. pulverulentum.* Using kraft lignin agar plates, it was found that cellulose was the best cosubstrate for degradation of kraft lignin by the wild-type. The order in which the tested carbohydrates stimulated kraft lignin degradation was as follows: cotton DP 2000 > cotton DP 500 > Walseth cellulose DP 150 > cellobiose or glucose. As an example, glucose and cotton DP 2000 gave 27 and 66% kraft lignin degradation, respectively.[6] Kirk recently reported that both cellulose and glucose support strong degradation of synthetic ^{14}C-lignins (dehydrogenative polymerizates [DHP]) (see Volume II, Chapter 4). The reason for this discrepancy in the observations may be that kraft lignin is a partially degraded lignin from the kraft cooking process with a lower molecular weight and with more phenolic hydroxyl groups than DHPs. Addition of glucose or maltose to wood has also been found to stimulate lignin degradation.[1]

The importance of CBQ in kraft lignin degradation was further supported by the finding that Cel 44, which does not produce CBQ, could degrade only 20% of the kraft lignin in agar plates in the presence of celulose, as compared to 66% for the wild-type. Some increase in the kraft lignin degradation (from 10 to 20%) by Cel 44 on addition of cellulose is believed to be due to impurities such as glucose and cellobiose in the different cellulose preparations.[6]

Further studies have shown that degradation of kraft lignin, lignin sulfonates, and milled wood lignin by *Pleurotus ostreatus* was stimulated by cellulose.[7] Lignin sulfonates were polymerized more in a medium without cellulose compared to a cellulose-containing medium.[8] This effect is certainly due to the CBQ activity.

At this point, one question seemed particularly important to answer, namely, is CBQ so necessary that a specific degradation of lignin in wood cannot take place in its absence? This question was studied by cultivating Cel 44 on birch, pine, and spruce wood.[6] As seen in Table 1, lignin was degraded in all three wood species and in birch 31% of the lignin was degraded in 10 weeks without loss of cellulose. The xylan, however, was degraded to a great extent mainly following lignin degradation. Table 1 further shows that the wild-type degrades lignin better than Cel 44 and that birch is easier to degrade than pine and spruce.

All these results indicate that CBQ may be important but not entirely necessary in lignin degradation. CBQ is probably ubiquitous among white rot fungi,[9] but brown rot fungi do not seem to have this enzyme at all, as investigated with five different species of brown rot fungi.[1] The exact function of CBQ is not known at present. It may be part of an extracellular electron transport chain, thereby reducing quinones to phenols which may be used as substrate for ring-cleaving enzymes. The activity of ring-cleaving enzymes in whte rot fungi is low,[1] however, and low-molecular-weight phenols may be degraded through other pathways, as described below.

B. The Importance of Phenol Oxidase Activity in Lignin Degradation

By the use of similar methods as in the production of Cel 44, we were able to isolate one phenol oxidaseless mutant (Phe 3) and one phenol oxidase-positive revertant (Rev 9) from *S. pulverulentum*.[10] As shown in Table 2, the mutant Phe 3 could not degrade lignin or the other wood components. Rev 9, however, degraded all wood components to the same extent as the wild-type. Kraft lignin also was not degraded by Phe 3, unless purified laccase was added to the medium. In that case, degradation of kraft lignin increased to near normal.[10]

The lack of wood degradation by Phe 3 may be due to the fact that cellulase and xylanase production was strongly inhibited by the presence of wood phenols (Table 3). By addition of laccase to the Phe 3 cultures, the mutant again produced normal amounts of cellulases. These results may indicate that phenol oxidases have a regulatory role, since lignin as well as polysaccharide degradation is affected by the absence of phenol oxidase production.[10] A phenol oxidaseless (peroxidaseless) mutant of *Phanerochaete chrysosporium* has now also been found (see Volume II, Chapter 5). The mutant does not release $^{14}CO_2$ from labeled DHP.[28] These results can all be considered as strong evidence for the necessity of phenol oxidase activity in lignin degradation.

C. Screening of White Rot Fungi for Selective Lignin Degradation

The use of white rot fungi which preferentially degrade lignin is probably necessary if biological delignification processes are to be realized. In order to find such fungi, we cultivated 25 different species of white rot fungi on kraft lignin agar plates with and withoat cellulose. The result can be seen in Figure 1. Ten of the fungi had a stronger phenol oxidase reaction in the presence of cellulose (Group 1 fungi), while 15 fungi (Group 2) had the strongest phenol oxidase reaction in the absence of cellulose. Table 4 shows that the average kraft lignin degradation in the absence of cellulose is

TABLE 1

Weight Losses,ᵃ Moisture Content,ᵇ and Relative Losses in Wood Components,ᶜ Caused by WT and Cel 44 after Different Decay Times

Wood	Incubation period (weeks)	Wild type						Cel 44					
		Weight loss (%)	Moisture content (%)	Lignin (%)	Glucan (%)	Xylan (%)	Mannan (%)	Weight loss (%)	Moisture content (%)	Lignin (%)	Glucan (%)	Xylan (%)	Mannan (%)
Birch	4	12.1	118	17	7	13	NT	4.6	86	9	0	4	NT
	6	15.7	120	24	10	23	NT	4.5	76	13	−2	15	NT
	8	16.8	122	30	9	34	NT	7.7	110	23	1	35	NT
	10	18.3	125	29	12	22	NT	11.3	123	31	0	32	NT
Pine	4	6.2	42	14	8	NT	22	1.8	57	5	−1	NT	10
	6	6.3	40	9	5	NT	22	2.7	56	8	−3	NT	3
	8	9.7	100	15	7	NT	23	1.5	98	10	−2	NT	2
	10	11.8	88	22	4	NT	18	3.4	84	8	0	NT	14
Spruce	4	8.2	68	8	4	NT	15	0.6	65	2	−5	NT	10
	6	7.0	56	11	5	NT	15	1.5	43	4	−3	NT	21
	8	5.9	55	7	5	NT	25	0.9	61	4	−2	NT	24
	10	8.8	74	15	2	NT	30	1.3	73	5	−3	NT	12

Note: NT = not tested; minus (−) indicates increase.

ᵃ   Based on original dry weight.
ᵇ   Based on the dry weight of the decayed blocks.
ᶜ   Based on original amount of component, dry weight basis.
ᵈ   Corrected for glucose released from glucomannan and galacto-glucomannan.

From Ander, P. and Eriksson, K.-E., *Sven. Papperstidn.*, 78, 643, 1975. With permission.

TABLE 2

Weight Losses[a] and Losses of Wood Components[b] caused by WT, Phe 3, and Rev 9 after 46 days

	Fungal strain	Birch	Pine	Spruce
Weight loss	WT	15.9	7.1	6.9
(%)	Phe 3	0.6	0.9	1.2
	Rev 9	11.0	10.0	6.8
Lignin loss	WT	25	10	10
(%)	Phe 3	−6	−3	4
	Rev 9	33	17	7
Glucan loss[c]	WT	10	5	5
(%)	Phe 3	1	3	−2
	Rev 9	3	13	10
Xylan loss	WT	25	NT	NT
(%)	Phe 3	−1	NT	NT
	Rev 9	29	NT	NT
Mannan loss	WT	NT	23	17
(%)	Phe 3	NT	2	4
	Rev 9	NT	17	29

Note: NT = not tested.

a Based on original dry weight.
b Based on original amount of component.
c Represents cellulose loss.

From Ander, P. and Eriksson, K.-E., *Arch. Microbiol.*, 109, 1, 1976. With permission.

TABLE 3

Endo-1,4-β-Glucanase Production by WT, Phe 3(and Rev 9 in Cellulose Shake Flasks with or without 0.25% Kraft Lignin and 10^{-3} M Phenols.

Addition	Endo-1,4-β-glucanase production (units/ml)		
	WT	Phe 3	Rev 9
Ethanol and/or boiled laccase	0.47	0.92	0.56
Vanillic acid	0.22	0.06	0.64
Vanillic acid + 2 μg laccase	NT	0.14	NT
Vanillic acid + 8 μg laccase	NT	0.44	NT
p-Hydroxybenzoic acid	0.18	0.05	0.52
Ferulic acid	0.03	0.02	0.08
Kraft lignin	1.08	0.02	1.30
Kraft lignin + 2 μg laccase	NT	0.03	NT
Kraft lignin + 8 μg laccase	NT	0.12	NT
Kraft lignin + 16 μg laccase	NT	0.20	NT

Note: NT = not tested. The Phenols were added 1 day after inoculation, whereas kraft lignin was present during sterilization. Laccase was added 6 hr after addition of vanillic acid, the addition of laccase to the kraft lignin flasks being carried out at the same time.

From Ander, P. and Eriksson, K.-E., *Arch. Microbiol.*, 109, 1, 1976. With permission.

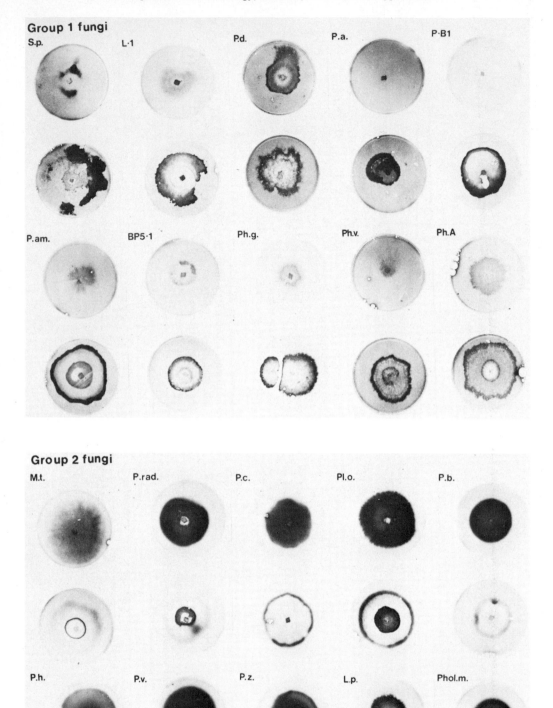

FIGURE 1.   Coloration and bleaching of kraft lignin only (top row) and kraft lignin plus cellulose (bottom row) by fungi from Groups 1 and 2. The fungi, their designations as well as their incubation times before photographing, are given in Reference 9. These photographs of representative kraft lignin plates were taken at different times during two separate cultivations. For this reason, some plates appear to have a darker background (e.g., P.d., P.a., and Ph.v.) compared to other plates (e.g., Ph.g., P.rad., and P.c.). (From Ander, P., and Eriksson, K.-E., *Physiol. Plant.,* 41, 239, 1977. With permission.)

Group 2 fungi

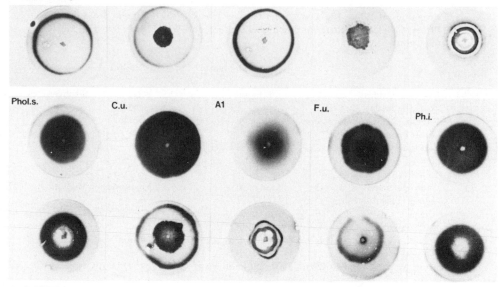

FIGURE 1

somewhat better by Group 2 fungi than by Group 1 fungi (19 vs. 10%). The Group 1 fungi are most dependent on cellulose to degrade kraft lignin, and consequently their kraft lignin degradation increases from 10 to 41% if cellulose is added.

Representative fungi from the two groups also differ in other respects. *S. pulverulentum* (S.p.), *Phanerochaete* sp L-1 (L-1), and *Polyporus dichrous* (P.d.) from Group 1 have a high endo-1,4-β-glucanase and CBQ activity in cellulose shake flasks, but low phenol oxidase activity in standing wood meal flasks (Table 4). The four fungi—*Merulius tremellosus* (M.t.), *Phlebia radiata* (P. rad.), *Pycnohorus cinnabarinus* (P.c.), and *Pleurotus ostreatus* (Pl. o.)— from Group 2, on the other hand, produce low levels of endo-glucanase and CBQ in the cellulose flasks and a high level of phenol oxidases in the wood meal medium.[9]

These results, as well as the finding that *P. cinnabarinus* in the presence of malt extract degrades only the lignin in pine wood, supports the conclusion that preferential lignin degraders are easier to find among Group 2 fungi producing large amounts of phenol oxidases than among Group 1 fungi having a strong endo-glucanase production. The results, however, also support earlier research,[1] indicating that lignin in wood cannot be degraded unless a more easily metabolized carbon source is also available. In the case of *P. cinnabarinus*, malt extract is probably used as an easily available carbon source instead of the wood polysaccharides.[9]

III. DEGRADATION OF VANILLIC ACID AND RELATED SUBSTANCES BY WHITE ROT FUNGI

One of the most commonly found products after fungal degradation of lignin is vanillic acid.[1,11,12] Reports in the literature indicate that white rot fungi metabolize vanillic acid through protocatechuic acid[13,14] or through methoxyhydroquinone.[15] There is, however, no definite proof for the degradation pathway(s) of vanillic acid in white rot fungi. For example, no pure enzyme preparation of a demethylating monooxygenase or a ring-cleaving dioxygenase from a truly lignin-degrading fungus has been obtained so far.[1] Studies on the degradation of vanillic acid by *S. pulverulentum* were therefore started to investigate the enzymes involved.

TABLE 4

Properties of White Rot Fungi from Groups 1 and 2 (Kraft Lignin: Solid Agar Plates, Cellulose Powder: Shaking Liquid Culture, and Wood Meal: Standing Liquid Culture)

Property	Medium	Group 1	Group 2
Phenol oxidase reaction, kraft lignin degrad. (%)	Kraft lignin	Weak (10)	Stronger (19)
Phenol oxidase reaction, kraft lignin degrad. (%)	Kraft lignin + cellulose	Strong (41)	Weaker (31)
Endo-glucanase and cellobiose dehydrogenase production	Cellulose powder	High	Low
Endo-glucanase and cellobiose dehydrogenase production	Wood meal	Medium	Medium
Phenol oxidase production	Wood meal	Low	High
Phenol oxidase production	Wood blocks	Normal	Normal

From Ander, P. and Eriksson, K.-E., *Physiol. Plant.*, 41, 239, 1977. With permission.

Eventually such studies will lead to isolation and characterization of lignin-degrading enzymes. Some experiments were also performed with veratric acid (vanillic acid methyl ether). The results from the above-mentioned preliminary studies are described below.

A. Vanillic Acid Degradation Products

S. pulverulentum was cultivated in shaking flasks containing 0.25% cellobiose and a nutrient salt medium modified from Kirk et al.,[16] so that it contained no phthalate buffer but low amounts of nitrogen. After 24 hr, vanillic acid was added to 1 or 3 mM. Degradation products after 1, 2, or 3 days were analyzed by thin layer and gas chromatography (TLC, GC). For the GC analyses, the culture solutions were extracted by ethyl acetate at pH 6.5 or pH 2.0. Derivatization was made by silylation. By comparison with authentic reference substances, the compounds listed in Figure 2 could be identified. In the case of 3 mM vanillic acid, an unidentified peak also appeared in the GC chromatograms. No protocatechuic acid was detected.

B. On the Formation of Protocathechuic Acid from Vanillic Acid

Demethylation of vanillic acid to form protocatechuic acid is a well-known oxygen-requiring step in the degradation of vanillic acid by bacteria.[17-19] For ring cleavage of protocatechuic acid, oxygen is also needed.[1] We therefore thought it possible to find out whether vanillic acid was degraded via protocatechuic acid by studying O_2 consumption in the presence of vanillic acid, using mycelia from *S. pulverulentum* which had been grown with and without vanillic acid. Oxygen electrode measurements (Yellow Spring Instruments, U.S.) were performed with both starved and nonstarved mycelia with and without cellobiose as cosubstrate, and with mycelium grown under induction conditions in both standing and shaking cultures. In no case was an increased oxygen consumption by vanillic acid-induced mycelia obtained, as compared to non-induced mycelia. Similar experiments with protocatechuic acid gave the same result. One reason for this may be that more mycelia are needed to detect an increased oxygen consumption by demethylation or ring cleavage in these short-time experiments. It is, however, impossible to use much more than 10 mg mycelia in our oxygen cell, since

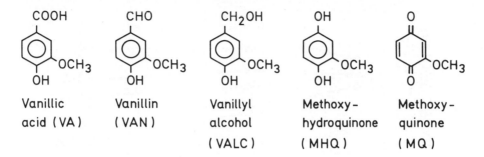

FIGURE 2. Products formed from vanillic acid by *S. pulverulentum*.

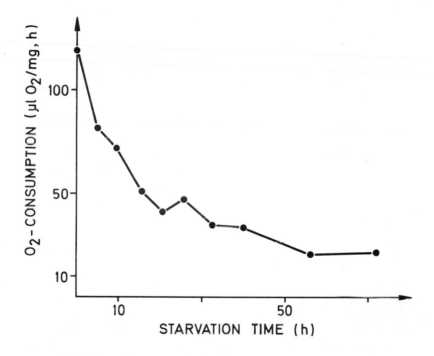

FIGURE 3. O₂ consumption by *S. pulverulentum* during starvation.

even after starvation (see Figure 3), there is a considerable endogenous respiration.

It may be pointed out that oxygen uptake measurements are not specific only for demethylation or ring cleavage. Other reactions such as oxidative decarboxylations and phenol oxidase reactions will also consume oxygen. Since no increased oxygen uptake could be detected with vanillic or protocatechuic acids under our conditions, it seems that oxygen uptake measurements using an oxygen electrode and limited amounts of mycelia (enzyme) is of limited value in this case.

Several attempts have also been made to accumulate protocatechuic acid by incubating fungal mycelium in the presence of 3 mM vanillic acid and the iron chelators, α,α'-bipyridyl, tiron or o-phenanthroline (mainly as described by Chapman and Hopper[20]). However, no protocatechuic acid has been detected in these experiments. Since the degradation of protocatechuic acid is rather slow (see below), it should be possible to detect this acid, if it is formed, even without the above-mentioned inhibitors.

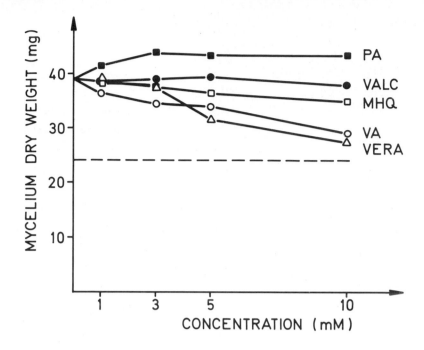

FIGURE 4.   Mycelium growth of *S. pulverulentum* on cellobiose, yeast extract, and different concentrations of vanillic acid and related substances, which were added after 24-hour cultivation. Growth was for 48 hr on VA = vanillic acid, VALC = vanillyl alcolhol, MHQ = methoxyhydroquinone, PA = protocatechuic acid, and VERA = veratric acid. (---) = mycelium growth at addition of these substances.

C. Cultivation of *S. pulverulentum* in the Presence of Vanillic Acid and Related Substances

1. Mycelium Growth Studies

In these experiments, *S. pulverulentum* was grown on 0.25% cellobiose, 0.05% yeast extract, and the medium according to Kirk et al.,[16] using phthalate as buffer and starting at a pH value of 4.7. The following substances were added to different flasks after 24-hr cultivation: vanillic acid, vanillyl alcohol, methoxyhydroquinone (MHQ), protocatechuic acid, and veratric acid. Since vanillic and veratric acids were dissolved in 0.1 *M* NaOH before addition, the initial pH was around 6 for these cultures. As shown in Figure 4, 10 m*M* vanillic and veratric acids are quite toxic, and after addition of such high amounts of these compounds, only a small increase in growth is obtained. In the presence of vanillyl alcohol or MHQ, the growth is nearly normal, while protocatechuic acid enhanced the mycelium growth somewhat as compared to the growth without addition of these compounds. Similar results with vanillic and protocatechuic acids were also obtained by Li et al.,[21] using two isolates of *Poria weirii*. Other results obtained by us indicate that vanillic and veratric acids are still more toxic if a lower starting pH is used. Adding these substances at inoculation also affects the germination rate, thereby decreasing the mycelium growth. At the end of the cultivation (48 hr after addition), dark particles were observed in the vanillyl alcohol cultures containing 3 and 5 m*M* vanillyl alcohol, indicating formation of polymeric products. These could be dissolved in ethanol and were only formed at pH 3 to 5.

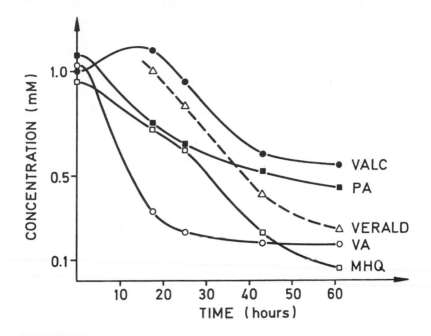

FIGURE 5. Degradation of vanillic acid and related substances by *S. pulver-ulentum* as measured by UV spectroscopy. Conditions and lignin models as in Figure 4. VERALD = veratrum aldehyde. The degradation rates may be considered as minimum degradation rates due to possible absorbance caused by metabolic products.

2. Degradation as Measured by UV and GC Analyses
a. UV Analyses

The preceding experiment (Section III.C.1) showed that the investigated substances were not very toxic at 1 mM concentration. Consequently, this concentration was used in a similar type of experiment where the degradation of the substances was followed by UV spectroscopy. The medium contained cellobiose and yeast extract but no phthalate, which disturbs both UV and GC analyses. Figure 5 shows that vanillic acid is degraded rather fast, and after 17.5 hr, the cultures smelled strongly from vanillin. Vanillin appears at 342 mm as a large peak, probably together with MHQ. MHQ shows a blue color at pH 11 in contrast to vanillin, which makes it possible to distinguish between these two compounds.

When *S. pulverulentum* was cultivated on MHQ, this compound was metabolized nearly completely but not as rapidly as vanillic acid (see Ref. 15). Protocatechuic acid was slowly converted to unknown compounds. No change in the UV spectrum indicating the formation of β-carboxy-*cis,cis*-muconic acid was observed. The strange form of the vanillyl alcohol curve is probably due to the formation of polymeric and other substances absorbing in the UV region. Veratric acid was almost quantitatively reduced to the aldehyde, which was metabolized rather fast as seen in Figure 5. An absorption maximum indicating monooxygenase-induced demethylation and formation of vanillic acid could not be detected.

The formation of unknown UV-absorbing metabolites may have affected these measurements.* The results were, however, supported by the following GC analyses.

* Note added in proof: The unknown compound mentioned in Section III. C. 2 is probably veratryl alcohol which is formed independent of added phenol.[29,30] Similarly, in the presence of vanillyl alcohol, not MHQ, but veratryl alcohol was detected by GC.

b. GC Analyses

When *S. pulverulentum* was cultivated on vanillic acid, the compound was rapidly converted to the same products as shown in Figure 2. After growth on MHQ, only the unknown compound formed during growth on vanillic acid could be detected. Cultivation on vanillyl alcohol and extraction at pH 2 gave the same unknown compound, while extraction at pH 6.5 gave MHQ. It is not known at present whether in this case MHQ is formed directly from vanillyl alcohol or if it is formed through methoxyquinone (MQ). No metabolic products were observed after growth in the presence of protocatechuic acid, and a rather slow degradation was indicated by the GC chromatogram. Veratric acid was reduced to the corresponding aldehyde and alcohol, while demethylation to vanillic acid was not detected. It must be pointed out here that these analyses are preliminary and quantification of the different metabolic steps remains to be done.

D. Degradation of Vanillic Acid by Three Other White Rot Fungi

The white rot fungi *Pycnoporus cinnabarinus, Polyporus versicolor,* and *Pleurotus ostreatus* were cultivated on 1 and 3 m*M* vanillic acid on the phthalate-deficient medium as with *S. pulverulentum.* By GC analyses, it was shown that from 3 m*M* vanillic acid, *P. cinnabarinus* formed vanillin, vanillyl alcohol, MHQ, possibly MQ, and also the above-mentioned unknown compound in the same way as *S. pulverulentum.* The other fungi, however, only formed the unknown compound from vanillic acid, and only a slow metabolism of the 3 m*M* vanillic acid took place.

Some variations in the enzymatic activity of the different cultures were also observed. On 1 m*M* vanillic acid, *Polyporus versicolor* and *Pleurotus ostreatus* both produced large amounts of phenol oxidases (28 and 3.5 units,* respectively) but no CBQ. *Pycnoporus cinnabarinus,* however, produced 24 CBQ units,* but only small amounts of phenol oxidases on this medium. This apparently low phenol oxidase activity may be the result of inhibition of phenolic coupling of the substrate due to the high CBQ activity (see Section II). In Figure 1, it is also shown that a weaker phenol oxidase reaction on kraft lignin agar plates containing cellulose is obtained by the Group 2 fungus, *Pycnoporus cinnabarinus* (P.c.), as compared to kraft lignin plates without cellulose.

IV. DISCUSSION

Based upon the results presented, the possible transformations of vanillic acid by *S. pulverulentum* and perhaps *P. cinnabarinus* are shown in Figure 6. Some of these reactions are known before in the literature but perhaps, except for the recent report by Nishida and Fukuzumi,[22] this is the first time where both formation of MHQ and reduction to vanillin and vanillyl alcohol are shown to be performed by two different white rot fungi. To our knowledge, very little has been reported about reduction of vanillic acid by white rot fungi. Ishikawa et al.[23] reported that *Polyporus versicolor* and *Fomes fomentarius* converted small amounts of vanillic acid to vanillin. Reduction of veratric acid by *P. versicolor*[24,25] and *Polystictus sanguineus*[26] has, however, been found.

It appears that phenol oxidases are of importance in the degradation of vanillic acid under our cultivation conditions (Figure 6). Their ability to demethoxylate phenolic substances has been reported by many authors.[1] Also their necessity in lignin degradation has now been established, using phenoloxidaseless mutants (see Section II).

---

*   For definition of units see Reference 9.

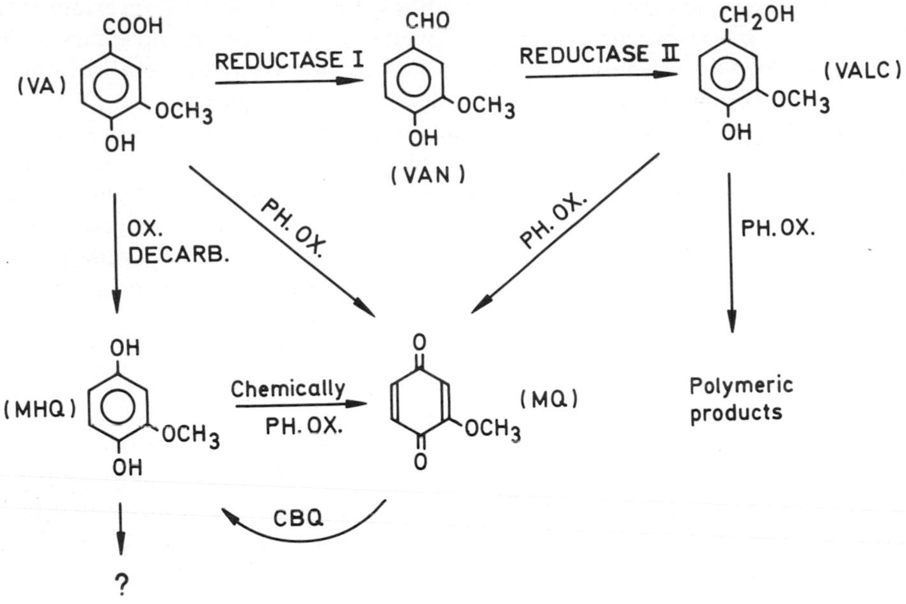

FIGURE 6. Possible transformations of vanillic acid by *S. pulverulentum*.

Work which should be done in the future is to investigate the delicate balance between phenol oxidase and CBQ activity, using pure enzyme preparations to elucidate the possible role of CBQ in the formation of MHQ and other reduced quinones. By comparative cultivations of *S. pulverulentum* and its phenol oxidaseless and CBQ-less mutants, Phe 3 and Cel 44, respectively, on the substances in Figure 6, we hope to find out whether the formation of MHQ goes via the reductive pathway and through MQ or if vanillic acid is oxidatively decarboxylated directly to MHQ, as Kirk and Lorenz[15] found using *Polyporus dichrous*. It is interesting to note that *S. pulverulentum* and *P. dichrous* both are placed in Group 1 in Table 4, since they both have high CBQ activity on cellulose, but low phenol oxidase activity on wood meal. The results reported here may indicate that they form MHQ in the same way.

Finally, our results indicate that *S. pulverulentum* does not transform vanillic acid into protocatechuic acid; the same result was obtained with *Polyporus dichrous*.[15] Neither have we been successful in obtaining cell-free activity against protocatechuic acid. Maybe other cultivation conditions, as well as lignin substrates of higher molecular weight, must be tested before such an activity can be found. Degradation of lignin models bound to nondegradable polymers may be one way to similate the conditions under which lignin is degraded in nature.[27]

V. SUMMARY

This paper summarizes both earlier and more recent results regarding degradation of lignin and lignin-related substances by white rot fungi.

Sporotrichum pulverulentum and *Polyporus versicolor* were found to produce a hitherto unknown enzyme now called cellobiose:quinone oxidoreductase. This enzyme reduces quinones produced from phenols by phenol oxidases and simultaneously oxidizes cellobiose to cellobiono-δ-lactone. It seems to be of importance both in cellulose and lignin degradation, as investigated using *Pleurotus ostreatus* and the wild type plus a cellulaseless mutant of *S. pulverulentum*.

Using a phenol oxidaseless mutant and a phenol oxidase-positive revertant from *S. pulverulentum*, it was shown that phenol-oxidizing enzymes are necessary in lignin degradation, since the phenol oxidaseless mutant did not degrade kraft lignin or lignin in wood. The revertant, however, degraded lignin as well as the wild type.

By cultivating white rot fungi on kraft lignin agar plates in the presence or absence of cellulose, it is possible to divide the fungi into two groups due to differences in phenol oxidase reactions in the lignin agar plates. The production of endo-1,4-β-glucanase, cellobiose:quinone oxidoreductase, and phenol-oxidizing enzymes by representative fungi from the two groups is described in relation to the degradation of different wood components. Fungi with high oxidase production seems to degrade lignin most selectively.

The metabolism of the lignin degradation product, vanillic acid, has been studied with *S. pulverulentum*. After growth of the fungus in the presence of cellobiose and vanillic acid, vanillin, vanillyl alcohol, MHQ, and MQ were detected in the culture solutions. The fungus did not grow on vanillic acid alone and no transformation into protocatechuic acid was detected. Veratric acid was not demethylated to vanillic acid. Some experiments using an oxygen electrode and fungal mycelia indicate that the level of oxidative enzymes is too low to detect an increased oxygen consumption at addition of vanillic acid. These results, as well as the metabolism of vanillic acid, are further discussed. Limited experiments with three other white rot fungi are also described.

ACKNOWLEDGMENTS

Research support for A. Hatakka by the Academy of Finland and by the Foundation for Research of Natural Resources in Finland is gratefully acknowledged.

REFERENCES

1. **Ander, P. and Eriksson, K.-E.,** Lignin degradation and utilization by micro-organisms, in *Progress in Industrial Microbiology,* Vol. 14, Bull, M. J., Ed., Elsevier, Amsterdam, 1978, 1.
2. **Eriksson, K.-E. and Goodell, E. W.,** Pleiotropic mutants of the wood-rotting fungus *Polyporus adustus* lacking cellulase, mannanase, and xylanase, *Can. J. Microbiol.,* 20, 371, 1974.
3. **Westermark, U. and Eriksson, K.-E.,** Carbohydrate-dependent enzyme quinone reduction during lignin degradation, *Acta Chem. Scand.,* B28, 204, 1974.
4. **Westermark, U. and Eriksson, K.-E.,** Cellobiose:quinone oxidoreductase, a new wood-degrading enzyme from white rot fungi, *Acta Chem. Scand.,* B28, 209, 1974.
5. **Westermark, U. and Eriksson, K.-E.,** Purification and properties of cellobiose:quinone oxidoreductase from *Sporotrichum pulverulentum, Acta Chem. Scand.,* B29, 419, 1975.
6. **Ander, P. and Eriksson, K.-E.,** Influence of carbohydrates on lignin degradation by the white-rot fungus *Sporotrichum pulverulentum, Sven. Papperstidn.,* 78, 643, 1975.
7. **Hiroi, T. and Eriksson, K.-E.,** Microbiological degradation of lignin. I. Influence of cellulose on the degradation of lignins by the white-rot fungus *Pleurotus ostreatus, Sven. Papperstidn.,* 79, 157, 1976.
8. **Hiroi, T., Eriksson, K.-E., and Stenlund, B.,** Microbiological degradation of lignin. II. Influence of cellulose upon the degradation of calcium lignosulfonate of various molecular sizes by the white-rot fungus *Pleurotus ostreatus, Sven. Papperstidn.,* 79, 162, 1976.
9. **Ander, P. and Eriksson, K.-E.,** Selective degradation of wood components by white-rot fungi, *Physiol. Plant.,* 41, 239, 1977.
10. **Ander, P. and Eriksson, K.-E.,** The importance of phenol oxidase activity in lignin degradation by the white-rot fungus *Sporotrichum pulverulentum, Arch. Microbiol.,* 109, 1, 1976.
11. **Higuchi, T.,** Formation and biological degradation of lignins, in *Advances in Enzymology,* Vol. 34, Nord, F. F., Ed., Interscience, New York, 1971, 207.
12. **Kirk, T. K.,** Effects of microorganisms on lignin, *Ann. Rev. Phytopathol.,* 9, 185, 1971.

13. **Flaig, W. and Haider, K.,** Die Verwertung phenolischer Verbindungen durch Weissfaulepilze, *Arch. Mikrobiol.,* 40, 212, 1961.

14. **Cain, R. B., Bilton, R. F., and Darrah, J. A.,** The metabolism of aromatic acids by micro-organisms. Metabolic pathways in the fungi, *Biochem. J.,* 108, 797, 1968.

15. **Kirk, T. K. and Lorenz, L. F.,** Methoxyhydroquinone, an intermediate of vanillate catabolism by *Polyporus dichrous, Appl. Microbiol.,* 26, 173, 1973.

16. **Kirk, T. K., Yang, H. H., and Keyser, P.,** The chemistry and physiology of the fungal degradation of lignin, in *Developments in Industrial Microbiology,* Vol. 19, Underkofler, L. A., Ed., American Institute of Biological Sciences, Washington, D. C., 1978, 51.

17. **Cartwright, N. J. and Buswell, J. A.,** The separation of vanillate O-demethylase from protocate-chuate 3,4-oxygenase by ultracentrifugation, *Biochem. J.,* 105, 767, 1967.

18. **Ribbons, D. W.,** Stoicheiometry of O-demethylase activity in *Pseudomonas aeruginosa, FEBS Lett.,* 8, 101, 1970.

19. **Kawakami, H.,** Bacterial degradation of lignin model compound. IV. On the aromatic ring cleavage of vanillic acid, *J. Jpn. Wood Res. Soc.,* 22, 246, 1976.

20. **Chapman, P. J. and Hopper, D. J.,** The bacterial metabolism of 2,4-xylenol, *Biochem. J.,* 110, 491, 1968.

21. **Li, C. Y., Lu, K. C., Nelson, E. E., Bollen, W. B., and Trappe, J. M.,** Effect of phenolic and other compounds on growth of *Poria weirii* in vitro, *Microbios,* 3, 305, 1969.

22. **Nishida, A. and Fukuzumi, T.,** Formation of coniferyl alcohol from ferulic acid by the white-rot fungus *Trametes, Phytochemistry,* 17, 417, 1978.

23. **Ishikawa, H., Schubert, W. J., and Nord, F. F.,** Investigations on lignin and lignification. The degradation by *Polyporus versicolor* and *Fomes fomentarius* of aromatic compounds structurally related to softwood lignin, *Arch. Biochem. Biophys.,* 100, 140, 1963.

24. **Farmer, V. C., Henderson, M. E. K., and Russel, J. D.,** Reduction of certain aromatic acids to aldehydes and alcohols by *Polystictus versicolor, Biochim. Biophys. Acta,* 35, 202, 1959.

25. **Zenk, M. H. and Gross, G. G.,** Reduktion von Veratrumsäure zu Veratrylaldehyd und Veratrylalkohol durch zellfreie Extrakte von *Polystictus versicolor* L., *Z. Pflanzenphysiol.,* 53, 356, 1965.

26. **Minami, K., Tsuchiya, M., and Fukuzumi, T.,** Metabolic products from aromatic compounds by the wood-rotting fungus *"Polystictus sanguineus (Trametes sanguinea)".* IV. Culturing condition for reduction and demethoxylation of veratric acid, *J. Jpn. Wood Res. Soc.,* 11, 179, 1965.

27. **Kirk, T. K., Connors, W. J., and Brunow, G. A.,** Fungal degradation of lignin-related aromatics attached to biologically inert polymers, *Tappi,* p. 163, 1977.

28. **Gold, M.,** personal communication, 1978.

29. **Lundquist, K. and Kirk, T. K.,** *Phytochemistry,* 17, 1676, 1978.

30. **Ander, P.,** recent results, 1979.

Chapter 2

THE ROLE OF LACCASE IN LIGNIN BIODEGRADATION

Tatsuo Ishihara

TABLE OF CONTENTS

I. INTRODUCTION

The relationship between ligninolytic activity and phenol oxidase activity of wood-rotting fungi has been a problem of interest for a long time, but a well defined conclusion has not yet been attained. Many years ago, Bavendamm showed that most white rot fungi, which are capable of degrading lignin, gave a colored zone around mycelium on agar plates containing tannin and that the phenomenon is due to phenol oxidases excreted by the fungi.[1] Kirk and Kelman[2] and Sundman and Näse[3] have reconfirmed that the ligninolytic activity of wood-rotting fungi and the positive Bavendamm's test are almost parallel. It seemed to be reasonable, therefore, to consider that phenol oxidases are related to lignin degradation. On the other hand, however, phenol oxidases are known to cause oxidative polymerization of phenols. Haider, Lim, and Flaig[4] claimed that enzymes other than phenol oxidases may cause lignin degradation. Konishi and Inoue reported recently that laccase of *Coriolus versicolor* decreased the amount of polymeric milled wood lignin (MWL) and also made it more water soluble.[5] Furthermore, Kirk, Harkin, and Cowling[6,7] showed that syringyl glycol-β-guaiacyl ether was decomposed by the same enzyme to give guaiacoxyacetaldehyde and 2,6-dimethoxy-p-benzoquinone.

In this paper, I should like to describe some experiments concerning the effects of laccase on MWL and lignin models and discuss the role of laccase in lignin biodegradation.

II. THE RELATIONSHIP BETWEEN LACCASE PRODUCTION AND THE CAPABILITY OF WOOD-ROTTING FUNGI TO DEGRADE LIGNIN

A. Bavendamm's Reaction

Fifty years ago, Bavendamm made an important observation that white rot fungi and brown rot fungi can be discriminated by coloration of agar medium around the mycelium caused by white rot fungi when certain phenolic compounds are present in the medium.[1] Later, Higuchi[8] showed that several phenolic compounds can be used for this color reaction. Kirk and Kelman[2] and Sundman and Näse[3] examined many Basidiomycetes for their ability to degrade lignin and their Bavendamm's reaction. Although they found that the ligninolytic activity and a positive Bavendamm's reaction are almost parallel, a few exceptions existed. Nevertheless, it seemed not unreasonable to consider that phenol-oxidizing enzymes play some role in lignin biodegradation.

B. Degradation of Wood Carbohydrates and Lignin

It is well known that white rot fungi decompose lignin as well as carbohydrates, while brown rot fungi decompose carbohydrates only, although they cause some alteration in lignin. It has been reported that white rot fungi can grow with kraft lignin as a sole carbon source, but it is also found that carbohydrates accelerate the decomposition of lignin.[9] An interaction between decomposition of the two components is quite probable, as is suggested for the action of cellobiose:quinone oxidoreductase.[10]

III. OXIDATION OF MILLED WOOD LIGNIN BY LACCASE

Although several phenol oxidases are known to be present in wood-rotting fungi, laccase (EC 1.10.3.2) is abundant and easy to purify. This enzyme was therefore prepared from the filtrate of tank cultures of *Coriolus versicolor* and used in most experiments reported here.

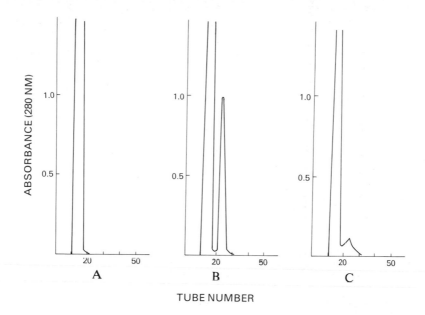

FIGURE 1. Elution curves of MWL on Sephadex® LH-20 (dioxane-water solvent). (A) Untreated, (B) incubated with laccase, and (C) incubated without enzyme. (Reproduced from Ishihara, T., and Miyazaki, M., *Mokuzai Gakkaishi,* 18, 415, 1972. With permission).

A. Change in Molecular Weight Distribution[11]

Although fungal degradation of lignin is thought to be very complicated, with many enzymes working together, the effect of a single enzyme on milled wood lignin (MWL) was examined to simplify the problem. MWL of maple wood (*Acer pictum* Thunb) was used as a substrate in order to lessen the chance of condensation at the 5 position in phenolic units. A suspension of MWL was incubated with laccase at 30°C for 4 days, and then an aqueous dioxane solution of the residual MWL was passed through a Sephadex® LH-20 column. As shown in Figure 1A the original MWL was completely excluded from the gel. On the other hand, incubated MWL gave a second peak in addition to the peak of the excluded portion (Figure 1B), showing the formation of a lower-molecular-weight portion. The sample treated without enzyme at pH 4.0 gave a very small second peak, indicating slight nonenzymatic decomposition (Figure 1C). In Figure 1, the lignin concentration was measured by $A_{280 \, nm}$. Hence the enzyme, which also has a UV maximum at 280 nm, may interfere in the determination. However, the absorption intensity of the latter is about 1/20 of the former, and, further, the amount of the enzyme used was also 1/20. Thus, interference was negligible. When the incubated MWL suspension was evaporated to dryness under reduced pressure and the residue dissolved in dioxane water (10:1 v/v) about one third of the MWL was found to become insoluble.

Since the LH-20 column chromatography mentioned above did not give any information on the actual molecular weight of the excluded portion, gel permeation chromatography (GPC) was conducted, using acetylated samples. The incubated MWL suspension was evaporated to dryness under reduced pressure and the residue was dissolved in pyridine. After centrifugation, about one third of MWL was again found to be insoluble. The soluble portion was acetylated in the usual manner with acetic anhydride and subjected to GPC columns (Styragel® 5.0×10^4 to 1.5×10^5, 5.0×10^3 to 1.5×10^4, 2000 to 5000, 900 to 2000, 350 to 900, 100 to 350 Å), using tetrahydrofuran as a solvent. As can be seen in Figure 2, the original MWL showed a pattern of a nearly monodisperse system, whereas the incubated one showed a newly formed shoul-

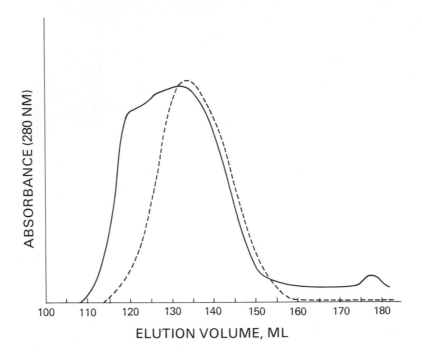

FIGURE 2.. GPC of acetylated MWL on Styragel® (tetrahydrofuran solvent). Broken line represents MWL; solid line, laccase-treated MWL. (Reproduced from Ishihara, T. and Miyazaki, M., *Mokuzai Gakkaishi*, 18, 415, 1972. With permission).

der on the higher molecular weight side of the original MWL. Although the molecular weight of lignin cannot be estimated accurately from the calibration curve prepared by the use of standard linear polymers of known chain length (Figure 3), the MWL treated with laccase was eluted at about 120 mℓ (position of the shoulder) and can tentatively be expressed in terms of chain length for comparison with the peak of the original MWL (peak at about 130 mℓ). The former was found to have a chain length three times longer than the latter. Another peak eluted at 175 mℓ indicates the presence of degradation products.

Thus, the molecular weight of the MWL increased after incubation with laccase, and in addition, about one third of the MWL became insoluble even in aqueous dioxane or pyridine. On the other hand, a small part of MWL was found to be degraded to low-molecular-weight products. Judging from the GPC elution curve, the occurrence of random scission seems to be unlikely.

B. Degradation Products

The incubated MWL suspension was centrifuged after addition of a small amount of sodium chloride, and the almost clear supernatant solution was concentrated under reduced pressure and subjected to gel filtration chromatography of Sephadex® G-25, using water as a solvent. Figure 4 shows the elution curve of the water-soluble portion of the incubated MWL. The peak fractions showing yellow color were extracted with chloroform. Evaporation of the solvent gave yellow needles (1.4 mg from 1 g of MWL). This compound was identified as 2,6-dimethoxy-*p*-benzoquinone (I) by mixed melting point determination with authentic sample and also by TLC, UV, and VIS

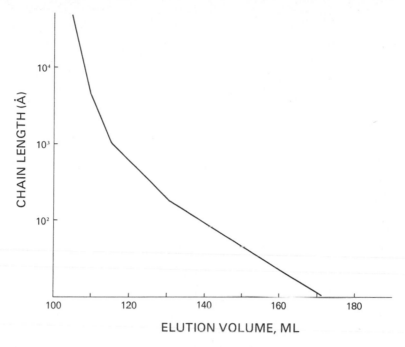

FIGURE 3. GPC calibration curve for Styragel® (tetrahydrofuran solvent); curve prepared with polystyrene standards. (Reproduced from Ishihara, T. and Miyazaki M., *Mokuzai Gakkaishi*, 8, 415, 1972. With permission).

spectra. In

addition to this, some other degradation products were observed on TLC, but their isolation has not been successful because of their low amounts.

Formation of 2,6-dimethoxy-*p*-benzoquinone from syringylglycol-β-guaiacyl ether by laccase has already been reported,[7] and the same quinone was also obtained from syringaresinol.[12] This is the first instance, however, of the isolation of the same quinone from lignin itself by the action of laccase.

IV. OXIDATION OF LIGNIN-RELATED AROMATICS BY LACCASE

A. Demethylation

It has long been known that the methoxyl content of lignin in decayed wood is generally lower than that in sound wood. Catechol moieties in "enzymatically liberated" (brown-rotted) lignin were proved by Kirk and Adler,[13] and these were attributed to the demethylating action of the fungi. Farmer, Henderson, and Russel[14] and Minami, Tsuchiya, and Fukuzumi[15] reported demethylation of methoxy aromatic acids by *Polyporus (Coriolus) versicolor* and *Polystictus sanguineus*, respectively. As enzymes related to demethylation of lignin and model compounds, peroxidase, laccase, and *O*-demethylase have been suggested. Trojanowski, Leonowicz, and Hampel reported on

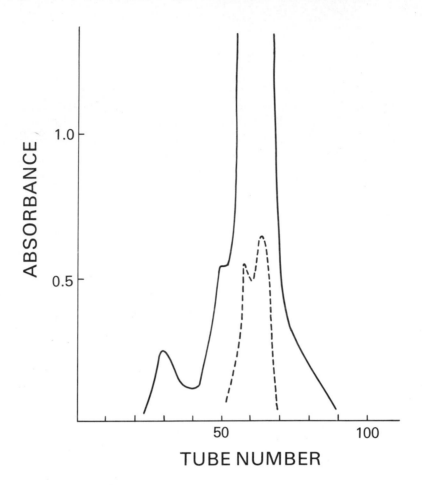

FIGURE 4.   Elution curve of water-soluble portion of laccase-MWL reaction mixture (Sephadex® G-25/H$_2$O). Solid line represents A$_{285nm}$ broken line, A$_{380nm}$ (Reproduced from Ishihara, T. and Miyazaki, M., *Mokuzai Gakkaishi*, 8, 415, 1972. With permission.)

the basis of inhibition experiments[16] that peroxidase (with H$_2$O$_2$) and laccase (with O$_2$) in *Pholiota mutabilis* cultures were responsible for demethylation of lignin and model compounds. Connors et al. also made experiments using horseradish peroxidase (with H$_2$O$_2$) and model methoxyphenols and proved demethylation and orthoquinone formation.[17] Fukuzumi, Takatsuka, and Minami[18] and Cartwright and Smith[19] mentioned O- demethylation in the presence of NADH or NADPH by enzymes of *Poria subacida* and *Pseudomonas fluorescens,* respectively. In these cases, formation of formaldehyde was demonstrated.

In our work, demethylation of lignin and model compounds by laccase was studied, and the liberation of methanol was observed.[20] Since laccase is a kind of oxidase and not a hydrolase, at least one of the fission products should be in an oxidized form. Since the methoxyl group is recovered as methanol, the counterpart, the aromatic ring, should become an o-quinone.

As substrates for these studies, vanillic acid, vanillyl alcohol, maple MWL, and its sulfonate were employed. These substrates were incubated with laccase at 30°C for 3 hr, and the distillate was used for methanol determination. As shown in Table 1, all these substrates gave methanol in yields ranging from about 2 to 10%, based on methoxyl. Biological or enzymatic demethylation of lignin and model compounds has been

TABLE 1

Liberation of Methanol and Formaldehyde
from Substrates on Oxidation with Laccase

| Substrate | CH$_3$OH | | HCHO |
100 mg	(mg)	(%)[a]	(mg)
Vanillic acid	0.40	2.2	0
	0.40	2.2	0
	0.48	2.6	0
Vanillyl alco-	2.60	12.9	2.4
hol	3.10	15.5	2.8
	1.80	8.9	3.0
MWL	0.26	1.3	0
	0.26	1.3	0
	0.28	1.4	0
MWL sulfo-	0.64	—	0
nate	0.68	—	0
	0.40	—	0

[a] Mole percent based on methoxyl.

Reproduced from Ishihara, T. and Miyazaki, M., *Mokuzai Gakkaishi,* 20, 39, 1974.
With permission.

discussed by many authors, but the products derived from methoxyl group are not identical. In case of *Pseudomonas* O-demethylase, the first product was reported to be formaldehyde.[19] Flaig studied the cleavage of the methyl ether in methoxy-labeled phenolic model compounds by fungal cultures and reported that 60% of the radioactivity was recovered as carbon dioxide.[21] Trojanowski and Leonowicz[22] mentioned that *Pholiota mutabilis* culture of lignin gave a product which showed a negative chromotropic acid reaction.[22] This means that the product was not formaldehyde.

Since the action of laccase is oxidative (as mentioned above), the aromatic ring after demethylation should become an o-quinone. The dark color of the reaction mixture is in accord with quinone formation. In order to verify the occurrence of o-quinone, estimation of these structures in reaction mixtures was carried out by the use of a specific color reaction, together with the determination of methanol liberated.[23] The color reaction for o-quinone determination was conducted by the use of Schaller's method, employing 3-methyl-2-benzothiazolinone hydrazone hydrochloride. The calibration curve was prepared using 3-methoxy-5-*tert*-butylbenzoquinone-(1,2).

The results of the estimation of methanol liberated and o-quinone formed from the methoxybenzoic acids by the action of laccase are shown in Table 2. The amounts of methanol and o-quinone are expressed in mole percent. It was observed that less than 10% of the methoxyl groups from vanillic acid and about twice that in the case of syringic acid were split during 3 hr incubation with laccase. (The experimental conditions are somewhat different in Tables 1 and 2, so the values are not identical on the same substrate.) Although syringic acid possesses two methoxyl groups, whether both of them are involved in demethylation is as yet unclear. Also, the ratio (B/A in Table 2) of methanol and o-quinone is higher with syringic acid than vanillic acid.

During the incubations with laccase, an evident difference was observed between the behavior of vanillic acid and syringic acid. Just after the addition of the enzyme, the solution of the former became brown and cloudy, but the latter solution turned to reddish brown without any turbidity. Apparently, the presence of two methoxyl groups

TABLE 2

Formation of *O*-Quinone Products and Methanol on Laccase-
Catalyzed Oxidation of Syringic and Vanillic Acids

Substrate	Orthoquinone (%) A	CH_3OH (%) B	B/A
Syringic acid	18	17	0.94
	15	15	1.00
	18	15	0.83
	13	13	1.00
Vanillic acid	9	7	0.77
	8	6	0.75
	8	6	0.75
	8	5	0.62

From Ishihara, T. and Ishihara, M., *Mokuzai Gakkaishi*, 21,
324, 1975. With permission.

at the position ortho to the phenolic hydroxyl group prevents polymerization of syringic acid.

B. Side Chain Splitting

Splitting of the side chains of model compounds and liberation of aliphatic aldehyde on laccase oxidation were shown by Kirk, Harkin, and Cowling.[7] In our experiments, using vanillyl alcohol as a substrate, formaldehyde was liberated, in addition to methanol, and this is supposed to be due to splitting of the side chain (Table 1).[20] In the case of vanillic acid and syringic acid, the *p*-quinone and *o*-quinone products were obtained and their structures elucidated.[24-26] The formation of the *p*-quinone is obviously a result of splitting of the carboxyl side chain. Liberation of carbon dioxide was demonstrated in manometric experiments.[24]

C. Oxidation Products

Several monomers and dimers have been reported as degradation products of lignin by various wood-rotting fungi. Some lignin-related aromatics have also been reported to be degraded by the organisms. Among the degradation products, vanillic acid has often been found. However, this compound is not an end product, and further microbiological alteration can be expected. Although Cartwright and Smith[19] reported the degradation of vanillic acid by bacterial enzymes, its degradation by enzymes of wood-rotting fungi has not yet been clarified (Volume II, Chapter 1).

In an attempt to know the mechanism of degradation of vanillic acid by a wood-rotting fungus, *Coriolus versicolor,* crude enzyme of the organism was applied to vanillic acid, and the reaction products (degradation or polymerization) were examined. From the reaction products, a yellow crystalline substance having mp 225 to 226° was isolated.[24,25] The crude enzyme of the fungus was incubated with 4 g vanillic acid in 5.5 *ℓ* of buffer solution (pH 4.0) at 30° for 2 hr and the mixture was kept overnight at room temperature. From the ethyl acetate extract, about 200 mg of the yellow crystals were obtained. The structure of this compound was elucidated by examining spectral and analytical data, preparation of derivatives, and characterizing products formed on hydrolysis (Figure 5). All these data support the structure 2-methoxy-6-(2′-methoxy-4′-carboxyphenoxy)-benzoquinone-(1,4) (Figures 5,II) for this compound. It cannot be said to be a "degradation" product of vanillic acid.

Fukuzumi et al.[27] observed the formation of 2-methoxy-*p*-benzoquinone from van-

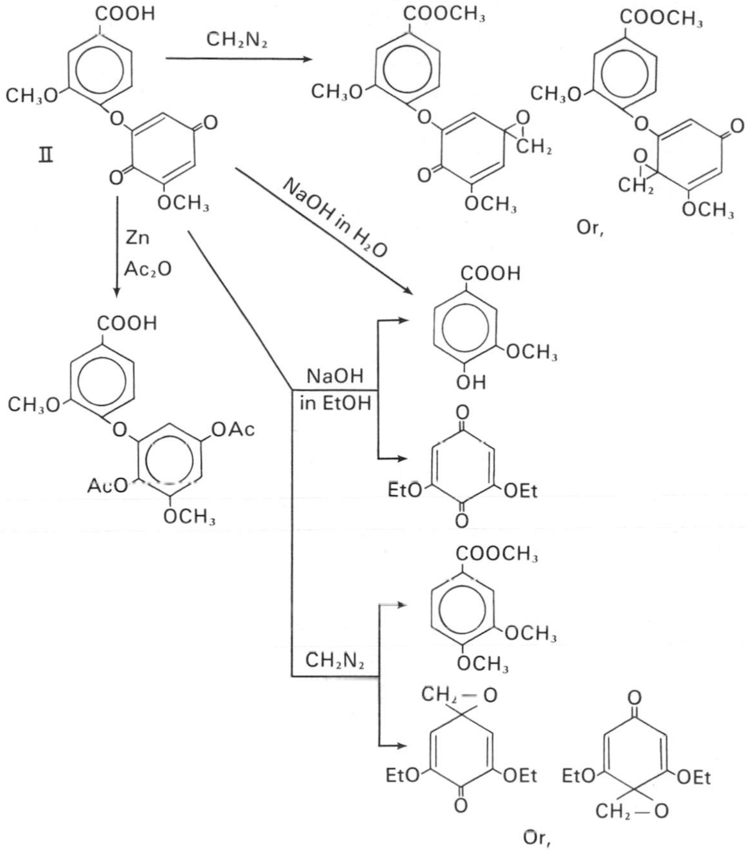

FIGURE 5. Structure and characterizing reactions of compound II. (Reproduced from Ishihara, T. and Miyazaki, M., *Mokuzai Gakkaishi*, 16, 185, 1970. With permission.)

illic acid by enzymes of the wood-rotting fungi, *Polystictus sanguineus* and *Poria subacida*, and discussed the mechanism of carbon dioxide liberation from vanillic acid by the action of laccase. The compound (II) is considered to be formed through a coupling reaction between a phenoxy radical of vanillic acid and a radical (at 5 position) of another molecule of vanillic acid, both formed by the action of laccase, as the first step. Rearomatization of the phenolic unit, followed by further oxidation and decarboxylation of the phenolic unit, affords the quinonoid substance (II).

Musso et al.[28] studied the reaction of *p*-benzoquinone and phenol and reported the nonenzymatic formation of phenoxy linkage at 2 position of the quinone.

Freudenberg and Chen[29] reported the presence of diphenyl ether-type carboxylic acids among the products of methylation and permanganate oxidation of lignin. Since lignin is a polymer formed from *p*-hydroxycinnamyl alcohols by enzymic abstraction of hydrogen from the phenolic hydroxyl group of the alcohols, there is a similarity in the mechanism of formation of such ether linkages in lignin and in compound (II).

Pew and Connors[30] obtained a "dioxepin" from lignin model compounds by the action of peroxidase in the presence of H_2O_2. This substance also contains a quinone moiety formed by the enzyme, but in this case, one carbonyl group of the quinone is opened to form another linkage. In this connection, the type of linkage is somewhat different from that in compound (II).

As shown in Table 2, on laccase oxidation of syringic acid, methanol was found to

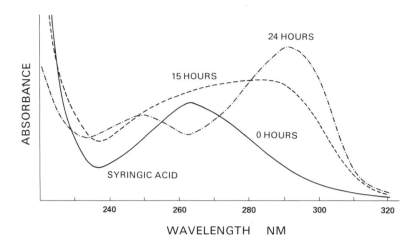

FIGURE 6.   UV spectra taken during reaction of syringic acid with laccase/O_2. (Reproduced from Ishihara, T. and Ishihara, M., *Mokuzai Gakkaishi,* 22, 371, 1976. With permission.)

be formed as a result of demethylation. The laccase oxidation products of syringic acid have been studied.[26] Syringic acid (400 mg) was incubated with 20 mg of laccase at pH 4.5 at 30°. After 24 hr, a color reaction with diazotized sulfanilic acid disappeared completely and appreciable change was observed in the UV spectrum (Figure 6). Examination by TLC indicated two compounds, yellow and red. The two substances were separated and purified by means of column chromatography and recrystallization. The yellow one was identified as 2,6-dimethoxy-*p*-benzoquinone (I) by comparison with an authentic sample (TLC, UV, VIS, m.p.). The structure of the red compound was elucidated by spectral (UV, VIS, IR, NMR, MS) and analytical data and by preparation of derivatives and was determined to be 6-methoxy-4-(2′,6′-dimethoxy-4′-carboxyphenoxy)-benzoquinone-(1,2) (Figure 7, III.). The amount of these two compounds in the reaction mixture was esimated by the use of the evalues of the compounds at two different wave lengths (Table 3) and was found to be about 2:1 in molar ratio (1:1 by weight). A compound having a similar structure as (III) was obtained from another model compound by peroxidase (with H_2O_2).[17] This type of compound again is not a true degradation product, just as (II) was not. These compounds are dimers of the starting materials linked together by ether bonds, although decarboxylation (in the case of II) and demethylation (in the case of III) occurred.

In the case of syringic acid oxidation (Figure 8) the molar ratio of the *p*-quinone and *o*-quinone products was found to be 2:1 and a corresponding amount of methanol was liberated. As has been described by many authors, the reaction is initiated by abstraction of hydrogen from the phenolic hydroxyl group, and the resulting phenoxy radical and its mesomeric forms proceed to couple. Some of the mesomeric structures are favored by the influence of the substituents (side chain and methoxyl), and therefore the frequency of the coupled intermediate radicals varies according to the compounds used. This may be one of the reasons for the varied ratio of *p*- and *o*-quinone products, depending on the substrate used. Another reason may be the stability of the products. Usually *o*-quinones are unstable, but some that are substituted, including *o*-quinone (III), are rather stable.

Concerning demethylation in this experiment, 1 mol of methanol was found to be liberated from 4 mol of the substrate. It is of interest to know whether one or both methoxyl groups in a given molecule are involved in demethylation. Judging from the structure of (III), the former seems most likely.

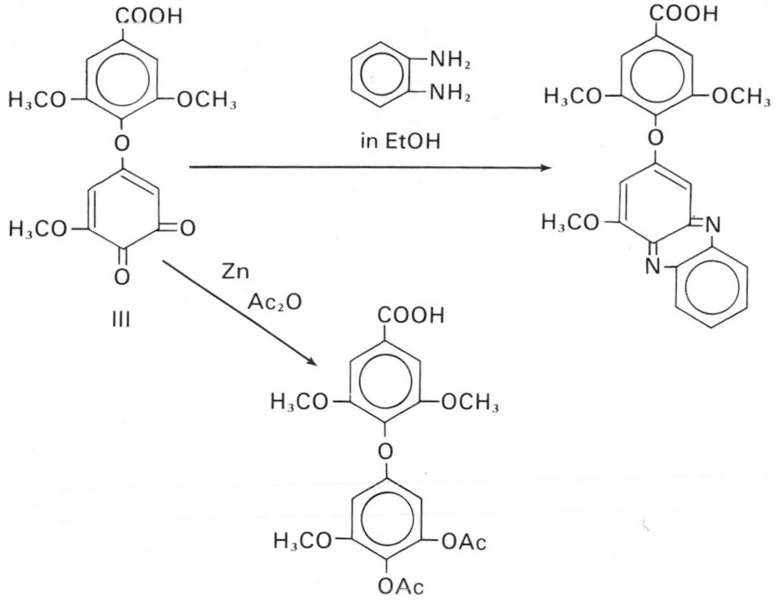

FIGURE 7.. Structure and characterizing reactions of compound III.

TABLE 3

UV Spectra of the Reaction Mixture Containing Yellow and Red
Compounds of Syringic Acid and Laccase

UV property	249 nm	290 nm
O.D. of reaction mixture (measured after 50-fold dilution)	0.58	1.00
ε of yellow compound (in 50% aqueous ethanol)	0.21×10^4	1.31×10^4 (max)
ε of red compound (in 50% aqueous ethanol)	1.47×10^4 (max)	0.54×10^4

Note: ε of syringic acid at 265 nm = 0.87×10^4.

From Ishihara, T. and Ishihara, M., *Mokuzai Gakkaishi,* 22, 374, 1976. With permission.

V. POSSIBILITY OF REDUCTION OF OXIDATION PRODUCTS

A. Cellobiose:Quinone Oxidoreductase

Westermark and Eriksson discovered a very interesting enzyme, cellobiose:quinone oxidoreductase, from wood-rotting fungi.[10] This enzyme catalyzes the reduction of a quinone and simultaneous oxidation of cellobiose and shows one possibility for an interaction between the degradation of lignin and that of carbohydrate. The substrate specificity for the quinone was found to be rather wide, so it is quite probable that these quinones formed from lignin and model compounds mentioned above by the action of laccase can be substrates. The reduction of an o-quinone yields a catechol group and that of a p-quinone, a hydroquinone group.

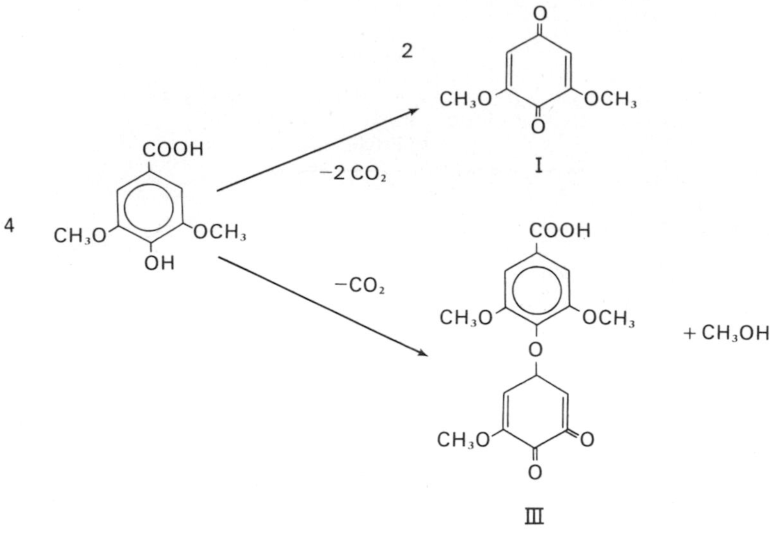

FIGURE 8.   Laccase-catalyzed reactions of syringic acid. (From Ishihara, T. and Ishihara, M., *Mokuzai Gaikkaishi,* 22, 371, 1976. With permission.)

B. Compatibility of Oxidation and Reduction

Despite its speculative nature, I want to mention something concerning the possible compatibility of oxidation and reduction in lignin biodegradation by wood-rotting fungi. The formation of *p*-quinones and *o*-quinones from lignin and lignin models by laccase was shown above, and the possibility of the reduction of quinones to catechol or hydroquinone compounds by cellobiose:quinone oxidoreductase has also been mentioned. Westermark and Eriksson[10] have shown that in the presence of laccase their enzyme reduces quinones if cellobiose is present. According to my speculation, therefore, lignin units carrying a free phenolic hydroxyl group can be oxidized by a phenol-oxidizing enzyme, and as a result of further reaction, lose methoxyl or side chains, yielding *o*-quinone and *p*-quinone. Next, these quinones can be reduced to catechol or hydroquinone moieties by cellobiose:quinone oxidoreductase or a similar enzyme. A question naturally arises of how both oxidation and reduction – obviously competing reactions – can be physically compatible.

There are two possibilities: (1) the oxidizing enzyme and the reducing enzyme (of course, in the case of oxidation, the counterpart is reduced and vice versa, so the term is used from the lignin side) appear at different growth stages. Formation of a double ring on the plate culture observed by Westermark and Eriksson[10] may show this possibility. (2) Differences in location or distribution of the two enzymes in the organism cells is another possibility. For instance, one may be excreted from the cells, while the other is cell-bound, but these are only speculations and no experimental evidence is yet available. It is said to be unlikely that lignin is first decomposed to monomers and the monomers disintegrated, as is the case with polysaccharides, but rather that lignin is decomposed "exo-wise", with retention of its high molecular backbone. If this is the case, penetration of cell walls and assimilation in the cell of monomers is not so plausible. Unfortunately, cellobiose:quinone oxidoreductase is reported to be an extracellular enzyme just as laccase, so stepwise action of the two enzymes in different locations is not likely in this case. There may be other enzymes, however, which work to reduce quinones in another location in the organisms, although we have no experimental evidence. At any rate, most white rot fungi decompose lignin and at the same time excrete laccase, so there should be some reductive mechanism which works to prevent polymerization.[10]

C. Aromatic Ring-Splitting of Reduced Quinones

The biodegradation of lignin cannot be achieved without breakdown of the aromatic ring. Ring-splitting of aromatic compounds has been extensively studied using *Pseudomonas,* and *o*-diphenols such as catechol and protocatechuic acid and *p*-diphenols such as homogentisic acid are known to be degraded to aliphatic acids.[32] Many aromatic compounds are decomposed via protocatechuic acid or catechol. Although many studies of aromatic ring cleavage have been conducted, the organisms used were mostly pseudomonads, and examples with wood-rotting fungi are very few. Two phenolic hydroxyl groups ortho or para to each other are necessary. In proto-lignin such groups are negligible. As mentioned in the above sections, the stepwise action of laccase and cellobiose:quinone oxidoreductase can afford such diphenols via *p*-quinone or *o*-quinone.

As seen in the structure of compound (III), the *o*-quinone from the syringyl group still carries one methoxyl group, so it is of interest to know whether such methoxy-*o*-quinone or methoxycatechol is decomposed or not. When pyrogallol mono-methylether (methoxycatechol) was added to a shaking culture of *Pleurotus ostreatus,* the color of the medium turned brown immediately by the action of laccase, but after 10 days or so, the medium became light in color, indicating the possibility of decomposition of the methoxydiphenol.[33]

VI. SUBSTRATES OF LACCASE

A. Amount of Nonetherified Phenolic Hydroxyl Groups in Lignin

Since the action of laccase is to abstract hydrogen from phenolic hydroxyl groups, etherified phenol naturally cannot be the substrate of laccase. The estimation of free phenolic hydroxyl groups in lignin has been attempted by several methods, and a value of about 20 phenolic units per 100 C_9 units was found for spruce MWL.[34] Accordingly, the remaining 80 units per 100 C_9 units cannot be substrates of laccase. It is evident, therefore, that laccase is not capable of oxidizing many of the units in lignin.

B. Enzymes which Liberate Phenolic Groups

As mentioned above, although laccase is considered to play some role in lignin biodegradation, its substrate is confined to units with free phenolic groups. Kirk, Harkin, and Cowling[7] stated that the glyceraldehyde-2-arylether moieties liberated from aromatic nuclei by the action of laccase should be easily hydrolyzed to yield new free phenolic groups. This, however, is not a direct action of laccase, so some ether-splitting enzymes would be needed.

VII. CONCLUSION

The role of laccase in lignin biodegradation has attracted the interest of many lignin chemists, plant pathologists, and microbiologists. Some have concluded that laccase contributes to the biodegradation of the polymer, and others have concluded that the enzyme plays no role or that it just detoxifies phenols and some other enzymes are concerned with the biodegradation (see Volume I, Chapter 10 and Volume II, Chapter 1).

As shown here in the experiments with MWL, the enzyme works to cause both polymerization and depolymerization. In nature where lignin is always present with carbohydrates, however, polymerization may be suppressed.

The demethylating action of laccase is also proved here, both in lignin and models, and it is thought to be an initial step of biodegradation. The *o*-quinone moieties left after demethylation can be reduced to catechol compounds, which are known to be substrates of the dioxygenases responsible for ring cleavage.

Side chain elimination is another important reaction caused by laccase and is a direct contribution in biodegradation. The *p*-quinone moieties left after side chain elimination can be reduced to hydroquinone-type compounds, which are also known to suffer oxygenase-catalyzed ring cleavage. Demethylation and side chain elimination mentioned above are, of course, initiated by hydrogen abstraction from phenolic hydroxyl groups, followed by various coupling of mesomeric forms of phenoxy radicals.

In conclusion, laccase plays an important role, but it cannot act alone. Other enzymes which are capable of splitting aryl-alkyl ether linkages must be sought for elucidation of the biodegradation of this complex polymer.

VIII. SUMMARY

Although most white rot fungi — potent lignin destroyers in nature — secrete laccase, the role of the enzyme in lignin biodegradation is still not clear. When treated with this enzyme in the presence of O_2, milled-wood lignin underwent an increase in molecular weight, but at the same time a small amount of 2,6-dimethoxybenzoquinone-(1,4) was liberated. When the enzyme was applied to syringic acid, the substrate was completely converted into orthoquinone and paraquinone derivatives as a result of demethylation and side chain splitting. These products presumably can be converted into the corresponding catechol and hydroquinone compounds, respectively, by cellobiose:quinone-oxidoreductase. Since aromatic ring-splitting by oxygenases usually occurs via catechol compounds or hydroquinone compounds, there may be a possibility of ring-splitting in the above-mentioned types of lignin degradation products formed by the action of laccase/O_2. If such a sequence of reactions is present in nature, the location of the enzymes (laccase, quinone reductase, and ring-splitting enzymes) in fungal cells or secretion of the enzymes in various growth stages must be balanced. A mechanism for preventing polymerization of lignin by laccase should be present in nature, but the mechanism is not yet clear. Another problem is that etherified phenols cannot be substrates for laccase, and 80% of the phenolic groups in lignin are etherified. Since laccase is not capable of yielding phenolic groups, it is clear that other enzymes which can split ether groups must play a big role in lignin biodegradation.

REFERENCES

1. **Bavendamm, W.,** Über des Vorkommen und den Nachweis von Oxydasen bei holzzerstorenden Pilzen, *Z. Pflanzenkr.*, 38, 257, 1928.
2. **Kirk, T. K. and Kelman, A.,** Lignin degradation as related to the phenol oxidases of selected wood-decaying basidiomycetes, *Phytopathology*, 55, 739, 1965.
3. **Sundman, V. and Näse, L.,** A simple plate test for direct visualization of biological lignin degradation, *Pap. Puu*, 53, 67, 1971.
4. **Haider, V. K., Lim, S., and Flaig, W.,** Experimente und Theorien über den Ligninabbau bei den Weissfaule des Holzes und bei der Verrottung Pflanzlicher Substanz in Boden, *Holzforschung*, 18, 81, 1964.
5. **Konishi, K. and Inoue, Y.,** Decomposition of lignin by *Coriolus versicolor*. I, II, and III, *Mokuzai Gakkaishi*, 17, 209 and 214 and 255, 1971.
6. **Kirk, T. K., Harkin, J. M., and Cowling, E. B.,** Oxidation of guaiacyl- and veratryl-glycol-β-guaiacyl ether by *Polyporus versicolor* and *Stereum frustulatum*, *Biochim. Biophys. Acta*, 165, 134, 1968.

7. **Kirk, T. K., Hardin, J. M., and Cowling, E. B.,** Degradation of the lignin model compound syrin-gylglycol-β-guaiacyl ether by *Polyporus versicolor* and *Stereum frustulatum, Biochim. Biophys. Acta,* 165, 145, 1968.
8. **Higuchi, T.,** Biochemical study of wood rotting fungi. I. Studies on the enzymes which cause Bavendamm's reaction, *J. Jpn. For. Soc.,* 35, 77, 1953.
9. **Hiroi, T. and Eriksson, K.-E.,** Microbiological degradation of lignin. I. Influence of cellulose on the degradation of lignin by the white-rot fungus *Pleurotus ostreatus, Sven. Papperstidn.,* 79, 157, 1976.
10. **Westermark, U. and Eriksson, K.-E.,** Carbohydrate dependent enzymic quinone reduction during lignin degradation, *Acta Chem. Scand.,* B28, 204, 1974; Cellobiose:quinone oxidoreductase, a new wood-degrading enzyme from white-rot fungi, *Acta Chem. Scand.,* B28, 209, 1974.
11. **Ishihara, T. and Miyazaki, M.,** Oxidation of milled wood lignin by fungal laccase, *Mokuzai Gakkaishi,* 18, 415, 1972.
12. **Freudenberg, K., Harkin, J. M., Reichert, M., and Fukuzumi, T.,** Die an der Verholzung beteiligten Enzyme, Die Dehydrierung des Sinapinalkohols, *Chem. Ber.,* 91, 581, 1958.
13. **Kirk, T. K. and Adler, E.,** Methoxy-deficient structural elements in lignin of sweetgum decayed by a brown-rot fungus, *Acta Chem. Scand.,* 24, 3379, 1970.
14. **Farmer, V. C., Henderson, M. E. K., and Russel, J. D.,** Reduction of certain aromatic acids to aldehydes and alcohols by *Polystictus versicolor, Biochim. Biophys. Acta,* 35, 203, 1959.
15. **Minami, K., Tsuchiya, M., and Fukuzumi, T.,** Metabolic products from aromatic compounds by wood-rotting fungus *Polystictus sanguineus.* IV. Culturing conditions for reduction and demethylation of veratric acid, *Mokuzai Gakkaishi,* 11, 179, 1965.
16. **Trojanowski, J., Leonowicz, A., and Hampel, B.,** Exoenzymes in fungi degrading lignin. II. Demethoxylation of lignin and vanillic acid, *Acta Microbiol. Pol.,* 15, 17, 1966.
17. **Connors, W. J., Aeyers, J. S., Sarkanen, K. V., and Gratzl, J. S.,** Demethylation and ortho-quinone formation in enzymic dehydrogenation of lignin model phenols, *Tappi,* 54, 1284, 1971.
18. **Fukuzumi, T., Takatsuka, H., and Minami, K.,** Enzymic degradation of lignin. V. The effect of NADH on the enzymic cleavage of aryl-alkyl ether bond in veratrylglycerol-β-guaiacyl ether as lignin model compound, *Arch. Biochem. Biophys.,* 129, 396, 1969.
19. **Cartwright, N. J. and Smith, A. R. W.,** Bacterial attack on phenolic ethers, an enzyme system demethylating vanillic acid, *Biochem. J.,* 102, 826, 1967.
20. **Ishihara, T. and Miyazaki, M.,** Demethylation of lignin and lignin models by fungal laccase, *Mokuzai Gakkaishi,* 20, 39, 1974.
21. **Flaig, W.,** Effects of microorganisms in the transformation of lignin to humic substances, *Geochim. Cosmochim. Acta,* 28, 1523, 1964.
22. **Trojanowski, J. and Leonowicz, A.,** Investigation on degradation of lignin by *Pholiota mutabilis, Ann. Univ. Mariae Curie-Sklodowska Sect. C,* 18, 441, 1963.
23. **Ishihara, T. and Ishihara, M.,** Estimation of ortho-quinone structure and methanol formed from lignin models by fungal laccase, *Mokuzai Gakkaishi,* 21, 323, 1975.
24. **Ishihara, T. and Miyazaki, M.,** Formation of a new yellow crystalline compound from vanillic acid by the action of crude enzyme of *Polyporus versicolor, Mokuzai Gakkaishi,* 16, 181, 1970.
25. **Ishihara, T., and Miyazaki, M.,** The structure of a new quinone derivative formed from vanillic acid by the action of crude enzyme of *Polyporus versicolor, Mokuzai Gakkaishi,* 16, 185, 1970.
26. **Ishihara, T. and Ishihara, M.,** Oxidation of syringic acid by fungal laccase, *Mokuzai Gakkaishi,* 22, 371, 1976.
27. **Fukuzumi, T., Uraushihara, S., Oohashi, T. and Shibamoto, T.,** Enzymic degradation of lignin. III. Oxidation accompanying carbon dioxide liberation of vanillic acid, vanillyl formic acid and guaiacyl pyruvic acid by enzyme of *Polystictus sanguineus* and *Poria subacida, Mokuzai Gakkaishi,* 10, 242, 1964.
28. **Musso, H., Gizycki, U. V., Zahorszky, U. I., and Borman, D.,** Formation of hydroxy aryl quinones by the addition of phenols to quinones, *Liebigs Ann. Chem.,* 676, 10, 1964.
29. **Freudenberg, K. and Chen, C. L.,** Weitere Oxydationsprodukte des Fichtenlignins, *Chem. Ber.,* 100, 3683, 1967.
30. **Pew, J. C. and Connors, W. J.,** New structures from the enzymic dehydrogenation of lignin model p-hydroxy-α-carbinols, *J. Org. Chem.,* 34, 580, 1969.
31. **Shimazono, H. and Nord, F. F.,** Transformation of anisic acid and methyl anisate by the mold *Polyporus versicolor, Arch. Biochem. Biophys.,* 87, 140, 1960.
32. **Hayaishi, O.,** *Oxygenases,* Academic Press, New York, 1962, 1.
33. **Ishihara, T.,** unpublished, 1977.
34. **Adler, E. and Hernestam, S.,** Estimation of phenolic hydroxyl groups in lignin. I. Periodate oxidation of guaiacol compounds, *Acta Chem. Scand.,* 9, 319, 1955.

Chapter 3

ENZYMATIC TRANSFORMATIONS OF LIGNIN

Philip L. Hall, Wolfgang G. Glasser, and Stephen W. Drew

TABLE OF CONTENTS

I. INTRODUCTION

The structure of lignin is complex. That in itself constitutes a formidable challenge to the investigator who would attempt to elucidate lignin biodegradation, and it is the challenge upon which primary consideration is focused in this chapter. In nature lignin is transformed ultimately to carbon dioxide, microbial biomass, and refractory soil components such as "humic acids". But what structural transformations occur along the way toward these ultimate products, especially early in the biodegradation pathways while the lignin still retains a complex polymeric structure? Those early transformations are what we seek to monitor analytically and characterize structurally.

The structure suggested in Figure 1[1] is for a softwood or conifer lignin in its native state. It is an attempt to provide, as accurately as possible given current knowledge of the problem, a representation of all structural variations of the C_9 repeating units in lignin with a statistically correct indication of their relative frequencies of occurrence. The lignin is represented as if it were one continuous molecule, without finite molecular weight. Although in natural wood the lignin may well have essentially infinite molecular weight, isolated lignins, such as Björkman's "milled wood lignin,"[2] or lignins recovered from commercial pulping operations such as kraft lignins or lignin sulfonates are broadly polydisperse materials with molecular weight ranges encompasssing everything from C_9 monomers and dimers to oligomers and polymers of undetermined degree of polymerization. More to the point of the present discussion, any isolated lignin incorporates the structural consequences of operations performed during the isolation process. Extensive ball milling and extraction with organic solvents results in minimal structural transformation (other than molecular weight reduction), so milled wood lignin is not misrepresented by Figure 1. Kraft pulping, on the other hand, results in certain structural transformations which will be considered very briefly here, since kraft lignin has been employed as the substrate in the enzymatic transformation studies discussed later in this chapter.

The chemistry and technology of alkaline delignification has an extensive literature which has been reviewed by Marton.[3] The predominant delignification reactions which take place during a kraft process cook of wood chips with hot alkaline sodium sulfide are shown in Figure 2. The hot alkali initially brings about a dehydration of free phenolic phenylpropanoid units to give highly electrophilic quinonemethide intermediates. These are attacked at the α-position by SH^- nucleophiles. The newly introduced thiol groups then bring about intramolecular nucleophilic displacement of phenoxide ions by attack at the adjacent β-position, thus accomplishing depolymerization and the creation of additional free phenolic groups, which in turn give rise to more quinonemethide intermediates, and so forth. Inasmuch as the predominant linkage between C_9 units in lignin is of the β-aryl ether type, depolymerization is extensive. If one reexamines Figure 1, then, to assess structural changes brought about by kraft pulping, the principal transformation is extensive diminution of glycerol-β-aryl ether linkages with concomitant increases in free phenolic hydroxyl groups. There is also some incorporation of thiol groups, loss of hydroxyl groups from α-positions on the side-chains, and (by side reactions not presented here) loss of hydroxymethyl groups from the γ-position in the form of formaldehyde.

In what follows, experiments with fungal biotransformations of kraft lignin by *Coriolus versicolor* are described, as are enzymatic transformation experiments employing cell-free soluble extracellular proteins produced in submerged-culture fermentations of *C. versicolor*. Analysis data on biodegraded kraft lignin are then presented and discussed.

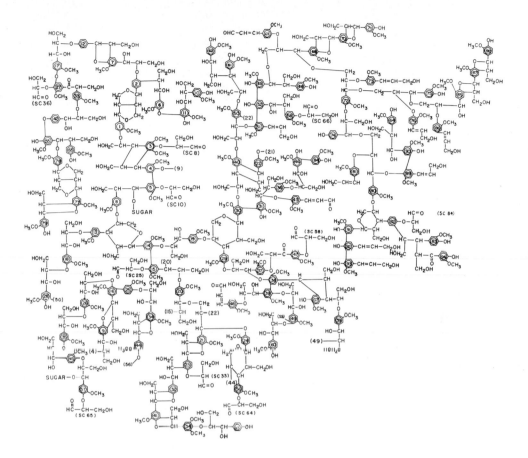

FIGURE 1.   Structural representation of a simulated softwood lignin.[1]

II. FERMENTATIONS AND ENZYME DIGESTIONS

A. *Coriolus versicolor* Fermentations

Although it has not been an objective of the work described here to screen numbers of wood-degrading microorganisms for their ability to metabolize or transform lignin, *C. versicolor* was selected for study from among three well-known strains of white-rot fungi on the basis of its apparently superior ability to metabolize kraft lignin in submerged culture. Data supporting this choice are presented in Table 1. The criterion of interest in these experiments was release of radioactive $^{14}CO_2$ from uniformly ^{14}C-labeled kraft lignin, prepared by the method of Crawford et al.[4] from loblolly pine seedlings, which had been allowed biosynthetically to transform U^{14}C-phenylalanine into lignin.

Also indicated in Table 1 are some effects of variations in the carbohydrate carbon source supplied along with the kraft lignin in these fermentations. Although all three white-rot fungi metabolize lignin at rates and to extents which reflect the nature of the carbohydrate cosubstrate (and fail to metabolize lignin at all in the absence of carbohydrate), *C. versicolor* is unique among the three in its response to Avicel® (a highly microcrystalline cellulose, which presumably is more difficult for the organism to metabolize than phosphoric acid swollen cellulose [PSC] or cellobiose). The data are not included in Table 1, but further experiments have shown that a cellulose product similar to Avicel®, Sigma Cell® (Sigma Chemical Company), and to a lesser extent, soluble starch, also stimulate lignin metabolism by *C. versicolor*.

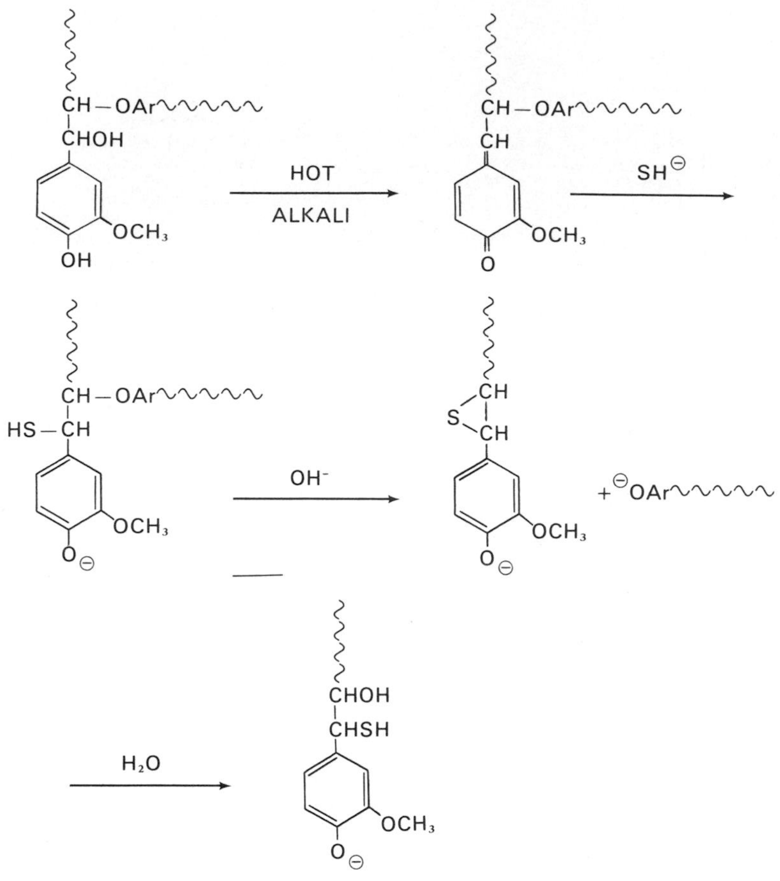

FIGURE 2.   Principal lignin depolymerization mechanism in kraft process wood pulping.

Conditions conducive to lignin metabolism by *C. versicolor* are distinctly not optimal for growth of the organism. In fact, little, if any growth* at all, is observed in fermentations of *C. versicolor* on kraft lignin plus Avicel® or Sigma Cell®. On the other hand, with PSC or cellobiose, growth rates, while low compared to the other white-rot fungi tested, are appreciable. Finally, when glucose is present in high concentration (>10g/ℓ), *C. versicolor* grows quite rapidly, but one can observe no measurable metabolism of lignin to CO_2. It seems not unlikely that a form of catabolite repression or other regulatory pattern controlling growth and lignin metabolism may be the result of rapid and facile carbohydrate metabolism (Volume II, Chapter 4). In any case, these findings emphasize what has often been observed in other investigations reported in the literature, that in white-rot fungi lignin metabolism is very much dependent upon the presence of carbohydrate.

Table 2 summarizes a series of large-scale submerged culture fermentations of *C. versicolor* carried out for the production of extracellular enzymes with lignin-transforming activities and for the production of biotransformed lignin preparations to be

---

*   In these fermentations with suspended lignin and cellulose present, it is not possible to monitor growth directly by dry cell weight. Thus, a method has been adopted for use in this work in which glucosamine, released by hydrolysis from the chitin of fungal cell walls, is determined and related by a linear calibration curve for the organism in question to dry cell weight.[5]

TABLE 1

Lignin Degradation by Three White-Rot Fungi with Variations in Carbohydrate Co-substrate

Organism	Carbohydrate	% of total radioactivity released as $^{14}CO_2$ after 15 days	% of total radioactivity released as $^{14}CO_2$ after 21 days
C. versicolor	Cellobiose	1.94	2.72
S. pulverulentum	Cellobiose	1.49	—
P. chrysosporium	Cellobiose	1.24	—
C. versicolor	Avicel®	5.91	8.87
S. pulverulentum	Avicel®	1.34	—
P. chrysosporium	Avicel®	1.37	—
C. versicolor	PSC[a]	2.94	4.08
S. pulverulentum	PSC	2.14	—
P. chrysosporium	PSC	1.98	—

Note: Submerged cultures in shake-flasks incubated at 28°C. Medium contained (in units of g/ℓ) lignin (1.0), carbohydrate (10.0), $(NH_4)H_2PO_4$ (2.0), K_2HPO_4 (0.6), KH_2PO_4 (0.4), $MgSO_4 \cdot 7H_2O$ (0.5), $CaCl_2 \cdot 2H_2O$ (0.75), plus trace salts ferrous citrate, $MnSO_4$, $ZnSO_4$, $CoCl_2$, $CuSO_4$.

[a] Phosphoric acid swollen cellulose.

structurally characterized. All of these fermentations were of about one-week duration and 8- to 10-ℓ volume in an agitated 14-ℓ fermenter at 28°C. Each was inoculated with a seed culture grown on malt broth, and each contained 1% by weight reprecipitated and then autoclave-sterilized kraft lignin (Indulin ATR-C1) and basal medium in addition to the supplemental carbon source noted. pH was controlled between 4.5 and 5.5. Foaming was controlled in fermentations CE-6 through CE-10 with Dow-Corning silicone-base antifoam type AF, and in CE-11 through CE-13 with Dow-Corning type FG-10 antifoam. CE-16, with only 0.1% lignin added, required no antifoam. In CE-6, CE-7, and CE-9, xylidine (2,5-dimethylaniline) was added to induce the production of laccases.[8] Each addition (1, 3, or 14 times at equally spaced intervals during the course of CE-6, 7, and 9, respectively) provided enough xylidine to give a net medium concentration of 2×10^{-4} M. In CE-13 and CE-16 no cellulosic material at all was present. Instead, glucose was added frequently throughout the course of the fermentation (41 mg every 84 min) in an attempt to maintain a very low steady-state concentration of glucose, such as might otherwise be provided by the action of β-glucosidases on cellulosic material. Furthermore, in CE-16, C. versicolor cell density was allowed to build up to many times inoculum level by growth on 1% glucose prior to introduction of the lignin and glucose limitation. It was hoped that this would allow production of extracellular protein in greater quantity than previous fermentations had provided.

Fermentations were harvested by a series of screenings on U.S. Standard Sieves to collect bulk fractions, followed by centrifugation or filtration on glass fiber filters to collect suspended fine solids. The cell-free, Millipore®-filtered fermentation broth was subjected to ultrafiltration on a membrane filter with nominal retention above mol wt 10,000. The protein-containing retentate was concentrated to 50 mℓ and dialyzed.

B. Extracellular Enzymes

White-rot fungi have long been known to produce extracellular cellulolytic enzymes[7]

TABLE 2

Summary Data on *C. versicolor* Fermentations

Fermentation identity	Carbohydrate Supplement	Xylidine induced	Protein recovered (mg)[a]	Phenol-oxidizing enzyme activity[b]	Solids recovered[c] (g)	Solids added initially (g lignin/g CH_2O)
CE-6	0.1% PSC	1X	81	1.67	—	100/10
CE-7	0.1% PSC	3X	82	0.58	15	100/10
CE-9	0.1% PSC	14X	80	7.80	68	100/10
CE-10	1.0% Avicel®	—	109	0.024	92	100/100
CE-11	0.1% Avicel®	—	78	1.13	54	100/10
CE-12	0.1% PSC	—	136	0.59	59	100/10
CE-13	Glucose	—	79	0.34	55	100/d
CE-16	Glucose	—	19	16.77	—	8/d

[a] Protein concentration determined by method of Bradford[6] using Coomassie Blue G-250. This method is much more sensitive and more reliable in the presence of phenolic contaminants than the standard Lowry assay for protein.

[b] Assayed spectrophotometrically at 480 nm with guaiacol as substrate in water at 25°C. Specific activity units are milligrams guaiacol oxidized per minute per mg protein.

[c] Weights refer to total lignin plus cellulose. Weights of fungal mycelia recovered are not included.

[d] Glucose added periodically in small portions. See text.

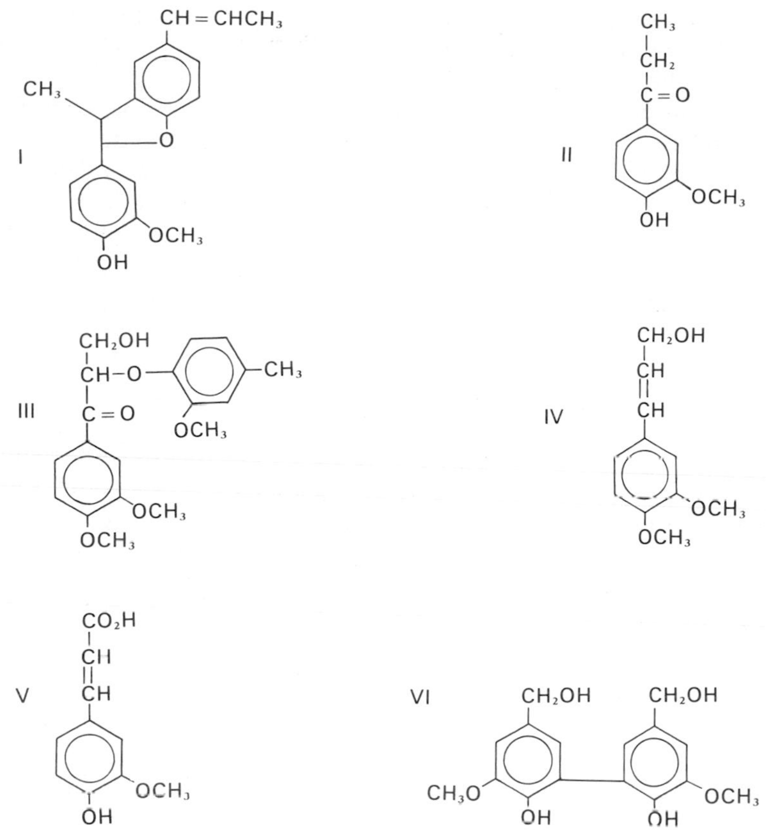

FIGURE 3. Lignin model compounds used in lignolytic enzyme assays.

and phenol-oxidizing enzymes,[8] and it would seem reasonable to expect (though the assumption has been questioned) to find lignin-transforming enzymes as well among the soluble extracellular proteins of these lignin-degrading basidiomycetes. Phenol-oxidizing enzymes seem in some not-yet-understood way to be essential to lignin degradation,[9] and another enzyme which too may be involved, cellobiose:quinone oxidoreductase, has been identified.[10]

Table 2 provides data on the amounts of soluble protein recovered from various *C. versicolor* fermentations which, as was stated earlier, were undertaken in large part for the purpose of producing proteins for enzymological studies. One is struck at once by how little protein there appears to be in these extracellular broths. The highest soluble protein concentration found in these fermentations, before concentration of the broth, was only 13.6 $\mu g/m\ell$ in CE-12. CE-16 contained about 2.3 $\mu g/m\ell$. With such small amounts of protein to work with, the investigation has concentrated not on protein fractionation and purification, but rather on examination of these extracellular protein mixtures for the enzymatic activities they contain, and for protein variety by electrophoresis and isoelectric focusing.

The enzyme activities looked for in the extracellular ultrafiltration retentates include the following: phenol-oxidizing enzymes, cellobiose:quinone oxidoreductase, *o*-pyrocatechuic acid carboxylyase, protocatechuic acid 3,4-dioxygenase, and protease enzymes. As indicators of possible lignin-transforming enzyme activity, the six lignin model compounds in Figure 3 were employed as substrates with each of the protein fractions of Table 2.

Phenol-oxidizing enzyme activity was detected in all of the retentates, as indicated in Table 2. Cellobiose:quinone oxidoreductase activity was definitely present in CE-6 and CE-9 only, with some indication of very low activity in CE-7 and CE-11. The other fermentations did not have this enzyme, and none of the other enzyme activities looked for could be observed in any of the *C. versicolor* retentates. Yet, as will be described, some of these retentates have indeed been shown to have lignin-transforming activity. It is not yet apparent what sorts of enzymes are responsible.

From a number of disc gel electrophoresis experiments at both normal (9.5) and low (3.8) pH, as well as slab gel isoelectric focusing experiments over the ampholine pH range 3.5 to 9.5, it was concluded that the extracellular proteins from *C. versicolor* fermentations are rich in variety, if not in amount of recognizable activity. The CE-10 extracellular retentate yielded over thirty resolved protein bands on isoelectric focusing. Some of these proteins are cellulolytic enzymes of course, but some may indeed be the sought-after lignolytic enzymes.

A finding of interest in the low pH disc gel electrophoresis experiments was that phenol-oxidizing enzyme activity is associated with an unusually acidic protein. Very few of the proteins in the *C. versicolor* extracellular retentates migrate into the separator gel at all in the pH 3.8 experiments, yet the phenol-oxidizing enzymes do (guaiacol staining to identify the bands), showing the greatest mobility toward the anode of all the proteins present. In isoelectric focusing gels, a phenol-oxidizing enzyme focuses directly beside the anode at pH 3.5. Also, chromatography of the CE-9 retentate on DEAE-Sephadex®, eluting with a triethylamine/acetate buffer of pH 5.0, demonstrated that the phenoloxidase activity is strongly retained on the column while other, less acidic proteins pass through.

A concluding word is in order concerning the discouragingly low protein yields of these *C. versicolor* fermentations. It is well known that phenolic materials and proteins interact strongly with each other through hydrogen bonding,[11] and that lignin in particular has a high affinity for protein.[12] Thus, it is not unlikely that the suspended lignin present in all of the fermentations served as an effective protein scavenger, reducing recoverable soluble protein yields to a level not at all representative of the actual production of extracellular protein by the organism. Experiments are now underway which show that suspended kraft lignin does indeed associate with enzymes (horseradish peroxidase as well as phenoloxidase from CE-9), inhibit their activity, and remove them from solution. Fortunately, it is possible to displace the adsorbed proteins once again from the lignin and recover their activities using water-soluble polyvinylpyrrolidone (PVP). These recent findings suggest ways of improving protein yields from future fermentations.

C. Lignolytic Enzyme Activity

Gross lignolytic activities of cell-free ultrafiltration retentates containing extracellular proteins from several of our large-scale fermentations were evaluated by incubating these retentates with U-^{14}C-labeled kraft lignin. The design of these experiments is indicated schematically in Figure 4, and Table 3 shows the results. In each experiment, 3 mℓ of ultrafiltration retentate from a particular large-scale submerged-culture fermentation (Table 2) were incubated with 30 mg of ^{14}C-lignin at 30°C for 7 days. The data show that the extracellular enzymes from CE-10 (1% of lignin plus 1% Avicel®) were the most effective as measured by "solubilization" of radioactive lignin, while CE-12 produced the most significant release of labeled carbon dioxide.

A second set of experiments with the CE-12 extracellular retentate is summarized in Table 4. The specific activity of the lignin in these experiments was higher than that of the lignin used in those presented in Table 3, and the conclusion that extracellular enzymes do indeed have lignolytic activity is more compelling here.

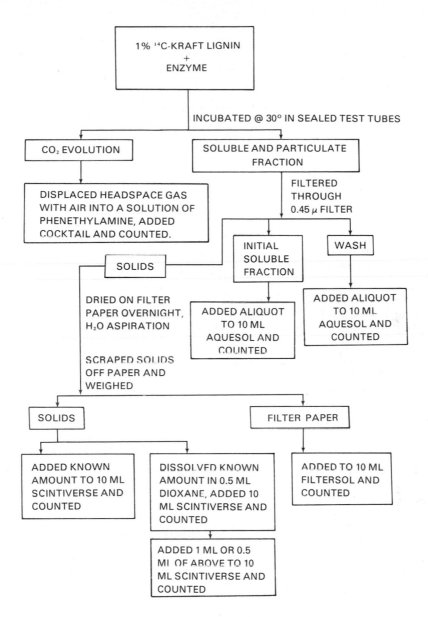

FIGURE 4. Schematic diagram of experiments designed to demonstrate lignolytic activity of cell-free extra cellular enzymes from *C. versicolor* on U-^{14}C-labeled kraft lignin.

TABLE 3

Distribution of Radioactivity (Percent) after Enzymic Reaction with ^{14}C-Lignin

Product Fraction	Control	CE-9	CE-10	CE-11	CE-12
CO₂	0.1	0.7	0.7	0.2	0.8
Solution	5	7	12	5	9
Solids	92	87	79	91	83
Filter	3	5	8	4	7

TABLE 4

Distribution of Radioactivity[a] in Various Fractions Following Incubation of Lignin with Extracellular Proteins from a *C. versicolor* Fermentation

Sample	CO_2	Soluble fraction	Particulate fraction	Total
Blank (H_2O)	0.01	5.5	95	101
Enzyme control (heat-inactivated)	0.01	6.4	94	100
Enzyme incubation (aerobic)	0.75	12.6	85	99
Enzyme incubation (anaerobic)	0.02	6.2	95	101

[a] Percent of total radioactivity (521,000 disintegrations/min) initially present in U-^{14}C-kraft lignin.

In parallel with these experiments with radioactive lignin, experiments on a larger scale (though with much lower protein concentrations) were carried out on "cold" kraft lignin. In these experiments 1-g samples of kraft lignin (reprecipitated Indulin ATR-C1) were incubated with 10-mℓ aliquots of the ultrafiltration retentates from each of fermentations CE-7, 9, and 10. Each sample run was accompanied by a control in which the retentate aliquot was denatured by heating at 100°C for 30 min. All flasks were degassed by aspiration and sealed prior to incubation. The incubations were carried out at 25°C in an incubator-shaker for 7 days. Then the solids were filtered off and dried prior to analysis.

Another experiment, employing just the CE-10 ultrafiltration retentate, used 9:1 (v/v) H_2O dimethyl formamide (DMF) as the incubation solvent. The DMF was present to enhance lignin solubility. The incubations were carried out at room temperature with stirring in parafilm-covered containers for 1 week prior to analysis for changes in lignin structure.

III. BIODEGRADED KRAFT LIGNINS

A. Lignins from *Coriolus versicolor* Fermentations

Biodegraded lignins (and other solid components) were recovered from large-scale fermentations CE-7, 9, 10, and 12 by a series of filtration steps. In every case the coarsest material, retained by sieve screening, was predominantly cell biomass along with the lignin particles which adhered to it. This fraction, referred to as the "retentate," was mildly sonicated to free the adhering lignin particles from the mycelia and then re-screened. The dislodged lignin now passed through the sieve. It was collected, dried, and, subsequently, repurified by dissolution in alkali, filtration, and reprecipitation by acidification. The final product is referred to herein as "retentate lignin." It is to be borne in mind that in each instance, this retentate lignin is the fraction which physically adhered to the fungal mycelium and was thus in intimate contact with cells.

The lignin which did not adhere to the fungal mycelia passed through the sieve in the original screening step, and is referred to herein as "filtrate lignin." In one case, (CE-10), there were two such filtrate lignins, one with finer particle size than the other. These filtrate lignins too, like the retentate lignins, were repurified prior to analysis. Table 5 indicates the results of these repurification steps. The alkali-insoluble material is mainly residual cellulose.

Each of the lignin fractions of Table 5 was carefully analyzed in order that structural

TABLE 5

Purification of Lignin Preparations Recovered from Fermentations

CE-Fermentation No.	Initial dry wt of fraction (g)	Lignin yield upon reprecip. (%)	Alkali insoluble material (%)	Ash content
7 (Unfractionated)	15.0	∿80—90	1.8	7.8%
9 Retentate	20.6	80.6	5.7[a]	3.5%
9 Filtrate	47.4	∿80—90	∿0	6.2%
10 Coarse filtrate	34.6	31.7	51.3[b]	12.8%
10 Fine filtrate	21.9	83.5	5.5	14.5%
10 Retentate	35.8	91.5	2.2	5.0%
12 Retentate	43.2	96.1	0.4	Incomplete
12 Filtrate	15.7	92.7	6.4	Incomplete

[a]   Methoxyl content of 1.04%.
[b]   This represents recovered cellulose. CE-10 initially contained 100 g of Avicel®.

transformations brought about by the biodegradation process might be assessed. Elemental analyses (C, H, N, S) and OCH_3 determinations provided empirical formulas for average C_9 units.[13] (The raw data were first corrected for contamination by ash and carbohydrate, each of which was individually determined for each sample.) ^{1}H-NMR spectra of acetylated samples revealed hydrogen and hydroxyl group distributions in each lignin fraction.[14] Carbonyl content was determined by volumetric determination of H_2 evolved upon treatment with sodium borohydride. Molecular weights were estimated by vapor pressure osmometry (VPO) on solutions of lignins in dimethylformamide. Finally, each lignin fraction was subjected to oxidative degradation according to the methods of Miksche et al.,[15] with alkaline potassium permanganate. Degradation products were identified and quantitated by GC/MS using authentic synthesized materials as reference standards.

Table 6 presents the results of analyses of the most extensively transformed lignin fraction from each of the four fermentations mentioned in Table 5. In each case this most transformed lignin fraction was "retentate" lignin. This finding tends to lend credence to the hypothesis that actual physical contact between fungal mycelia and lignin is required for optimal biodegradation. The "filtrate" lignins of Table 5, in all cases, were hardly different from the original Indulin®.

Table 6 shows that the biodegradative transformations brought about in these lignin fermentations are definitely oxidative. Total oxygen per C_9 increases, and with it the C_9 unit weight. But beyond this general (and not very surprising) observation, it is difficult from these data to assess the nature of the chemical changes brought about in the biodegraded lignins.

Where is the extra oxygen? The answer to that question varies from one fermentation to the next. But in the case of CE-12, which produced the most highly oxidized lignin, virtually all of it is in the "other oxygen" category. Although we have not yet settled upon a satisfactory analysis for carboxyl groups in lignin and have no reliable data on them as yet, we assume that carboxyl groups would have to account for most of this other oxygen.

If one does conclude that the extra oxygen belongs to carboxyl groups, how can they be accounted for? Figure 5 suggests some conceivable degradation schemes leading to carboxyl groups (or esters). But in each of these schemes the oxidation is accompanied by other changes which should show up in the analytical data (loss of aliphatic or aromatic hydrogen, loss of aliphatic hydroxyl, gain in aromatic hydroxyl, loss of

TABLE 6

Analysis Data — Fermented Lignins

	Indulin	CE-7	CE-9	CE-10	CE-12
C_9-Unit Weight	181.60	188.41	196.85	187.86	203.03
Total H/C_9[a]	8.25	8.20	9.23	8.54	8.05
Aromatic H	2.45	2.08	2.16	2.49	2.37
Aliphatic H	4.22	4.40	5.18	4.53	4.03
Total OH	1.46	1.56	1.71	1.38	1.49
Total SH	0.12	0.16	0.18	0.14	0.16
Total O/C_9[a]	2.27	2.67	3.06	2.59	3.54
Aromatic OH	0.58	0.37	0.50	0.54	0.58
Aliphatic OH	0.88	1.19	1.21	0.84	0.91
Aromatic OR[b]	0.42	0.63	0.50	0.46	0.42
Carbonyl	0.26	—·—	0.45	0.57	0.21
Other oxygen	0.13	0.48[c]	0.40	0.18	1.42
Methoxyl/C_9	0.81	0.78	0.80	0.81	0.81
Molecular weight (VPO)					
Unacetylated	1201	1401	1879	1749	1791
Acetylated	2084	2205	2425	2015	——

[a]   Not including -OCH_3 groups.
[b]   By difference. Aromatic OH and OR must sum to 1.00/C_9.
[c]   Includes any carbonyl oxygen.

methoxyl). One sees no such accompanying changes in the CE-12 data. There is perhaps some suggestion of loss of aromatic hydrogens in CE-7 and CE-9, but in no case is there any loss of methoxyl groups.

It is conceivable that route 2 in Figure 5 could produce a diester product in which the vinyl protons were grouped with aromatic protons in the NMR. But there is no known precedent for this particular sort of ring opening among the metabolic pathways of microorganisms.

Route 4 of Figure 5 (see discussion by T. Higuchi in Volume I, Chapters 9 and 10) results in oxidative sidechain cleavage with eventual loss of two carbons from a C_9 unit. It has been pointed out that if the data of Table 6 were recalculated for an average unit of fewer than nine carbons in the degraded lignin, there would indeed appear to be losses of hydrogen, hydroxyl groups, and methoxyl groups accompanying degradation.

In summary, the data of Table 6 seem to raise more questions than they answer. Answers to these questions await more definitive analyses, most particularly for carboxyl groups although the whole matter of oxygen distribution in lignin is currently very difficult to assess.

B. Lignins from Enzyme Incubations

The enzyme incubation mixtures were either filtered to recover suspended lignin or, in the case of the DMF-containing incubations, freeze-dried. The lignin was then purified, in the manner previously described for the fermentation lignins, prior to analysis. The analytical results for the enzyme-digested lignins appear in Table 7. The control lignins, in all cases, were essentially unchanged Indulin.

In accord with the observations with fermentation lignins, cell-free, enzyme-digested lignins exhibit O/C_9 to be the most significant change. The fact that CE-7 enzymes appear to be more efficient than CE-10 enzymes may be attributed to the experimental setup which possibly has seen CE-10 oxygen limited. Some changes are too small and the uncertainties in the data too great to draw definite conclusions on this point. Mo-

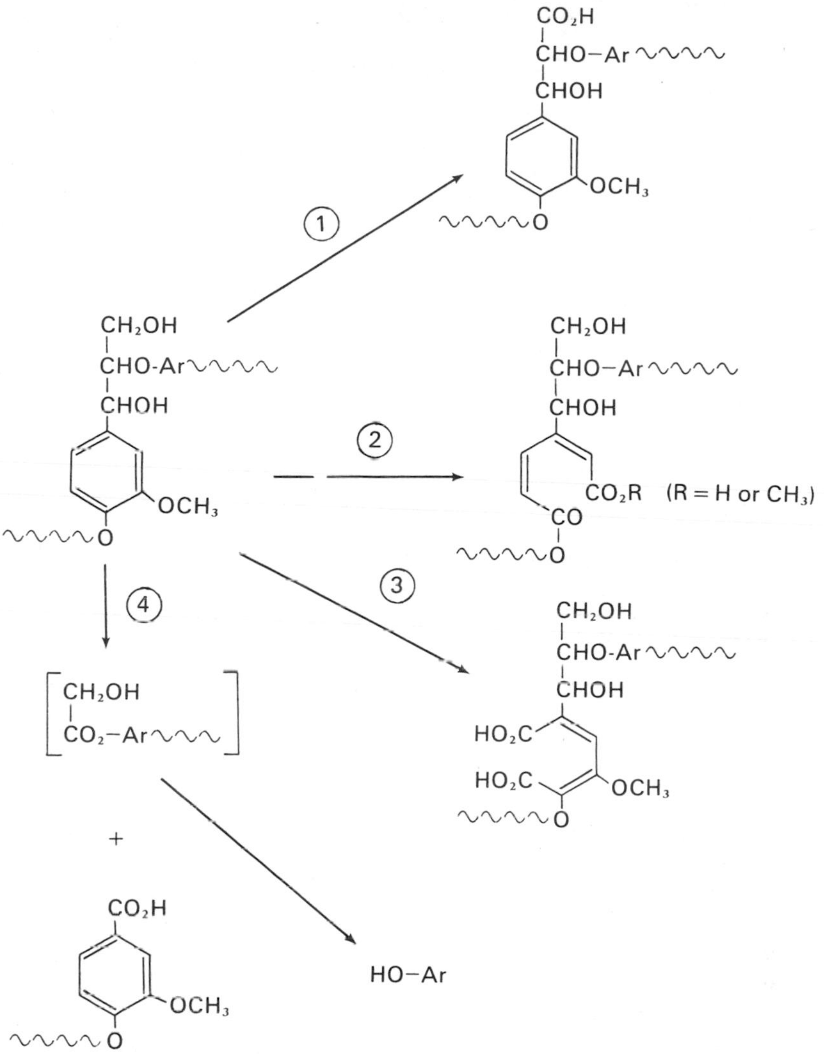

FIGURE 5. Conceivable oxidative biodegradation routes leading to carboxyl groups in lignin.

lecular weight increases of enzyme-incubated lignins are less significant than those of fermentations, possibly because smaller fragments are being efficiently metabolized and depleted in fermentations, leading to the reflection of higher molecular weights of the remaining substance.

The main conclusion one is compelled to draw concerning these enzyme-digested lignins is that the apparent transformation has been very small, albeit significant. One suspects that limited concentrations of the lignin-transforming enzymes, or limited oxygen availability, or both, may have contributed to this situation and experiments are underway to correct both limitations.

C. Permanganate Oxidation Results

Three of the lignins recovered from fermentations and one of the enzyme-digested lignins have been subjected to permanganate oxidative degradation.[15] The object of this analytical procedure is to gain information about the relative frequency of the

TABLE 7

Analysis Data — Enzymatically Degraded Lignins

	Indulin®	CE-7	CE-9	CE-10	CE-10 (DMF)[a]
C_9-unit weight	181.60	208.87	180.74	191.04	185.21
Total H/C_9[b]	8.25	8.16	8.66	7.13	10.78
Aromatic H	2.45	2.71	2.12	2.34	3.08
Aliphatic H	4.22	3.67	4.70	3.14	5.83
Total OH	1.46	1.53	1.49	1.40	1.67
Total SH	0.12	0.25	0.35	0.25	0.20
Total O/C_9[b]	2.27	3.66	2.17	2.61	2.29
Aromatic OH	0.58	0.67	0.53	0.55	0.59
Aliphatic OH	0.88	0.86	0.96	0.85	1.08
Aromatic OR[c]	0.42	0.33	0.47	0.45	0.41
Carbonyl	0.26	—	—	—	—
Other oxygen	0.13	1.80[d]	0.21[d]	0.76[d]	0.21[d]
Methoxyl/C_9	0.81	0.84	0.58	0.83	0.74
Molecular weight (VPO)	1201	1481	1682	1323	—

[a]   Incubation solvent was 9:1 H_2O/DMF. See text.
[b]   Not including -OCH_3 groups.
[c]   By difference. Aromatic OH and OR must sum to 1.00/C_9.
[d]   Includes any carbonyl oxygen.

various kinds of C_9-C_9 interunit linkages in a lignin sample. If biodegradation or enzymatic transformation results in attack upon some sorts of C_9 units more extensively than others, this should be reflected in the relative yields of the various permanganate oxidation products.

Figure 6 shows the structures of the six most prominent products of the (methylated) permanganate oxidation of methylated lignins derived from softwood species. By far the major product is the methyl ester of veratric acid VII. It comes from free phenolic guaiacyl units in lignin. Ester VIII (isohemipinic acid methyl ester) comes from phenylcoumaran structures — the result of β-5 coupling between coniferyl alcohol structures in lignin biosynthesis (Volume I, Chapter 1). The other "monomer" IX is from somewhat less common α-6 or β-6 linked C_9 units. The "dimers" X, XI, and XII come from 5-5, 1-5, and 4-0-5-linked units. Among these the 5-5 linkage is the most frequently encountered.

Table 8 presents the results of permanganate oxidation of Indulin and the aforementioned transformed lignins. It is difficult from these data to draw really compelling conclusions because there are few large differences from one lignin to the next. Furthermore, the total yields of identifiable monomers and dimers are not high with kraft lignins. However, attention can be directed to certain trends in the data.

Perhaps the most conspicuous feature is the apparent loss in biodegraded lignins of phenylcoumaran units and α-6- or β-6-linked aromatic rings. This results in relatively low yields of monomers VIII and IX. Somewhat surprisingly, however, the apparent depletions in these two monomers do not result in any change in the overall proportion of monomers to dimers among the biodegradation products. And there is no indication that any of the Ar-Ar, or Ar-O-Ar-linked, units in lignin is attacked preferentially by the lignolytic enzymes.

In all cases, the total yields of identifiable monomers and dimers are lower for biodegraded kraft lignins than for Indulin itself. The yields for the enzyme-digested sample are especially low, which may be cause to question conclusions based on these particular data. The reason for the low yields is not known.

The permanganate oxidation data obtained so far and reported here are intriguing,

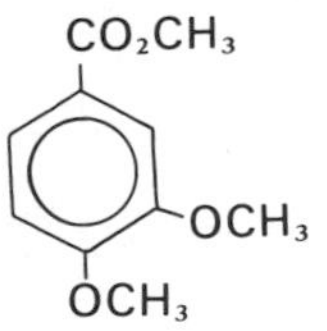

FIGURE 6.   Principal products from permanganate oxidation of methylated softwood lignins.

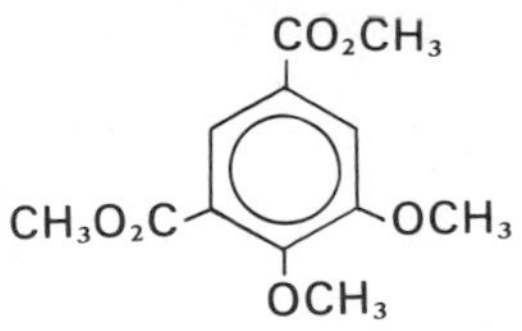

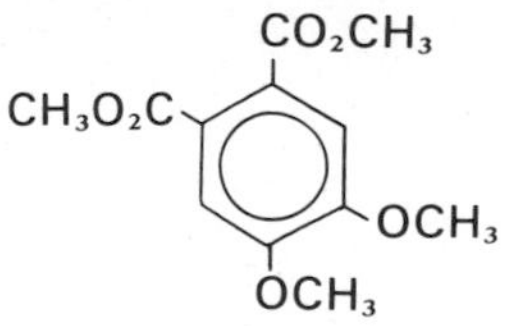

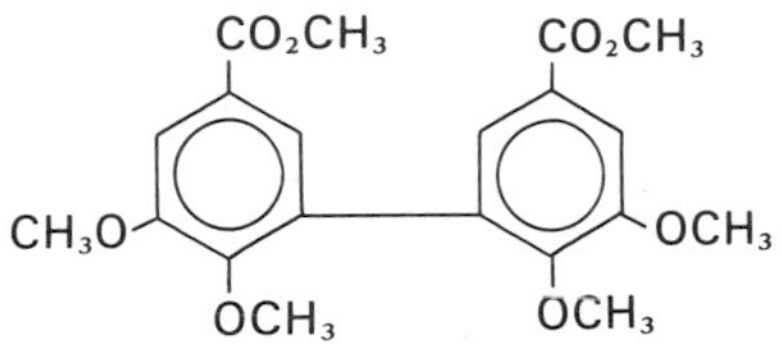

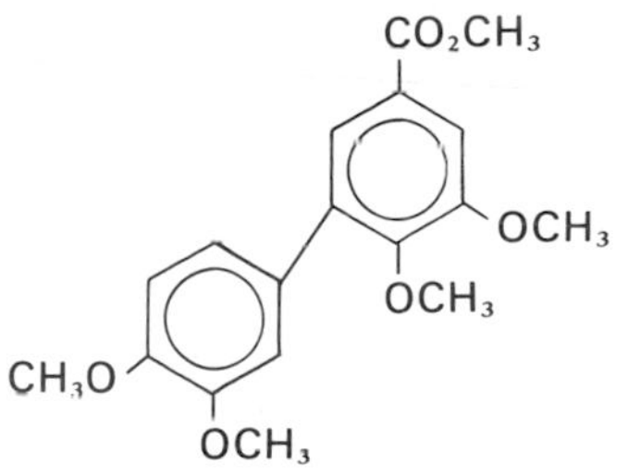

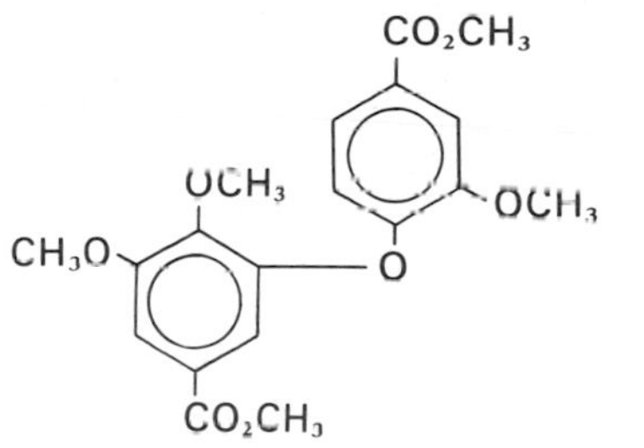

but many more experiments are needed before we can progress from the mere suggestions made here to firm conclusions.

IV. CONCLUSIONS AND DIRECTIONS FOR FUTURE WORK

Much of what is now known or surmised about lignin biodegradation has been deduced more or less indirectly in studies with model compounds as substrates or in observations of gross lignin metabolism by means of monitoring carbon dioxide evolution or disappearance of lignin from wood substrates. These approaches to the problem have been, and will continue to be, very powerful and useful, as many chapters in this volume amply demonstrate. However, elucidation of the chemical and biochemical details of lignin biodegradation will require examination of the structures of biodegraded lignin. The data presented in this chapter represent early and somewhat tentative steps in this latter direction.

It is clear that oxidative structural changes are perceptible and determinable in biodegraded lignins, though it is not yet clear what those changes actually are. It is also clear that cell-free extracellular enzymes are capable to some degree of lignin structural transformation, though again the nature and extent of the transformations remain to be brought to light. And finally, it is clear that the analytical techniques routinely used

TABLE 8

Permanganate Oxidation Yields[a]

	Indulin	CE-9	CE-10	CE-12	CE-10 (DMF)[b]
Monomer VII[c]	14.4(51)	12.1(57)	11.3(59)	10.2(57)	6.2(64)
Monomer VIII[c]	4.5(16)	2.0(9)	1.8(9)	1.6(9)	1.0(10)
Monomer IX[c]	1.2(4)	0.2(1)	0.1(1)	0.2(1)	0.1(1)
Total monomers[d]	23.9(79)	16.0(75)	14.7(77)	13.6(76)	7.7(79)
Dimer X[c]	3.6(12)	3.2(15)	2.5(13)	2.6(14)	1.2(12)
Dimer XI[c]	0.2(1)	0.1(1)	0.1(1)	0.1(1)	0.04(1)
Dimer XII[c]	1.4(5)	1.2(6)	1.1(6)	0.9(5)	0.6(6)
Total dimers[d]	6.0(21)	5.3(25)	4.4(23)	4.4(24)	2.1(21)
Total yield	28.8(100)	21.3(100)	19.0(100)	18.0(100)	9.7(100)

[a]   Yields given as mg methyl esters found per 100 mg of lignin oxidized. Values in parentheses are percent of total recovered yield.
[b]   Recovered from the 9:1 H_2O/DMF incubation of Indulin® with the protein concentrate from CE-10. See Table 7 and text.
[c]   See Figure 6.
[d]   Includes a number of minor products, all known and positively identified but not included in the table.

in characterizing lignin must be supplemented by additional and/or improved techniques if substantial progress is to be made in this difficult area. ^{13}C- and ^{19}F-NMR, as well as spin labeling, fluorescence tagging, and use of isotopic tracers are under investigation in this context.

Concerning the microbiology and enzymology of lignin biodegradation, investigation is plagued by very slow rates of biodegradation by microorganisms known to attack lignin, by poor growth, or no growth at all of these microorganisms on lignin as a substrate, and by discouragingly low levels of enzyme protein production. Wide and intensive searches for better lignin-degrading microorganisms have been undertaken by several laboratories (see reports in this volume). It is to be hoped that these searches will be successful. Meanwhile efforts will continue, through fermentation optimization and microbiological genetics, to improve the performance of the widely investigated white-rot fungi as lignin degraders and enzyme producers.

V. SUMMARY

Kraft lignin is partially biodegraded over a period of a week or two when suspended, along with an appropriate carbohydrate as a supplemental carbon source, in submerged culture fermentations of the white-rot fungus, *C. versicolor*. Some lignin is metabolized to carbon dioxide, very little is "solubilized," and the remainder is recovered as a distinctly transformed material. The recovered lignin has a higher average molecular weight than untransformed kraft lignin, and a dramatic increase in oxygen content. Our kraft lignin contains 2.27 oxygen atoms per C_9 unit, and we have found as much as 3.54 oxygen per C_9 in lignin recovered from *C. versicolor* fermentations. Hydroxyl group content per C_9 unit changes little. Surprisingly, even phenolic hydroxyl groups are scarcely depleted, and OCH_3/C_9 also remains constant. Hence, one concludes that the additional oxygen is probably associated with introduction of carbonyl and/or carboxyl groups on sidechains. Our data are consistent with some dearomatization, but it is not extensive.

Cell-free, soluble extracellular enzymes from *C. versicolor* fermentations are also capable of transforming lignin. The chemical nature of the transformations is not unlike that described above for the fermentations themselves. Electrophoresis and isoelectric focusing experiments show a variety of different proteins in our cell-free extracts. However, the amounts of protein recovered from fermentations are exceedingly small, and we have not progressed very far with our efforts to isolate and characterize ligninase enzymes. In keeping with previous reports, we observe phenol oxidase activity, as well as cellobiose:quinone oxidoreductase and various cellulases.

REFERENCES

1. Glasser, W. G. and Glasser, H. M., unpublished results of recent simulation studies; see also Glasser, W. G., and Glasser, H. M., Simulation of reactions with lignin by computer (Simrel) II. A model for softwood lignin, *Holzforschung*, 28, 5, 1974.
2. Björkman, A., Studies on finely divided wood. I. Extraction of lignincarbohydrate complexes with neutral solvents, *Sven. Papperstidn.*, 60, 243, 1957.
3. Marton, J., Reactions in alkaline pulping, in *Lignins: Occurrence, formation, structure and reactions*, Sarkanen, K. V. and Ludwig, C. H., Eds., Interscience, New York, 1971, Chap. 16.
4. Crawford, D. L., Crawford, R. L., and Pometto, A. L., III, Preparation of specifically labelled ^{14}C-(lignin)- and ^{14}C-(cellulose)-lignocelluloses and decomposition by microflora in the soil, *Appl. Environ. Microbiol.*, 33, 1247, 1977.
5. Cochran, T. W. and Vercellotti, J. R., Hexosamine biosynthesis and accumulation by fungi in liquid and solid media, *Carbohy. Res.*, 61, 529, 1978.
6. Bradford, M. M., A rapid and sensitive method for the quantitation of microgram quantities of protein utilizing the principal of protein-dye binding, *Anal. Biochem.*, 72, 248, 1976.
7. Streamer, M., Eriksson, K.-E., and Pettersson, B., Extracellular enzyme system utilized by the fungus *Sporotrichum pulverulentum (Chrysosporium lignorum)* for the breakdown of cellulose, *Eur. J. Biochem.*, 59, 607, 1975.
8. Malmström, B., Fahraens, G., and Mocbach, R., Purification of laccase, *Biochim. Biophys. Acta*, 28, 652, 1958.
9. Ander, P. and Eriksson, K.-E., The importance of phenol oxidase activity in lignin degradation by the white rot fungus *Sporotrichum pulverulentum*, *Arch. Microbiol.*, 109, 1, 1976.
10. Westermark, U. and Eriksson, K.-E., Purification and properties of cellobiose:quinone oxidoreductase from *Sporotrichum pulverulentum*, *Acta Chem. Scand. Ser. B*, 29, 419, 1975.
11. Loomis, W. D., Removal of phenolic compounds during isolation of plant enzymes, *Methods Enzymol.*, 13, 555, 1969.
12. Nakayama, H., Matijević, E., and Shinoda, K., Precipitation and microflotation of kraft lignin and hemoglobin, *J. Colloid Interface Sci.*, 61, 590, 1977.
13. Kirk, T. K. and Chang, H. M., Decomposition of lignin by white-rot fungi. II. Characterization of heavily degraded lignins from decayed spruce, *Holzforschung*, 29, 56, 1975.
14. Glasser, W. G. and Glasser, H. M., Simulation of reactions of lignin by computer (Simrel). III. The distribution of hydrogen in lignin, *Cellul. Chem. Technol.*, 10, 23, 1976.
15. Erickson, M., Larsson, S., and Miksche, G. E., Gaschromatographische Analyse von Ligninoxydationsprodukten. VII. Ein verbessertes Verfahren zur Charakterisierung von Ligninen durch Methylierung und oxydativen Abbau, *Acta Chem. Scand.*, 27, 127, 1973.

Chapter 4

PHYSIOLOGY OF LIGNIN METABOLISM BY WHITE-ROT FUNGI

T. Kent Kirk

TABLE OF CONTENTS

I. INTRODUCTION

Our research on lignin metabolism by white-rot fungi is aimed primarily at elucidating the chemical reactions that comprise decomposition of the polymer. Recent investigations[1-4] (summarized here) have been directed first at defining the culture parameters important for lignin metabolism, and second toward describing some of the physiological events that accompany appearance of ligninolytic activity in cultures. This work has now made it possible to consistently produce active ligninolytic cultures for study. Current investigations are directed at identifying the chemical changes catalyzed by the ligninolytic enzyme system in lignin-related aromatics. These studies are intended to complement continuing efforts using another approach, the chemical characterization of degraded lignin isolated from white-rotted wood (Volume I, Chapter 11).

II. CULTURE PARAMETERS IMPORTANT IN LIGNIN METABOLISM BY *PHANEROCHAETE CHRYSOSPORIUM*

A. Methodology

The first job in defining the culture conditions for lignin metabolism was to devise a sensitive, efficient, unequivocal, and quantitative assay for lignin catabolism. The heterogeneous nature of lignin dictates that a method be used which is not dependent on specific functional groups. Consequently, radioisotopic procedures are the methods of choice (see Volume I, Chapter 3). We have described the preparation of synthetic ^{14}C-lignins (DHPs) which were prepared from ^{14}C-coniferyl alcohol with peroxidase and H_2O_2. The lignins were labeled in the methoxyl, aromatic ring, or side chain carbons.[1] Determinations of $^{14}CO_2$ produced by microbial metabolism of these model lignins provided the required assay for the work described here.

The next task in physiological investigations was to choose one fungus for detailed study. After consultation with mycologists at the Center for Forest Mycology Research (Forest Products Laboratory, Forest Service, U.S. Department of Agriculture), we selected *Phanerochaete chrysosporium* Burds. This fungus causes a typical white rot of wood. It is a serious agent of deterioration in stored wood, particularly stored pulpwood chips. Salient taxonomic features have been described by Burdsall and Eslyn.[5] For physiological investigations, this species has several advantages over most white-rot fungi: (1) it produces copious conidia ("aleuriospores"), which greatly simplifies handling, (2) its optimum temperature for growth is 39 to 40°C, which, together with its pH optimum of 4.5 to 5, means that contamination is easily avoided, (3) it grows rapidly, and (4) examined strains produce only barely detectable amounts of phenol-oxidizing enzymes, which minimizes unwanted oxidative-coupling reactions in phenolic compounds. *P. chrysosporium* degrades lignin at least as rapidly as other examined white-rot fungi. Burdsall and Eslyn[5] have presented evidence that *P. chrysosporium* (which in our laboratory was initially called *Peniophora* "G") is synonymous with *Sporotrichum pulverulentum, Chrysosporium lignorum*, and *C. pruinosum*. We have used Strain ME-446 in the physiological investigations described here. This strain was isolated from beech wood chips collected in Waterville, Maine in 1964.

Results obtained with *P. chrysosporium* are probably representative for white-rot fungi in general. Limited experiments, noted here, with *Coriolus versicolor* (L. ex Fr.) Quel. (Mad. 697) have given similar results. *C. versicolor* and *P. chrysosporium* are taxonomically distinct.

Our first culture experiments with the radiolabeled lignins were conducted with a medium comprised of mixtures of wood meal and synthetic lignins moistened with a mineral solution and inoculated with various fungi, including *P. chrysosporium*.[1] As

TABLE 1

Basal Medium for Culturing *Phanerochaete chrysosporium* in Lignin Metabolism Experiments[a]

Component	Amount per liter
KH_2PO_4	200 mg
$MgSO_4 \cdot 7H_2O$	50 mg
$CaCl_2$	80 mg
Vitamin solution[b]	0.5 ml
Mineral solution[c]	1.0 ml
Water	to 1 l

[a] Additions to this medium included growth substrate, nutrient nitrogen, buffer, and synthetic ^{14}C-lignins (see text.)

[b] Vitamin solution contained (per liter of H_2O): biotin, 2 mg; folic acid, 2 mg; thiamine·HCl, 5 mg; riboflavin, 5 mg; pyridoxine·HCl, 10 mg; cyanocobalamine, 0.1 mg; nicotinic acid, 5 mg; Dl-calcium pantothenate, 5 mg; p-aminobenzoic acid, 5 mg; and thioctic acid, 5 mg.

[c] Mineral solution contained (per liter of H_2O): nitrilotriacetate, 1.5 g; $MnSO_4 \cdot H_2O$, 0.5 g; NaCl, 1.0 g; $FeSO_4 7H_2O$, 0.1 g; $CoSO_4$, 0.1 g; $ZnSO_4$, 0.1 g; $CuSO_4 \cdot 5H_2O$, 10 mg; $AlK(SO_4)_2$, 10 mg; H_3BO_3, 10 mg; and $NaMoO_4$, 10 mg.

the wood decayed, [14]CO_2 was evolved, indicating complete oxidation of the various carbon atoms in the polymer. Subsequent experiments were conducted with chemically defined liquid culture media and were directed at describing the important culture parameters favoring lignin metabolism.

The basal medium employed in all the work summarized here is described in Table 1. Additions to this medium included growth substrate (1% glucose, unless stated otherwise), nutrient nitrogen (equimolar L-asparagine and NH_4NO_3; 2.4 mM total N, unless stated otherwise), and synthetic ^{14}C-lignin. Various amounts of ^{14}C-lignins were used; 2.5 to 3.0 × 10[4] dpm per culture usually permitted satisfactory statistical treatment of the data from scintillation spectrometry. Cultures (10 ml total volume) in 125-ml Erlenmeyer flasks were maintained at 38 to 40°C. Flasks were closed with rubber stoppers fitted with glass tube ports which permitted periodic flushing to exchange gases and to allow trapping of [14]CO_2. Detailed descriptions of procedures and materials have been published.[2-4]

B. Requirement for a Growth Substrate

Experiments with cultures (*P. chrysosporium* and *C. versicolor*) containing ^{14}C-lignin plus various carbon compounds demonstrated that a readily utilizable growth substrate is required for lignin metabolism. Growth and metabolism of ^{14}C-lignin with spruce milled wood lignin as sole carbon addition to an otherwise complete medium were negligible.[2] The efficacies of various growth substrates in supporting lignin metabolism were found to vary, but a range of compounds supported lignin metabolism:

Glycerol	(Sodium) succinate
D-xylose	(Potassium) D-gluconate
D-glucose	Xylan (wood gum)
D-mannose	Cellulose (Whatman®)
D-glucuronolactone	Cellulose "Solka-Floc"
	(bleached wood pulp)
D-cellobiose	Cellulose (avicel)

For some unexplained reason, glucose supported relatively poor lignin metabolism in our initial experiments,[2] but has since been found to be excellent. The wood pulp cellulose and glucose are the best of the examined growth substrates in supporting lignin metabolism. Limited studies indicate an absence of growth on aromatic compounds such as vanillate.

The basis for the growth substrate requirement may be a complex one. Dagley[6] has interpreted our results to indicate that the initial degradation of lignin consumes rather than liberates energy. We have also offered this as a possible basis,[3] and further experiments may bear it out. However, our recent research has shown that ligninolytic activity does not appear until after a primary growth phase is complete, which would obviously preclude primary growth at the expense of lignin. This is discussed in Section III. We also suspect that the rate of lignin metabolism, once it does get under way, might be too slow to support growth.[3,4] Further investigation of this question is needed.

Following the observation that a growth substrate is required, other culture parameters were examined with glucose as growth substrate. Experiments showed that the influences of the various culture parameters described below were similar with cellulose, xylose, or glucose as growth substrate.

C. Effect of O_2 Concentration

Lignin metabolism by white-rot fungi has been shown to be largely an oxidative process,[7] which suggests that the O_2 concentration should be an important rate-determining factor. This proved to be the case. Metabolism of lignin in cultures maintained under atmospheres of 5, 21, and 100% O_2 was studied.[3]

Results (Table 2) showed that the rate and extent of conversion of lignin to CO_2 was two- to threefold greater under 100% O_2 than under 21% O_2(air), despite the fact that in air growth was somewhat better and glucose utilization was faster. Growth but not lignin metabolism proceeded at 5% O_2 in N_2.

Culture agitation is used frequently to increase the rate of O_2 exchange between atmosphere and culture medium. It was considered likely that agitation would enhance lignin metabolism. However, mycelial fungi, including *P. chrysosporium,* usually grow in the form of pellets in agitated cultures, and it was found that agitation resulting in pellet formation greatly slowed lignin metabolism under both 21 and 100% O_2 (Table 3). Growth in agitated cultures was similar under 21 and 100% O_2. If pellet formation was prevented by pregrowth in nonagitated cultures, subsequent agitation was slightly beneficial at 21% O_2, but almost completely stopped lignin metabolism at 100% O_2. Lignin metabolism under 5% O_2 did not occur with agitation, even with a stationary pregrowth period.

The basis for the negative effect of pellet formation on lignin metabolism appears to be related to the O_2 effect. We suspect that the O_2 concentration is simply too low in pellet interiors to support lignin metabolism. Degradation by the pellet surface mycelium is low because the surfaces represent only a small percentage of total hyphal

TABLE 2

Effect of Atmospheric O_2 Concentration on Glucose Depletion and Lignin Metabolism by *Phanerochaete chrysosporium* during 20 days

O_2 concentration[a] (% by vol)	Glucose remaining[b] (% of original)	Lignin metabolized to $^{14}CO_2$[c] (% of total ^{14}C)
5	17	0
21	5	24 ± 3
100	8	50 ± 3

[a] Culture flasks (125 m*l*) were flushed every third day to exchange gases and allow trapping of $^{14}CO_2$. Stationary cultures were grown at 39°C under 100% O_2.
[b] Originally 100 mg/10 m*l* culture.
[c] Each culture originally contained 2.6×10^4 dpm (280 μg) of synthetic [ring - ^{14}C]-lignin. Values are means ± SD for three replicate cultures.

TABLE 3

Lignin Metabolism by *Phanerochaete chrysosporium* as Affected by Culture Agitation[a]

Treatment[b]	Lignin metabolized to $^{14}CO_2$[c] (% of total ^{14}C)
Incubation under 21% O_2	
Continuous agitation	3 ± 2
No agitation	12 ± 3
Incubation under 100% O_2	
Continuous agitation	0
No agitation	42 ± 5

[a] Cultures were incubated for 23 days at 39°C.
[b] Agitation was on a rotary shaker at 150 rpm (2.5 cm amplitude).
[c] Each culture originally contained 3×10^4 dpm (320 μg) of synthetic [ring-^{14}C]-lignin. Values are means ± SD for four replicate cultures.

surface area. Why agitation with or without pellet formation was detrimental under 100% O_2 is not clear.

The strong influence of O_2 itself seems to be at the ligninolytic enzyme level. Cultures grown under 21 and 100% O_2 exhibit the same level of ligninolytic activity when assayed at 100% O_2 (or at 21% O_2), but activity is always highest when assayed at 100% O_2. The lack of lignin metabolism at 5% O_2 despite growth suggests that one or more components of the ligninolytic enzyme system has a relatively low affinity for O_2; 100% O_2 creates the maximum dissolved O_2 concentration and consequently favors this component(s). Gel permeation chromatographic examination of the ^{14}C-lignin recovered from cultures maintained for 35 days at 5% O_2 indicated that the lignin had not been attacked significantly, suggesting that the putative low-O_2-affinity enzyme(s) catalyzes an early step in metabolism of the polymer.

The effect of O_2 concentration was essentially identical for *P. chrysosporium* and *C. versicolor*.

TABLE 4

Effect of Medium pH on Growth and Lignin
Metabolism by *Phanerochaete chrysosporium*
during 9 days

pH[a]	Growth (mg dry mycelium)	Lignin metabolized to $^{14}CO_2$[b] (% of total ^{14}C)
3.0	8	<1
3.5	11	8 ± 6
4.5	14	38 ± 4
5.5	14	8 ± 1

[a] Cultures were buffered with 0.01 *M* (sodium) aconitate.
[b] Each culture originally contained 3 × 10⁴ dpm (320 μg) of synthetic [ring-^{14}C]-lignin. Values are means ± SD for five replicate cultures. Stationary cultures were grown at 39°C under 100% O_2.

D. Other Culture Parameters

Culture pH was found to be an important factor.[3] *P. chrysosporium* metabolized lignin optimally when grown at approximately pH 4.5. Cultures maintained at pH 3.5 or 5.5 metabolized lignin only about 20% as rapidly as those at pH 4.5 (Table 4). Similar results were obtained with cultures grown at pH 4.5 and then assayed for ligninolytic activity at various pHs,[17] indicating that the pH effect was not simply a growth response.

Several buffers were examined.[3] Salts of aliphatic nonconjugated carboxylic acids such as acetic, succinic, and aconitic have pK_as of approximately pH 4.5, but suffer from other deficiencies. Acetate is toxic, succinate is metabolized, and aconitate for some reason interferes with lignin metabolism — although not growth. Sodium *o*-phthalate has been found to give satisfactory pH control, even though its buffering capacity at pH 4.5 (pK_a = 5.5) is lower than the others mentioned. It is not metabolized to any detectable extent. An extensive survey of possible buffers has not been made. Recent examination of other buffers has shown 2,2-dimethylsuccinate (pH 4.5) to be superior to *o*-phthalate. The latter, in fact, is inhibitory to lignin metabolism (about 50% at 0.01 *M* in comparison to dimethylsuccinate, also at 0.01 *M*).[17]

Our initial studies[2,3] employed a chemically defined but dilute culture solution that contained basal medium (Table 1), 0.01 *M* phthalate buffer (pH 4.5), NH₄NO₃ and L-asparagine (both 0.6 m*M*N) as N sources, 1% carbohydrate as growth substrate, and ^{14}C-lignin. It was shown[3] that increasing the basal medium concentration tenfold had no effect on the onset, rate, or extent of oxidation of ^{14}C-lignin to $^{14}CO_2$. This result suggests that neither osmolality nor concentration of specific basal medium components (other than nitrogen) is a limiting factor in lignin metabolism in cultures grown as described here.

The source of nutrient nitrogen also was without significant direct influence on lignin metabolism by *P. chrysosporium*.[3] The ultimate extent of decomposition of ^{14}C-lignins was similar with (NH₄)₂SO₄, NH₄NO₃ + L-asparagine, urea, casamino acids, L-asparagine alone, ammonium tartrate, and NaNO₃ (total N in all cases = 2.4 m*M*). *P. chrysosporium* was shown to grow only slowly from spores with nitrate as the sole N source (spores were from malt agar slants), but to grow much more rapidly from nitrate-grown mycelial inoculum. Consequently, the onset of lignin metabolism was delayed with the spore inoculum on nitrate medium, in comparison with media containing reduced nitrogen.

Although the source of N had little influence on lignin metabolism, the concentration was found to be critical.[3] The optimum concentration for lignin metabolism under our experimental conditions (1% glucose or cellulose in buffered basal medium, 100% O_2 atmosphere) was 2 to 3 mM N. At 20 to 30 mM N, growth was much more extensive, but lignin metabolism was severely suppressed. This suppression at the higher levels of nitrogen was exacerbated by pH changes. The strong influence of nutrient nitrogen concentration was observed with *Coriolus versicolor* as well as *P. chrysosporium*.

The adverse influence of "high" nitrogen is due in part to a role of nitrogen depletion in regulation of synthesis of the ligninolytic system (or essential component of it) as discussed below.[4] A second influence of high nitrogen may be in its promotion of rapid depletion of the growth substrate known to be required[2] for lignin metabolism.

Experiments were conducted to determine the influence of lignin concentration and of growth substrate (glucose) concentration on the metabolism of lignin. In cultures containing 1% glucose, total lignin converted to CO_2 was increased by increasing the lignin concentration from 0.003% (the concentration used routinely) to approximately 0.05%, as determined by conversion of side chain- and ring-labeled lignins to $^{14}CO_2$. When assessed with methoxyl-labeled lignin, however, metabolism was greatest at the highest concentration examined — 0.1%. In cultures containing 0.05% lignin, total lignin decomposition was increased by increasing glucose concentration up to 1%. It was also found that 2% glucose supported the same rate and extent of lignin decomposition as 1%.

The influence of various minerals and trace elements on lignin metabolism was not systematically examined. A complex minerals solution was used in all experiments. Similarly, a complex mixture of vitamins was employed. However, most white-rot fungi require only thiamine for growth, and we found that thiamine alone can be substituted for the complex vitamin mixture (Table 1) with *P. chrysosporium*. Thiamine is required.

III. APPEARANCE OF LIGNINOLYTIC ACTIVITY IN CULTURES

After the culture parameters influencing lignin metabolism had been examined as described above, it was necessary to learn how to produce (reproducibly) maximum ligninolytic activity.[4] The first step was to develop a short-term assay that could be used to assess the relative ligninolytic activity of cultures. Polarographic and respirometric measurements of O_2 consumption by cultures known to be competent were found to be not sensitive enough, due to the low levels of ligninolytic activity present. (Attempts to increase the rate of lignin metabolism by various additions to the growing cultures were without notable success.)[8] Therefore, we reverted to the use of the assay based on $^{14}CO_2$ produced from radioactive lignins. To increase the sensitivity of this technique, we prepared synthetic [ring-^{14}C]-lignin with a specific activity of approximately 1×10^6 dpm/mg. With this preparation, sufficient $^{14}CO_2$ for reliable measurement was produced within an hour after addition of the lignin to competent cultures. The assay used in obtaining the results described in the following was based on the $^{14}CO_2$ produced during 6 hr following lignin addition to cultures; "units" are expressed as dpm/hr per culture.

Lignin metabolism in cultures did not coincide with growth. In fact, the first $^{14}CO_2$ produced by cultures containing ^{14}C-lignins appeared near the end of a linear primary growth phase. Cessation of linear growth and the onset of lignin metabolism did not

TABLE 5

Effect of Cycloheximide and Culture Age on the Ligninolytic Activity of *Phanerochaete chrysosporium*

Culture age (days)	Ligninolytic activity[a] ($^{14}CO_2$, dpm/hr per culture)	
	Cycloheximide-treated[b]	Controls
3	0	0
4	3 ± 2	13 ± 6
5	83 ± 17	92 ± 27
6	83 ± 50	123 ± 18

[a] $^{14}CO_2$ evolved during 6 hr at 39°C and under 100% O_2 following addition of 5×10^4 dpm (50 μg) per culture of synthetic [ring-^{14}C]-lignin to cultures of the indicated ages, grown without lignin. Values are means ± SD for six replicate cultures.

[b] Cycloheximide (1 mg per culture) was added 1 hr before ligninolytic assay.

reflect depletion of growth substrate glucose, but instead appeared to follow depletion of nutrient nitrogen.

Our first experiments with the short-term $^{14}CO_2$ assay examined the time after inoculation when cultures became competent to metabolize lignin. Under the conditions used, this was between 3½ and 4 days. Addition of ^{14}C-lignin (ring-, side chain- , or methoxyl-labeled) to cultures 4 days old or older, which had been grown in the absence of lignin, resulted in almost immediate evolution of $^{14}CO_2$, with maximum activity being reached in 2 to 3 hr. The question then arose whether induction of the putatively complex ligninolytic system by lignin was occurring very rapidly, or whether induction by lignin was in fact not necessary.

This question was examined in experiments with the protein synthesis inhibitor cycloheximide, using cultures grown in a lignin-free medium. Addition of the inhibitor to cultures up to 3½ days old prevented subsequent appearance of activity. When added to older, lignin-free cultures, however, the cycloheximide had no qualitative effect on ligninolytic activity. Activity was present in 4-day-old cultures even though lignin had not yet been introduced (Table 5). These results demonstrated that lignin was not necessary for appearance of ligninolytic activity, and that protein synthesis after 3½ days was necessary for appearance of the activity. (Following addition of cycloheximide to competent cultures, ligninolytic activity against [ring-^{14}C]- , [side chain-^{14}C]- , and [methoxyl-^{14}C]-lignins decreased by approximately 60, 25, and 40%, respectively, in 24 hr, in comparison to cultures which received no cycloheximide.)

Experiments were then conducted to determine whether lignin-induced ligninolytic activity could be demonstrated in addition to the existing activity. Surprisingly, results showed that the presence or absence of lignin had no influence on the time of appearance or on the level of activity. Thus, in *P. chrysosporium,* the ligninolytic system (or an essential component of it) is produced at the end of an initial linear growth phase in response to an intrinsic shift in metabolism and not in response to lignin. The next question was what caused this metabolic shift. We turned our attention to nutrient nitrogen because its depletion appeared to be associated with the appearance of ligninolytic activity.

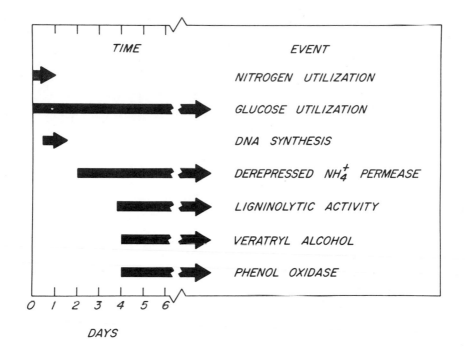

FIGURE 1. Schematic depiction of the timing of physiological events in cultures of *Phanerochaete chrysosporium* grown under conditions optimized for lignin decomposition.

The relationship between nitrogen metabolism and lignin metabolism was examined in some detail in cultures containing ammonium tartrate as nitrogen source. This source of nutrient nitrogen was selected because ammonium ion could be readily determined, and the tartrate salt did not lead to a lowering of pH as observed with $(NH_4)_2$ SO_4 and NH_4Cl. Tartrate was not assimilated. Figure 1 summarizes the results by showing the time relationship between primary growth (DNA synthesis), residual glucose, extracellular NH^+_4, ammonium permease activity, and ligninolytic activity. (Culture acidity remained consistant at pH 4.5 in the phthalate-buffered cultures.) Ammonium ion was depleted within 24 hr after inoculation of the medium. Growth (DNA increase) was discernable approximately 12 hr after inoculation and was linear until approximately 30 hr.[4] Ammonium permease activity was measured during growth to give an indication of nitrogen starvation. Others[9] have shown that this activity becomes derepressed under conditions of N starvation. The permease activity was not detected in 24-hr cultures, but was maximal (completely derepressed) in 2-day-old cultures, indicating that the organism was starved for N after 2 days. Ligninolytic activity appeared between days 3 and 4. Glucose was approximately 60% depleted after 7 days on this medium.

Thus the observed sequence of physiological events following inoculation with spores was as follows: 0 to 24 hr, germination, primary growth (as measured by DNA synthesis), and depletion of nutrient N; 24 to 48 hr, cessation of primary growth and derepression of ammonium permease activity, demonstrating N starvation; and 72 to 96 hr, appearance of ligninolytic activity (^{14}C-lignin $\rightarrow$ ^{14}CO$_2$).

Experiments were then conducted to assess more fully the relationship between NH^+_4 and the appearance of ligninolytic activity.[4] Addition of NH^+_4 in an amount equal to the original amount present (2.4 mM N) to 72-hr cultures, i.e., immediately prior to appearance of ligninolytic activity, delayed appearance of the activity approximately 48 hr (Figure 2). Addition of NH^+_4 to older, ligninolytic cultures caused no reproduc-

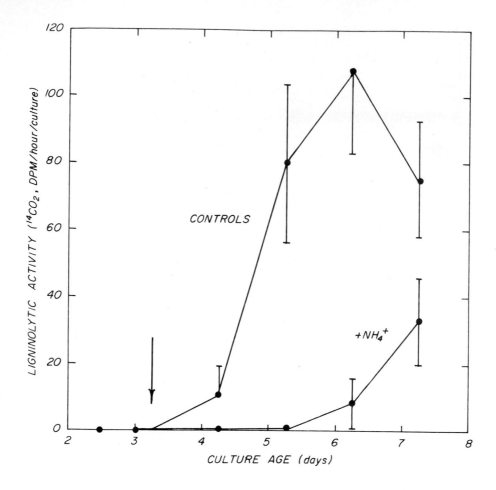

M 146 194

FIGURE 2.   Effect of NH$_4^+$ addition on appearance of ligninolytic activity in cultures of *Phanerochaete chrysosporium*. Immediately after inoculation each culture contained, in a total of 10 ml of basal medium, 1.2 mM ammonium tartrate, 100 mg glucose, and 0.01 M sodium *o*-phthalate as buffer (pH 4.5). Stationary cultures were grown at 39°C under 100% O$_2$. 78 hr after inoculation (arrow), 12 μ mol of ammonium tartrate in buffer were added to 6 of 12 replicate cultures. Control cultures received sodium tartrate in buffer.

ibly detectable decrease in activity for 6 hr. However, within 16 hr, a substantial suppression was evident, and this persisted for 40 to 50 hr. Analyses disclosed that the NH$^+_4$ added to the 3-day and older cultures was depleted within 16 hr.

Our interpretation of these results is that the ligninolytic enzyme system, or an essential proteinaceous component of it, is synthesized in response to N starvation. A role of N starvation is indicated not only by the sequential relationships observed between N depletion, ammonium permease derepression, and appearance of ligninolytic activity, but also by the effect that adding NH$^+_4$ has on activity and the demonstrated adverse effects of excess nutrient nitrogen demonstrated earlier. An indirect rather than a direct role of ammonium ion is indicated by the length of elapsed time between the onset of nitrogen starvation and the appearance of ligninolytic activity.

Two other physiological events apparently are initiated by nitrogen starvation in *P. chrysosporium*. First, phenol-oxidizing enzyme activity, which is weak in Strain ME-

446 grown as described here, appeared between the third and fourth day of incubation, just as ligninolytic activity did. Second, a secondary metabolite, veratryl alcohol, was discovered in cultures 4 days old or older. The compound was shown to be synthesized *de novo* from glucose.[10]

$$CH_2OH$$

—OCH$_3$

OCH$_3$

VERATRYL ALCOHOL, I

A direct relationship, if any, between the appearance of ligninolytic activity and synthesis of veratryl alcohol is not apparent. Addition of the alcohol to new cultures did not alter the time of appearance of the ligninolytic system. Interestingly, [U - ^{14}C]-veratryl alcohol was slowly decomposed to $^{14}CO_2$ by cultures that were 4 days old or older (perhaps by the ligninolytic system).

Phenol-oxidizing enzymes are considered to play a role in lignin metabolism by white-rot fungi[11] (see Volume II, Chapters 1 and 14). If this is the case, then it is to be expected that the oxidase activity and ligninolytic activity should appear simultaneously. We have not yet examined the effect of cycloheximide or added NH^+_4 on appearance of phenol-oxidizing activity or of veratryl alcohol. (It would also be interesting to examine low concentrations of cycloheximide, since this inhibitor has been shown to stimulate laccase secretion in *Neurospora*[12]).

The apparently simultaneous appearance of the ligninolytic system, veratryl alcohol, and phenol-oxidizing activity in response to nitrogen starvation suggests that all are part of a shift in physiology to an "idiophasic" or secondary metabolic phase. Nitrogen-mediated shifts to secondary metabolism have been reported in other filamentous fungi. For example, synthesis of the secondary metabolites (bikaverin by *Giberella fijukuroi*,[13] nigeran by *Aspergillus aculeatus*[9] and two phenolic compounds by *A. fumigatus*,[14] occur in response to depletion of nutrient nitrogen.

The fact that the ligninolytic system is not induced by lignin suggests that it has other substrates and is a relatively nonspecific enzyme system. Nonspecificity seems almost certain when one simply considers the diversity of lignin subunit structures that are metabolized. Another experimental indication of nonspecificity is the observation that kraft lignin and lignin sulfonates, both heavily modified industrial pulping by-product lignins, are still substantially degradable by white-rot fungi.[15,16] The relatively low rates of lignin metabolism — in comparison to glucose metabolism, for example — probably reflect low specificity and noninducibility, and the low rates, as well as the timing of synthesis of the system, probably help explain the fact that lignin is not a growth substrate.

IV. CONCLUSIONS

These investigations show that lignin metabolism by *P. chrysosporium* occurs in a simple, defined medium, but that control of several culture parameters — particularly O_2, nutrient N, and pH — is essential for maximum rates. Synthesis of the ligninolytic enzyme system, or an essential proteinaceous component of it, is not induced by lignin, but accompanies a physiological shift to secondary metabolism, a shift which is initiated by nutrient nitrogen starvation. These findings have obvious practical implications for any scheme purporting to produce ligninolytic enzymes or to use ligninolytic fungi.

Many unknowns remain. The research described here begs the question of why lignin metabolism is associated with "secondary metabolism," and what the mechanism is whereby nitrogen nutrition mediates appearance of the ligninolytic system. Other unknowns that await further research include specific reactions catalyzed by the ligninolytic enzymes, physical characteristics and cellular locations of the specific enzymes, and the possibility of enhancing activity by genetic or physiological means.

V. SUMMARY

Investigations were directed at defining the culture parameters important for lignin metabolism and describing some of the physiological events that accompany appearance of ligninolytic activity in white-rot fungi. The studies have been conducted with shallow batch cultures in a chemically defined liquid medium, using as the assay for lignin metabolism conversion of synthetic ^{14}C-lignins to ^{14}CO$_2$. *Phanerochaete chrysosporium* has been used in all experiments. (Limited experiments with *Coriolus versicolor* have given similar results.) Results have shown that lignin metabolism is promoted in cultures maintained under an atmosphere of 100% O$_2$, in cultures grown with low levels of nutrient nitrogen, in the presence of a required suitable growth substrate such as glucose or cellulose, and in cultures buffered at approximately pH 4.5. Under optimized conditions, ligninolytic activity appears in cultures following cessation of linear growth and the onset of nitrogen starvation. Experiments with cyclohexamide demonstrated that protein synthesis is involved in the appearance of ligninolytic activity and activity occurs irrespective of the presence of lignin. Lignin does not induce additional activity. These studies have (a) made it possible consistently to produce active ligninolytic cultures in a simple, defined medium, provided that key culture parameters are controlled, and (b) shown that the ligninolytic system or an essential component of it is synthesized as part of a series of secondary metabolic events initiated by nutrient nitrogen starvation.

ACKNOWLEDGMENTS

Research summarized here was supported in part by NSF Grant No. PCM 76-11144 and has involved input by several colleagues. The technical assistance of L. F. Lorenz and Kathleen Moore is gratefully acknowledged.

REFERENCES

1. **Kirk, T. K., Connors, W. J., Bleam, R. D., Hackett, W. F., and Zeikus, J. G.,** Preparation and microbial decomposition of synthetic [^{14}C]-lignins, *Proc. Natl. Acad. Sci. U.S.A.*, 72, 2515, 1975.
2. **Kirk, T. K., Connors, W. J., and Zeikus, J. G.,** Requirement for a growth substrate during lignin decomposition by two wood-rotting fungi, *Appl. Environ. Microbiol.*, 32, 192, 1976.
3. **Kirk, T. K., Schultz, E., Connors, W. J., Lorenz, L. F., and Zeikus, J. G.,** Influence of culture parameters on lignin metabolism by *Phanerochaete chrysosporium*, *Arch. Microbiol.*, in press.
4. **Keyser, P., Kirk, T. K., and Zeikus, J. G.,** The ligninolytic enzyme system of *Phanerochaete chrysosporium:* synthesized in the absence of lignin in response to nitrogen starvation, *J. Bacteriol.*, 135, 790, 1978.
5. **Burdsall, H. H. and Eslyn, W. E.,** A new *Phanerochaete* with a *chrysosporium* imperfect stage, *Mycotaxon*, 1, 123, 1974.
6. **Dagley, S.,** Microbial catabolism, the carbon cycle and environmental pollution, *Naturwissenschaften*, 65, 85, 1978.

7. **Kirk, T. K. and Chang, H.-M.,** Decomposition of lignin by white-rot fungi. II. Characterization of heavily degraded lignins from decayed spruce, *Holzforschung,* 29, 56, 1975.
8. **Kirk, T. K., Yang, H. H., and Keyser, P.,** The chemistry and physiology of the fungal degradation of lignin, in *Developments in Industrial Microbiology,* Vol. 19, Underkofler, L. A., Ed., American Institute of Biological Sciences, Washington, D. C., 1978, 51.
9. **Gold, M. H., Mitzel, D. L., and Segel, I. H.,** Regulation of nigeran accumulation by *Aspergillus aculeatus, J. Bacteriol.,* 113, 856, 1973.
10. **Lundquist, K. and Kirk, T. K.,** *De novo* synthesis and decomposition of veratryl alcohol by a lignin-degrading Basidiomycete, *Phytochemistry,* 17, 1676, 1978.
11. **Ander, P. and Eriksson, K.-E.,** The importance of phenol oxidase activity in lignin degradation by the white-rot fungus *Sporotrichum pulverulentum, Arch. Microbiol.,* 109, 1, 1976.
12. **Froehner, S. C. and Eriksson, K.-E.,** Induction of *Neurospora crassa* laccase with protein synthesis inhibitors, *J. Bacteriol.,* 120, 450, 1974.
13. **Bu'Lock, J. D., Detroy, R. W., Hostalek, A., and Munum-Al-Shakarchi, A.,** Regulation of secondary biosynthesis in *Giberella fujikuroi, Trans. Br. Mycol. Soc.,* 62, 377, 1974.
14. **Ward, A. C. and Packter, N. M.,** Relationship between fatty-acid and phenol synthesis in *Aspergillus fumigatus, Eur. J. Biochem.,* 46, 323, 1974.
15. **Hiroi, T. and Eriksson, K.-E.,** Microbiological degradation of lignin. 1. Influence of cellulose on the degradation of lignins by the white-rot fungus *Pleurotus osteatus, Sven. Papperstidn.,* 79, 157, 1976.
16. **Lundquist, K., Kirk, T. K., and Connors, W. J.,** Fungal degradation of kraft lignin and lignin sulfonates prepared from synthetic ^{14}C-lignins, *Arch. Microbiol.,* 112, 291, 1977.
17. **Fenn, P. and Kirk, T. K.,** Lignolytic system of *Phanerochaete chrysosporium:* inhibition by *o*-phthalate, *Arch. Microbiol.,* in press.

Chapter 5

GENETIC AND BIOCHEMICAL STUDIES ON *PHANEROCHAETE CHRYSOSPORIUM* AND THEIR RELATION TO LIGNIN DEGRADATION

Michael H. Gold, Therese M. Cheng, Kanit Krisnangkura,* Mary B. Mayfield, and Lawrence M. Smith**

TABLE OF CONTENTS

*   Present address: Department of Chemistry, Chiang Mai University, Chiang Mai, Thailand.
** Present address: Fish Pesticide Research Lab, U.S. Department of Interior, Columbia, Mo. 65201.

I. INTRODUCTION

The fungus *Phanerochaete chrysosporium (Sporotrichum* in its imperfect state)[1] and other white-rot species have potential for application in a variety of schemes for the commercial processing of lignocellulose. These organisms are especially adapted for removing lignin from wood, and *P. chrysosporium* has been used in numerous studies concerned with lignin degradation[2,3] and cellulose degradation.[4] The realization of these potential applications would be considerably enhanced, however, if genetic methods for selecting strains with superior capacities were available.

Although several mutants of *Sporotrichum pulverulentum* have been isolated,[3,5] little is known about the genetic aspects of its life cycle except for that described in a single morphological study.[1] Conspicuously lacking are reports on the isolation of suitable genetic marker strains, as well as reports of complementation, incompatibility, or recombination with this potentially valuable organism. Careful elucidation of these phenomena is required before any sophisticated fungal biochemical genetic studies can be undertaken.[6] Of the many hundreds of white-rot fungi, *P. chrysosporium* was chosen as the object of this study because it efficiently degrades both lignin and cellulose, because it produces asexual spores (conidia) prolifically — an advantage for genetic manipulation — and because it has been shown to form sexual fruiting structures in culture.[1]

II. TECHNIQUES FOR THE SELECTION OF MUTANTS

P. chrysosporium is filamentous, i.e., it has diffuse, unrestricted growth on ordinary agar medium. Our first task was to find a relatively nontoxic paramorphogenic agent which would yield compact colonies, permitting high plating densities, high viability of germinating conidia, and little inhibition of conidiation.[7] Figure 1 shows that in the presence of 4% sorbose and low levels of deoxycholate, induction of colonies could be attained after plating a dilute suspension of conidia. Germination was approximately 50% with this medium and conidiation was excellent. In addition, Figure 1 shows that this master copy could be replica plated onto supplemented and unsupplemented media, using velveteen discs[7] on wooden blocks, as described for bacteria by Lederberg and Lederberg.[8]

III. AUXOTROPHIC MUTANTS

Using these techniques of colony isolation and replica plating, we have mutagenized conidia and isolated several auxotrophic marker strains. Work with several of our auxotrophic strains is described below.

IV. COMPLEMENTATION AND HYMENIUM FORMATION

Sophisticated biochemical genetic studies with this organism require a much more detailed knowledge of the genetic aspects of its life cycle. The life cycle of a typical basidiomycete[9] falls into three stages: homokaryotic haploid, dikaryotic haploid, and homokaryotic diploid.

Basidiomycete sexual systems are grouped into two classes — heterothallic and homothallic — depending upon whether single basidiospores are self-sterile or self-fertile. Single basidiospores from heterothallic fungi will usually not give rise to a mycelium bearing fruit bodies. If mixtures of these spores are cultured, some of the combinations will lead to a dikaryotic mycelium after cell fusion and eventually to fruiting structures. In the fruiting structure, nuclear fusion and ultimately meiosis regenerate haploid bas-

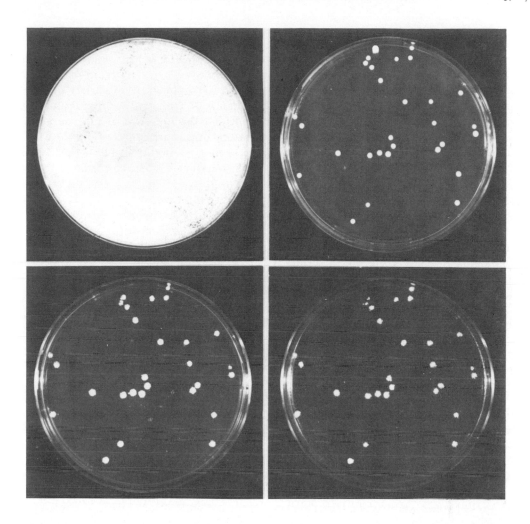

FIGURE 1. Upper Left: Growth of *P. chrysosporium* on medium containing inorganic salts, glucose, hydrolyzed casein, and vitamins. Upper Right: Growth from a conidial inoculum on the above medium plus 4% sorbose and 0.01% deoxycholate. Lower Left: Replica of the plate on upper right on supplemented medium. Lower Right: Replica of the plate on upper right on minimal medium.

idiospores. In some species, both monokaryotic and dikaryotic mycelia are able to form asexual spores (conidia). The compatibility factors controlling cell fusion in heterokaryotic fungi are termed dipolar or tetrapolar, depending on whether one or two sets of alleles are involved.

It is our intent in these studies to determine what aspects of this life cycle are exhibited by *P. chrysosporium*. Using riboflavin- and cysteine-auxotrophic marker strains isolated in our laboratory, we have attempted to demonstrate some of these events with this fungus. The two auxotrophic strains were plated on unsupplemented medium, and the heterokaryon was apparently produced because wild-type growth ensued.[10] The heterokaryon was then allowed to form conidia. When these conidia were plated on a colony-inducing medium, colonies with wild-type phenotype were produced. Colonies with wild-type phenotype were not produced, however, if conidia from the two auxotrophic strains were merely mixed and plated under colony-inducing conditions. These results indicate that heterokaryon formation rather than mere cross-feeding took place. It also indicates that the organism exhibits dikaryotic mycelial and conidial states. The dikaryotic mycelia do not make clamp connections, confirming an earlier observation.[1]

We have also attempted to define the conditions required for fruiting. Hymenium formation has been found to occur in as little as 10 days when the organism is plated on a medium consisting of 0.02% yeast extract and 1% glucose.[11] A variety of other starvation conditions will also lead to the induction of fruit-body formation. These include media containing a low level of nitrogen in the presence of glucose as the carbon source, or media containing an alternate carbon source such as cellulose. We are now investigating the role of 3′ to 5′-cyclic adenosine monophosphate (cAMP) in fruit-body formation. Low levels of cAMP will also induce hymenium formation when the organism is growing on Vogel's medium[12] plus glucose. This suggests that starvation conditions may lead to an increase in the intracellular levels of cAMP in *P. chrysosporium*. Details of this study have been published elsewhere.[11] cAMP has been shown to play a role in fruit body formation in *Coprinus macrorhizus*.[13] Studies are now in progress to elucidate the genetics of sexuality in this organism. One question remaining is whether the sexual system is heterothallic or hemothallic. Crosses made from colonies produced from individual basidiospores should provide an answer to this question.

V. LIGNIN DEGRADATION

In order to assay for the capacity of various strains for biodelignification, we have utilized previously published methods[2] to measure the evolution of $^{14}CO_2$ from *P. chrysosporium* growing on specifically ^{14}C-methoxyl- , side chain- , and ring-labeled synthetic lignins (dehydrogenative polymerizates, DHPs). Each of these substrates was prepared from the corresponding ^{14}C-coniferyl alcohol synthesized in our laboratory. Table 1 shows the results of an experiment where both DHPs and monomeric phenols were used as substrates. *P. chrysosporium* can oxidize the methoxyl carbon of 3-^{14}C-methoxyl vanillic acid methyester and 3-^{14}C-methoxyl veratric acid methyl ester, as shown by previous workers using other fungi.[14] It is also evident that *P. chrysosporium* is able to degrade synthetic DHPs. The finding of Kirk et al.[15] (see Chapter 4, Volume II) that nitrogen and oxygen levels are critical for delignification were also confirmed. Table 1 shows that in the presence of 12 mM NH_4NO_3, the amount of $^{14}CO_2$ evolved from either methoxyl- or sidechain-labeled DHP is only approximately 15% of that evolved in the presence of 1.2 mM NH_4NO_3. The data are not shown, but 12 mM NH_4NO_3 also effectively inhibited $^{14}CO_2$ evolution when 3-^{14}C methoxyl vanillic acid methyl ester was the ^{14}C-substrate. Maximum $^{14}CO_2$ evolution occurred in our experiments when the reaction flasks were purged with 100% O_2.

LH-20 gel filtration profiles of the ^{14}C-DHP substrates indicate that the DHPs elute, as does blue dextran, in the void volume and approximately twice as fast as coenzyme B_{12} (mol wt = 1580). This suggests that the organism is not merely degrading a low molecular weight fraction of the ^{14}C-DHP substrates.

VI. PHENOLOXIDASE

A. Enzyme Activity

The phenoloxidase enzyme activity of *P. chrysosporium* was assayed after growing the organism in standing cultures. *o*-Anisidine and H_2O_2 were used as the substrates. Assay mixtures were acidified with HCl after incubation[16] and measured at 540 nm. The highest levels of peroxidase were found when the organism was cultured on glucose, yeast extract, and wood meal as previously shown.[5] Under these conditions, as much as 0.1 μmol of product was oxidized at 30°C per minute by 1 mℓ of crude filtrate from 1-week-old cultures.

TABLE 1

$^{14}CO_2$ Evolution after Incubation of Wild-Type *P. chrysosporium* with ^{14}C-Substrates for 20 Days[a]

Substrate	$^{14}CO_2$ Trapped (% of total radioactivity)
3-^{14}C methoxyl vanillic acid methyl ester 1.2 mM NH$_4$NO$_3$	36
3-^{14}C methoxyl veratric acid methyl ester 1.2 mM NH$_4$NO$_3$	34
^{14}C-side chain-labeled DHP 1.2 mM NH$_4$NO$_3$	20[b]
^{14}C-side chain-labeled DHP 12 mM NH$_4$NO$_3$	3
^{14}C-methyl-labeled DHP 1.2 mM NH$_4$NO$_3$	30[b]
^{14}C-methyl-labeled DHP 12 mM NH$_4$NO$_3$	5

[a] Incubation was at 38°C. All flasks contained 55 mM glucose as a primary carbon source, inorganic salts, vitamins, and 10 mM Phthalate buffer pH 4.5. Nitrogen concentration was as indicated.

[b] The conditions for lignin degradation where improved in later experiments.

B. Mutants

A number of studies have implicated phenoloxidases (PO) in lignin degradation, although the only biochemical genetic report is that of Ander and Eriksson.[5] We thought it worthwhile to extend their interesting study using the $^{14}CO_2$ assay instead of the less powerful chlorine consumption assay they used. Using specifically labeled DHPs, we hoped to be able to determine which step(s), if any, in the DHP degradation pathway were catalyzed by peroxidase. Mutant colonies were visualized after mutagenizing and germinating conidia on plates containing o-anisidine, sorbose, and inorganic salts. Several PO⁻ mutants were isolated. They were not able to oxidize o-anisidine in our plate assay even after 3 weeks at 28°C. When these mutants were assayed as described above for peroxidase in standing cultures, no activity was detectable. Details of these studies will be published later.

The PO⁻ mutant which Ander and Eriksson[5] isolated could not degrade lignin, but their experiments were conducted with relatively high concentrations of kraft lignin present (0.2%). In addition, cellulose was used as the main carbon source. Those authors concluded that POs may function in regulating the production of lignin and polysaccharide-degrading enzymes by oxidation of phenols. In our present experiments, glucose is the main carbon source, thereby eliminating the need for the production of polysaccharide degrading enzymes. In addition, only low concentrations of ^{14}C-DHPs were used.

One of our peroxidase mutants, Prx⁻−1 showed less than 5% of the activity of the wild type against ring-, side chain-, and methoxyl-labeled DHPs in the assay described above. A phenotypic revertant produced by remutagenesis of this mutant, however, had normal activity against all three substrates.[17] While these studies are not complete, they suggest that peroxidase may play some role other than a regulatory one in lignin degradation by *P. chrysosporium*.

VII. FUTURE PERSPECTIVES

Although this symposium indicates a considerable amount of activity in the area of biodelignification, the field is still in its infancy. Needed are more efficient native strains of fungi and bacteria capable of delignification. An organism which can use lignin as a sole source of carbon would be especially helpful for genetic studies. The elucidation of the genetic life cycle of *P. chrysosporium* is essential if we are to construct superior mutant strains with that organism. Finally, the general availability of ^{14}C-labeled DHPs should help our endeavors toward the elucidation of the lignin degradative pathway and the characterization of the enzymes involved.

VIII. SUMMARY

The induction of colonial growth of *P. chrysosporium* on a medium containing L-sorbose and deoxycholate allowed replica plating and the selection of several auxotrophic mutants. Complementation studies were conducted with these mutants and heterokaryons were produced. These studies indicated that both the mycelium and conidia can be dikaryotic. A medium inducing fruiting structures and basidiospore production was also devised. Alternate carbon sources, low levels of nitrogen, and cAMP all induce fruiting.

As an assay for lignin degradation by *P. chrysosporium*, $^{14}CO_2$ evolution from ^{14}C-DHPs was measured. High levels of nitrogen were found to repress lignin degradation by this fungus. A peroxidase (Prx$^-$) mutant and a phenotypic revertant of *P. chrysosporium* were isolated. The Prx$^-$ mutant was not able to degrade lignin; the revertant regained this capacity.

ACKNOWLEDGMENTS

This work was supported by the Crown-Zellerbach, International Paper and Weyerhaeuser companies and by the Gottesman Foundation.

REFERENCES

1. **Burdsall, H. H. and Eslyn, W. E.**, A new Phanerochaete with a *chrysosporium* imperfect state, *Mycotaxon,* 109, 123, 1974.
2. **Kirk, T. K., Connors, W. J., Bleam, R. D., Hackett, W. F., and Zeikus, J. G.**, Preparation and microbial decomposition of synthetic [14C] lignins, *Proc. Natl. Acad. Sci. U.S.A.,* 72, 2515, 1975.
3. **Ander, P. and Eriksson, K.-E.**, Influence of carbohydrates on lignin degradation by the white rot fungus *Sporotrichum pulverulentum, Sven. Papperstidn.,* 78, 643, 1975.
4. **Eriksson, K.-E. and Petterson, B.**, Extracellular enzyme system utilized by the fungus *Sporotricum pulverulentum* for the breakdown of cellulose, *Eur. J. Biochem.,* 5, 193, 1975.
5. **Ander, P. and Eriksson, K.-E.**, The importance of phenol oxidase activity in lignin degradation by the white rot fungus *Sporotrichum pulverulentum, Arch. Microbiol.,* 109, 1, 1976.
6. **Fincham, J. R. S. and Day, P. R.**, *Fungal Genetics,* 3rd ed., Blackwell Scientific, Oxford, 1971, 19.
7. **Gold. M. H. and Cheng, T. M.**, Induction of colonial growth and replica plating of the white rot basidiomycete *Phanaerochaete chrysosporium, Appl. Env. Microbiol.,* 35, 1223, 1978.

8. **Lederberg, J. and Lederberg, E. M.**, Replica plating and indirect selection of bacterial mutants, *J. Bacteriol.*, 63, 399, 1952.
9. **Raper, J. R.**, *Genetics of Sexuality in Higher Fungi*, Ronald Press, New York, 1966, 9.
10. **Gold, M. H. and Cheng, T. M.**, Isolation of auxotrophic mutants and complementation studies with *P. chrysosporium*, in preparation.
11. **Gold, M. H. and Cheng, T. M.**, Conditions for fruit body formation in the wood rotting basidiomycete *Phanerochaete chrysosporium*, *Arch. Microbiol.*, 121, 37, 1979.
12. **Vogel, H. J.**, Distribution of lysine pathways among fungi: evolutionary implications, *Am. Nat.*, 98, 435, 1964.
13. **Uno, I. and Ishikawa, T.**, Chemical and genetical control of induction of monokaryotic fruiting bodies in *Coprinus macrorhizus*, *Mol. Gen. Genet.*, 113, 228, 1971.
14. **Haider, K. and Trojanowski, J.**, Decomposition of specifically ^{14}C-labelled phenols and dehydropolymers of coniferyl alcohol as models for lignin degradation by soft and white rot fungi, *Arch. Microbiol.*, 105, 33, 1975.
15. **Kirk, T. K.**, Personal communication, 1978.
16. **Gascon, S. and Lampen, J. O.**, Purification of internal invertase of yeast, *J. Biol. Chem.*, 243, 1567, 1968.
17. **Gold, M. H., Cheng, T. M., and Mayfield, M.**, Biochemical genetic studies on the role of *P. chrysosporium* peroxidase in lignin degradation, in preparation.

Chapter 6

MICROBIAL METABOLISM OF LIGNIN-RELATED AROMATICS

Toshio Fukuzumi

TABLE OF CONTENTS

I. METABOLISM OF LIGNIN-RELATED AROMATICS BY WHITE-ROT FUNGI

A. Introduction

Microbial lignin degradation is significant not only for humus formation, but also for the utilization of lignins found in pulping waste liquors. Therefore, white wood-rotting fungi, which degrade lignin, have attracted considerable attention.

White-rot fungi can be differentiated from brown-rot fungi by the Bavendamm reaction, which indicates the presence of phenol-oxidizing enzymes. Since laccase from the same fungi also polymerizes coniferyl alcohol to lignin, we became interested in examining this apparent contradiction. We selected fungi of rapid growth: *Polystictus sanguineus* and *Poria subacida* in early screenings, and *Trametes* sp. and *Tinctoporia borbonica* in more recent ones.

In my early study on the degradation of lignin by *P. subacida*, guaiaclypyruvic acid, vanillic acid, and guaiacylglycerol-β-coniferyl ether were detected in cultures containing Braun's conifer lignin and decayed conifer wood meal.[1] At the time, it was difficult to study the mechanism of degradation and metabolism of lignin by the fungus insofar as lignin is a high molecular weight polymer of random structure and weight. Thus, our work has been carried out primarily with a simple and "dimeric" aromatics structurally related to lignin.

B. Decarboxylative Oxidation of Aromatic Acids by Laccase-Type Enzymes[2]

1. Enzyme Preparation

Crude enzyme was prepared by extraction of wood meal (*Picea jezoensis*) decayed by *Poria subacida* or *Polystictus sanguineus* for 2 months. The 25 to 80% fraction from ammonium sulfate precipitation of the extract was used.

Partially purified enzyme was also prepared from culture filtrates of *P. sanguineus*, grown on a medium which initially contained an inducer, such as 2,5-xylidine or guaiacylpyruvic acid. The enzyme precipitate obtained from 50 to 80% ammonium sulfate saturation was eluted through a column of DEAE cellulose, and the fraction showing catechol oxidase activity was used.

2. Oxidation of Lignin-Related Aromatic Acids

Oxygen uptake and carbon dioxide liberation from vanillic acid, vanilloylformic acid, and guaiacylpyruvic acid with each enzyme preparation were measured using Warburg's manometric technique (Table 1). Oxidation was accompanied by carbon dioxide liberation, probably indicating that decarboxylation took place. The partially purified enzyme preparations, G80 (induced by guaiaclypyruvic acid) and X80 (induced by 2,5-xylidine), both showed approximately 0.7 equivalent moles of oxygen consumption and of carbon dioxide liberation in the reactions with vanillic acid, providing further support for a decarboxylation scheme (see also Volume II, Chapter 1). As a third check, the reaction products were studied.

The reaction mixtures of vanillic acid with G80 and with X80 were extracted with chloroform. The extracts were condensed and applied to thin-layer chromatography (TLC) (silica gel; solvent, $CHCl_3$:benzene = 4:1). Methoxy-*p*-benzo-quinone, very conceivably a decarboxylation product from vanillic acid, and vanillin, a product of an as yet unexplained reduction of vanillic acid, were detected in both extracts. The ESR spectrum of the reaction mixture of vanillic acid with crude enzyme from *Poria subacida* (50 to 80% fraction) showed a signal (g = 2.005) of spin resonance due to a radical, suggesting a radical coupling-type mechanism; we proposed the reaction scheme shown in Figure 1.

TABLE 1

Amount of Oxygen Uptake and Carbon Dioxide Liberation per Mole of Substrate

	Substrate					
	Guaiacylpyruvic acid (mol)		Vanilloylformic acid (mol)		Vanillic acid (mol)	
Enzyme	O_2	CO_2	O_2	CO_2	O_2	CO_2
Mixed enzyme of Poria Subacida	—	—	1.1	0.60	0.89	0.56
Mixed enzyme of Polystictus sanguineus	1.05	0.60	0.81	0.43	0.57	0.44
G80 enzyme	1.0	0.43	0.60	0.23	0.69	0.71
X80 enzyme	1.0	0.45	weak	weak	0.67	0.67

Substrate

Guaiacylpyruvic acid   Vanilloylformic acid   Vanillic Acid

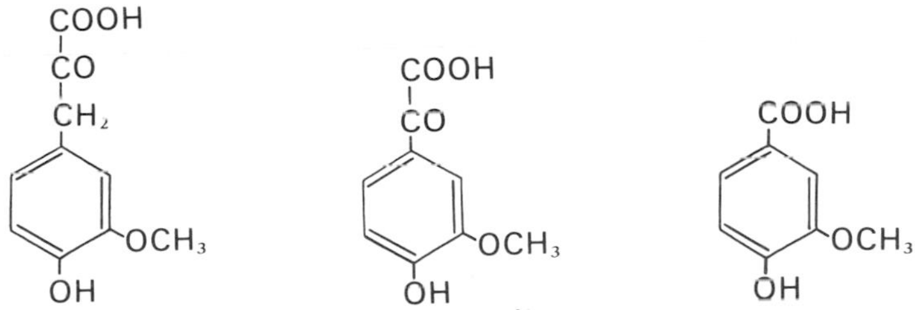

Reproduced from Fukuzumi, T., Uraushihara, S., Ohashi, T., and Shibamoto, T., *Mokuzai Gakkaishi*, 10, 242, 1964. With permission.

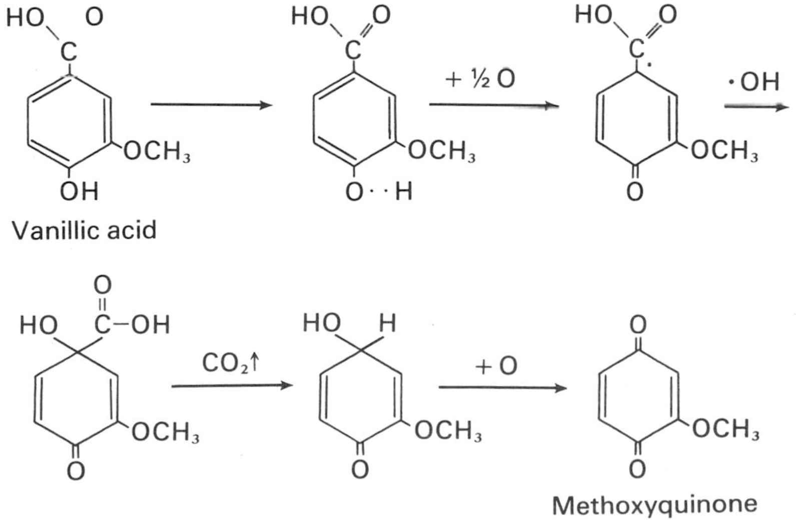

FIGURE 1. A scheme for oxidation of vanillic acid to methoxy-*p*-benzequinone by the enzyme of a wood-rotting fungus. (Reproduced from Fukuzumi, T., Uraushihara, S., Ohashi, T., and Shibamoto, T., *Mokuzai Gakkaishi*, 10, 242, 1964. With permission.)

C. Reduction of Aromatic Acids by White-Rot Fungi

In early studies of metabolic products from aromatics by *Polystictus sanguineus*, we found reduced, demethylated, and demethoxylated products, namely veratraldehyde, vanillic acid, and *p*-hydroxybenzoic acid, in veratric acid cultures.[3,4] Veratraldehyde was the main product. Recently, the reduction of aromatic acids, this time by *Trametes* sp., has been studied again in connection with the decolorization of kraft pulping waste liquor.

1. Formation of Coniferyl Alcohol from Ferulic Acid in Culture[5]

The fate of ferulic acid and its metabolic intermediates were followed through cultivation of *Trametes* sp. in 2 l of medium using a mini-fermenter. The best culture condition was a ferulic acid concentration of 3.3 mM with 0.5% glucose and 0.5% ethanol in a medium with an initial pH of 5.5. A small sample of the culture (50 ml) was periodically removed and analyzed by gas-liquid chromatography (GLC) after trimethylsilation. Coniferyl alcohol appeared after 48 hr of cultivation and reached a maximum concentration with approximately 40% yield at 90 hr of cultivation. After 120 hr, vanillyl alcohol became the predominant product, being formed from vanillic acid, which was also identified. Methoxyhydroquinone appeared before the transformation of vanillic acid to vanillyl alcohol. Vanillyl alcohol accumulated as an end product, while methoxyhydroquinone diminished after long incubation. Figure 2 illustrates these changes, showing the concentration of metabolites during incubation. The metabolic pathways of ferulic acid are shown in Figure 3.

2. Reduction of 3,4,5-Trimethoxybenzoic Acid by an Enzyme System[6]

a. Enzyme Preparation

Enzyme solution I: mycelia of *Trametes* sp. (120 g, frozen), were homogenized in a buffer solution of 0.2 M tris-HCl containing cysteine, mercaptethanol, and EDTA, each at 10 mM (pH 8.0), with 100 g of alumina powder using a Polytron® homogenizer. The homogenate (readjusted to pH 8.0) was centrifuged at 12,000 × g to collect an enzyme-containing supernatant free of cell debris. This crude solution (I) contained 10.6 mg of protein per milliliter.

Enzyme solution II: 290 ml of crude enzyme solution (I) were treated with 13 ml of 2% protamine solution. After 30 min, the precipitate was removed by centrifugation (12,000×g). The supernatant (solution II 300 ml) contained 10.6 mg of protein per milliliter.

Enzyme solution III: The 33 to 65% fraction from ammonium sulfate precipitation of solution II was desalted with Sephadex G-25® and concentrated to 53 ml with Sephadex G-50®. This solution (III) contained 13.7 mg of protein per milliliter.

Enzyme solution IV: solution III (30 ml) was fractionated by a column of DEAE cellulose (0 to 0.4 M NaCl in 0.01 M phosphate buffer, pH 7.5). No reductase activity was found in the eluted fractions; a fraction (6 ml) having aldehyde oxidase activity was retained (solution IV, 0.25 mg of protein per milliliter).

b. Cofactor Requirements for Reduction

As shown in Table 2, NADPH, ATP, and Mg^{2+} were necessary for the enzymatic reduction of the trimethoxybenzoic acid. The addition of CoA was inhibitory to the production of 3,4,5-trimethyxybenzyl alcohol, and the addition of NaF accelerated enzyme activity because NaF is an inhibitor of ATP pyrophosphatase. The enzyme system did not reduce other aromatic acids: ferulic acid, cinnamic acid, vanillic acid, veratric acid, and syringic acid.

Cofactor requirements for the enzymatic reduction of 3,4,5-trimethoxybenzaldehyde are shown in Table 3. NADPH was necessary for this reduction, and NADH could substitute for NADPH. GSH suppressed the formation of alcohol.

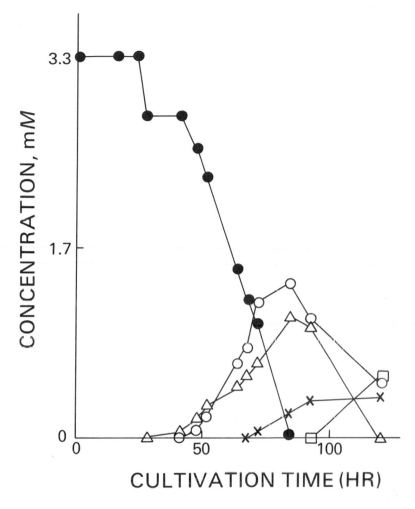

FIGURE 2. Changes in concentrations of ferulic acid and its products during incubation with *Trametes* sp. O ferulic acid, Δ vanillic acid, ⊙ coniferyl alcohol, □ vanillyl alcohol, × methoxyhydroquinone.

c. Cofactor Requirement for Reverse Oxidation[7]

The enzyme solution III, which was used in the above reduction, also oxidized aromatic aldehydes (Table 3). Study of the oxidase was carried out using enzyme solution IV, which had no reductase activity. Compared with reduction, oxidation of aromatic aldehydes by the enzyme occurred with little substrate specificity (Table 4). Cofactors required for this enzymatic reaction were $NADP^+$ or NAD^+. Addition of ATP to the reaction mixture was inhibitory, resulting in a 94% decrease of oxidase activity (Table 5).

d. Schemes for Reduction and Oxidation

Schemes for the enzymatic reduction of 3,4,5-trimethoxybenzoic acid to its benzyl alcohol and oxidation of 3,4,5-trimethoxybenzaldehyde to its acid are shown in Figure 4. The reduction proceeds with the same equation as the enzyme system of *Neurospora* sp. studied by Zenk and Gross[8], but the substrate specificity is different. The enzyme of *Neurospora* sp. showed no activity toward 3,4,5-trimethoxybenzoic acid.[8]

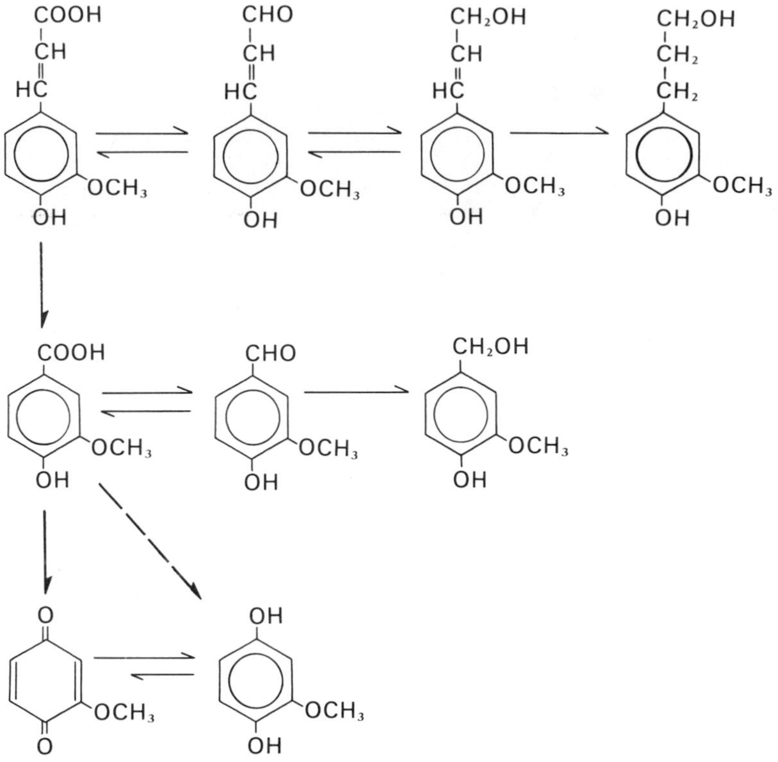

FIGURE 3.   Metabolism of ferulic acid by *Trametes.* sp.

TABLE 2

Cofactor Requirements for the Reduction of 3,4,5-Trimethoxybenzoic Acid by a Partially Purified Enzyme System from *Trametes* sp.

Reaction mixture	Alcohol formed (nmol/hr/mg protein)	Relative activity (%)
Complete system[a]	24.6	100
+ CoA[a]	3.67	15
+ NADH,[a] − NADPH	0.00	0
− NADPH	0.00	0
− ATP	0.00	0
− MgCl₂	0.00	0
− NaF	0.37	26
− GSH	8.67	35
Boiled enzyme	0.00	0

[a]   Complete system: 1 μmol 3,4,5-trimethoxybenzoic acid, 1 μmol NADPH, 10 μmol ATP, 10 μmol MgCl₂, 10 μmol NaF, 2 μmol GSH, 0.4 mℓ enzyme solution (III), and enough phosphate buffer (200 μM, pH 7.5) to make total vol 2.0 mℓ. Incubated at 30°C for 3 hr. + CoA: 0.5 μmol CoA added, + NADH: 1 μmol NADH added.

D. Degradation of Lignin-Related Aromatics[7]

1. Degradation of "Dimers" in Culture

Tinctoporia borbonica was cultivated in media which contained "dimer" aromatics (3.3 mM), together with 0.5% glucose and 0.5% ethanol. After 5 days of cultivation with shaking, the cultures were extracted with 0.1 M NaOH solution, and the extracts

TABLE 3

Cofactor Requirements for the Reduction of 3,4,5-Trimethoxybenzaldehyde by a Partially Purified Enzyme System from _Trametes_ sp.

Reaction mixture	Alcohol formed (nmol/hr/mg protein)	Relative activity (%)	Acid formed (nmol/hr/mg protein)
Complete system[a]	6.87	100	273
+ NADH, − NADPH	0.00	0	387
− NADPH	0.00	0	42.7
− GSH	32.50	473	278
+ MgCl₂, + NAF	5.93	86	232
Boiled enzyme	0.00	0	0.0

[a] Complete system: 1 μmol 3,4,5-trimethoxybenzaldehyde, 1 μmol NADPH, 2 μmol GSH, 0.4 m*l* enzyme (III) solution, and enough phosphate buffer (200 μ*M*, pH 7.5) to make the total vol 2 m*l*. Conditions of incubation and amounts of addition MgCl₂ and NaF are the same as in Table 2.

TABLE 4

Acid Formation from Aldehydes by an Enzyme System Partially Purified from _Trametes_ sp.

Substrate	Alcohol formed	Acid formed (nmol/hr/mg protein)
Cinnamyl aldehyde	0	143
Coniferyl aldehyde	0	13.3
Benzaldehyde	0	31.7
p-Hydroxybenzaldehyde	0	30.0
p-Methoxybenzaldehyde	0	233
Vanillin	0	57.3
Veratraldehyde	0	93.3
Syringaldehyde	0	71.0

Note: Reaction mixture: 1 μmol of substrate, 1 μmol NADPH, 2 μmol GSH, 0.4 m*l* enzyme solution (III), and enough phosphate buffer (200 μ*M*, pH 7.5) to make a total volume of 2.0 m*l*. Conditions of incubation are the same as in Table 2.

TABLE 5

Cofactor Requirements for the Oxidation of 3,4,5-Trimethoxybenzaldehyde by a Partially Purified Enzyme from _Trametes_ sp.

Reaction mixture	Acid formed (nmol/hr/mg protein)	Relative activity (%)
Complete system[a]	7.07	100
+ NAD, − NADP	2.70	38
− NADP	0.00	0
+ ATP	0.43	6
+ ATP, + NAD, −NADP	0.00	0
Boiled enzyme	0.00	0

[a] Complete system: 1 μmol 3,4,5-trimethoxybenzaldehyde, 1 μmol NADP, 2 μmol GSH, 0.4 m*l* enzyme solution (IV) and enough phosphate buffer (200 μ*M*, pH 7.5) to make the total 2.0 m*l*. Incubated at 30°C for 3 hr. + ATP: 10 μmol ATP was added in complete system.

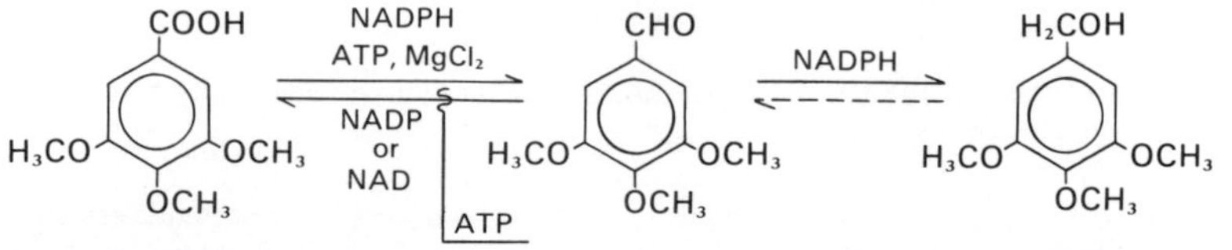

FIGURE 4. Cofactor requirements for the reduction of 3,4,5,-trimethoxy benzoic acid to its alcohol by a partially purified enzyme system of *Trametes* sp.

TABLE 6

Degradation of Lignin-Related Dimers by the Fungus, *Tinctoporia borbonica*

Substrate	λ_{max}[a] (nm)	Residual substrate[b] after 5 days (%)
Dehydrodivanillic acid	250	10
Guaiacylglycerol-β-coniferyl-ether	255	50
Dehydrodiisoeugenol	265	30
Dehydrodiferulic acid	350	28
Dehydrodiconiferyl alcohol	270	13
Pinoresinol	260	37
(4-Methoxy-benzyl)-ferulic acid	280	23
α-Conidendrin	245	18
3,3'-Dimethoxy-4,4'-dihydroxy-chalcone	430	7

[a] Measured in aq 0.1 *N* NaOH.
[b] Amount of substrate was calculated from the absorbance at λ_{max} and was set at 100% before cultivation.

measured for UV absorption (Table 6). Although all of the dimers were degraded to some degree, dehydrodivanillic acid and 3,3'-dimethoxy-4,4'-dihydroxychalcone were degraded exceptionally well. Metabolic products from these two dimers were investigated by GC-MS (gas chromatography and mass spectrometry) as TMSi (trimethylsilyl) derivatives. Vanillic acid was detected as a metabolite from the latter dimer. The other metabolic products have not yet been identified.

Trametes sp. was similarly cultivated in media containing dimer aromatics with a β-ether bond (3.3 m*M*) with glucose and ethanol (0.5% each). The metabolic products from each dimer were surveyed by GC-MS as TMSi derivatives. Syringic acid was produced from syringoylmethyl syringic acid ether (A), 3,4,5-trimethoxybenzoylmethyl syringic acid ether (B), and from α-syringoxy-β-hydroxypropioreratrone (C).

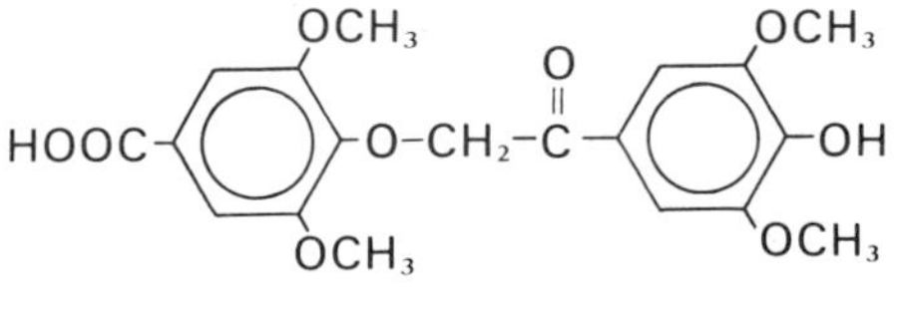

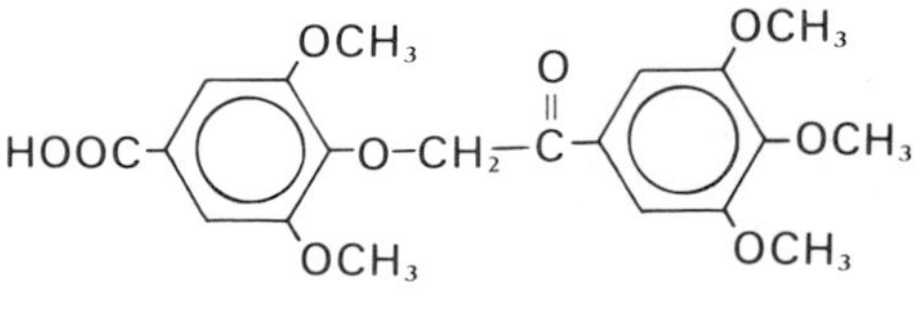

(C)   HOOC—⟨benzene ring with OCH₃ top and OCH₃ bottom⟩—O—CH—C—⟨benzene ring with OCH₃ top and OCH₃ right⟩
                                                    | ‖
                                                 CH₂OH   O

The structure (C) shows: $HOOC$-substituted aromatic ring bearing two OCH_3 groups, linked via $O-CH(CH_2OH)-C(=O)-$ to a second aromatic ring bearing two OCH_3 groups.

2. Degradation of Ferulic Acid by an Enzyme System[6]

Enzymatic degradation of ferulic acid was investigated using a crude enzyme preparation from *Trametes* sp., prepared as described in section, *Reduction of 3,4,5-Trimethoxybenzoic Acid by An Enzyme System*, in the presence of dithiothreitol to protect the SH groups. Vanillic acid was detected in the extract of the reaction mixture with CoA, ATP, and NAD^+ were added. Without these cofactors, little or no ferulic acid degradation occurred. Thus, degradation probably proceeds by the pathway proposed by Zenk[9] involving CoA, rather than by the pathway reported by French et al.[10] for *Polyporus hispidus*, which does not involve CoA.

3. Cleavage of Ether Bonds by an Enzyme System

An enzyme system that cleaves aryl-alkyl ether bonds in lignin was studied by isolating enzymes from mycelia of the white-rot fungus *Poria subacida*. The fungus grew well in a medium containing Braun's native lignin as the sole carbon source, as reported by Van Vliet.[11]

We first found that cleavage of the β-ether bond of veratrylglycerol-β-guaiacyl ether, indicated by the liberation of guaiacol, by an enzyme from *P. subacida* mycelia occurred only when NADH was added to the mixture.[12,13] In further studies concerning this enzymatic cleavage, guaiacylglycerol was formed repeatedly. For example, this compound was detected as its TMSi by GC-MS in the extract of the enzymatic reaction: veratrylglycerol-β-coniferyl ether, NADH, and enzyme.[14]

We also suspected that demethylation of the *p*-methoxyl group of the veratrylglycerol moiety of the substrate occurred. If an oxygenase demethylated the methoxyl group, formaldehyde should have been formed; the results of the formaldehyde determination shown in Table 7 confirm that this demethylation probably took place. Is demethylation preferred to cleavage of the β-ether bond, and are the enzymes participating in these two reactions separable? For elucidation of these questions, the following experiments were carried out.[15]

Fresh mycelia of *P. subacida* obtained from cultures containing lignin (1 g/ℓ) were extracted twice with cold 0.01 N NH₄OH solution. The first extract could not cleave the ether bonds. The second was divided into four fractions by ammonium sulfate saturation: a 20% fraction (E_5), a 40% (E_6), a 60% (E_7), and an 80% (E_8) with protein concentrations of 0.055 mg, 0.037 mg, 0.093 mg, and 0.062 mg/mℓ, respectively. Each enzyme solution had NADH-veratrylglycerol-β-guaiacyl ether oxidoreductase activity. Each enzyme and boiled enzyme E_6 were tested. The reaction mixture contained: enzyme solution, 2 mℓ; 10 mM veratrylglycerol-β-guaiacyl ether, 1 mℓ; 10 mM NADH, 1 mℓ; and 0.1 M acetic acid buffer solution of pH 4.5, 2 or 3 mℓ. The reaction mixtures were incubated at 30°C for 22 hr and then extracted with CHCl₃. The products of the reaction were determined by TLC and GC (Table 8). Enzyme fractions E_5, E_6, and E_7, in the presence of NADH, were able to cleave both the methoxyl ether and β-ether bonds. Enzyme fraction E_8 with NADH cleaved only the methoxyl ether bond. In each enzyme reaction with NADH, an unidentified carbonyl compound was detected on TLC. In the reactions without NADH, no products were detected except for E_5, which had the ability to cleave the β-ether bond even without the addition of NADH. This was probably due to the presence of an unknown cofactor in this enzyme solution or to a hydrolase with affinity for the β-ether bond structure.

TABLE 7

Determination of Formaldehyde and Guaiacol Produced in Reaction Mixture for Cleavages of Ether Bonds

Reaction system[a]	Formaldehyde (μg)	Guaiacol (μg)
VGE + GSH + NADH	6.46	1.73
VGE + GSH	1.94	trace
VGE + GSH + NADH + FeSO$_4$	7.39	trace
VGE + NADH + FeSO$_4$	8.69	trace
VGE + FeSO$_4$	7.44	trace
VGE	0.84	trace

[a] Reaction system: veratrylglycerol-β-guaiacylether, 8.25 mg; NADH, 10 mg; GSH, 0.3 mg; 0.01 M FeSO$_4$, 0.5 mℓ; enzyme (acetone precipitate from the buffer extract of mycelia of *Poria subacida*) solution, 1 mℓ; and acetic acid buffer (200 μM, pH 4.5) to a total vol of 4.3 mℓ. Incubation was at 26.5°C for 17.5 hr. Formaldehyde was determined by chromotropic acid reagent. Guaiacol was determined by GC.

TABLE 8

Activity of Fractionated Enzyme of *Poria subacida* for Cleavage of Methoxyl and Side Chain β-ether linkages[a]

Enzyme fraction	NADH	Guaiacol liberated (%) per substrate	Reaction products detected on TLC[b]	Ether bond cleaved
E$_5$	Without	0.4	Not detectable	β-Ether
E$_6$	Without	0	Nothing	
E$_7$	Without	0	Nothing	
E$_8$	Without	0	Nothing	
E$_5$	With	0.6	Guaiacylglycerol-β-guaiacylether	Methoxyl ether, β-ether
E$_6$	With	0.1	Guaiacylglycerol-β-guaiacylether, guaiacylglycerol	Methoxyl ether, β-ether
E$_7$	With	1.6	Guaiacol, guaiacylglycerol-β-guaiacylether	Methoxyl ether, β-ether
E$_8$	With	0	Guaiacylglycerol-β-guaiacylether	Methoxyl ether
E$_6$ boiled	With	0	None	None

[a] Substrate: 0.01 M veratrylglycerol-β-guaiacylether, 1 mℓ; 0.01 M NADH, 1 mℓ, or without NADH. Reaction: 22 hr at 30°C, pH 4.5
[b] Silica gel; spots detected by diazotized *p*-nitroaniline.

The enzyme activities of each fraction were not readily distinguishable, although fractions E$_5$ and E$_8$ were noticeably different. The fractions in between, E$_6$ and E$_7$, probably represent overlapping of fractions E$_5$ and E$_8$ to some extent, but they also seem to contain unique enzymes. E$_7$ was the only fraction to produce guaiacol in detectable amounts, and guaiacylglycerol was found only in the reaction mixture of fraction E$_6$. A hypothetical scheme of the cleavages made by each enzyme fraction, based on these results, is illustrated in Figure 5.

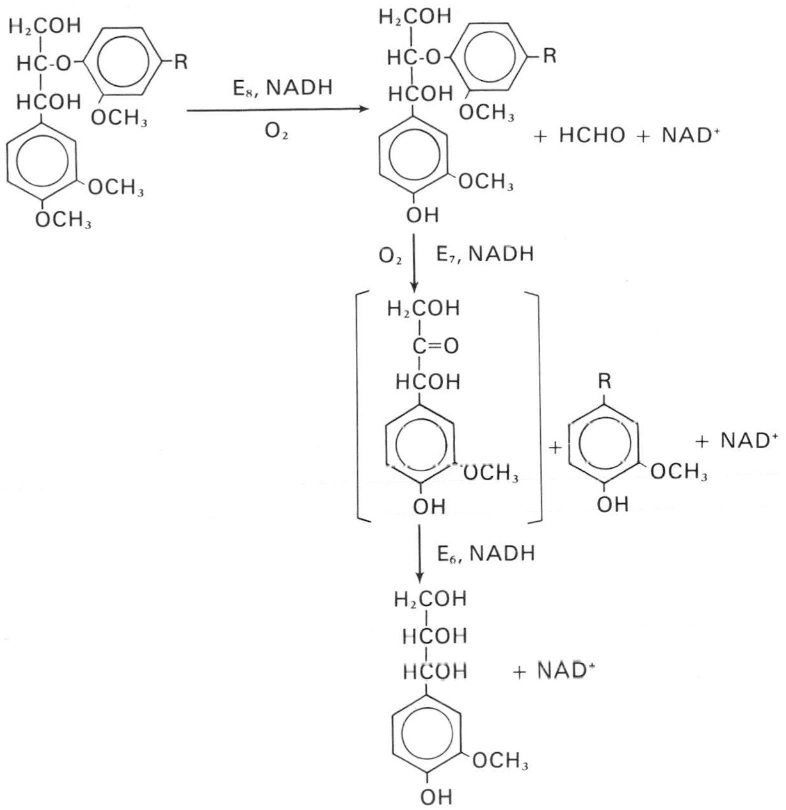

H₂COH
HC-O—⟨ ⟩—R   E₈, NADH   H₂COH
HCOH   OCH₃   ───────────→   HC-O—⟨ ⟩—R
                      O₂   HCOH   OCH₃   + HCHO + NAD⁺
OCH₃   OCH₃
OCH₃   OH

O₂ │ E₇, NADH

[H₂COH
  C=O
  HCOH   R
          ⟨ ⟩   ⟨ ⟩   + NAD⁺
          OCH₃   +   OCH₃
  OH]   OH

E₆, NADH

H₂COH
HCOH
HCOH   + NAD⁺
      ⟨ ⟩
      OCH₃
OH

FIGURE 5.   Hypothetical scheme of cleavage of ether bonds in veratryglycerol-β-ether compounds by an enzyme system of *Poria subacida*.

E. Discussion

There are two distinguishable types of wood-rotting fungi: white-rot and brown-rot. The latter type mainly decomposes cellulose and shows no Bavendamm color reaction (absence of extracellular phenol-oxidizing enzymes). The culture medium after growth of brown-rot fungi is strongly acidic, nearly pH 2, due to the accumulation of oxalic acid. The white-rot fungi primarily degrade lignin, give a positive Bavendamm reaction, and maintain a weakly acidic culture medium of around pH 4 with an enzyme capable of decomposing oxalic acid.[16]

It is supposed that extracellular and membrane-bound enzymes of white-rot fungi may work in this range of acidity near pH 4.5 to degrade high molecular weight constituents of cell walls in wood tissue. For example, hydrolases may cleave lignin-carbohydrate linkages and α-ether bonds in side chain structures of lignin.[17] A more important reaction must be the splitting of β-ether bonds in side chains of lignin attached to the mycelium of the fungus. The NADH-requiring enzyme for this cleavage can be prepared from mycelia which have been digested to some degree by autolysis, showing that the enzyme is probably membrane bound.

The role of laccase, an extracellular enzyme, remains obscure. In vitro, it causes radical coupling of lignins resulting in the formation of methoxyquinones on the one hand and of lignin polymers on the other.[18] Westermark and Eriksson[19] have recently found a cellobiose-quinone oxidoreductase in the white-rot fungus, *Polyporus (Coriolus) versicolor*, which prevents the above polymerization. We suggest that an NADH- or NADPH-oxidoreductase may also exist since hydroquinone was always one of the

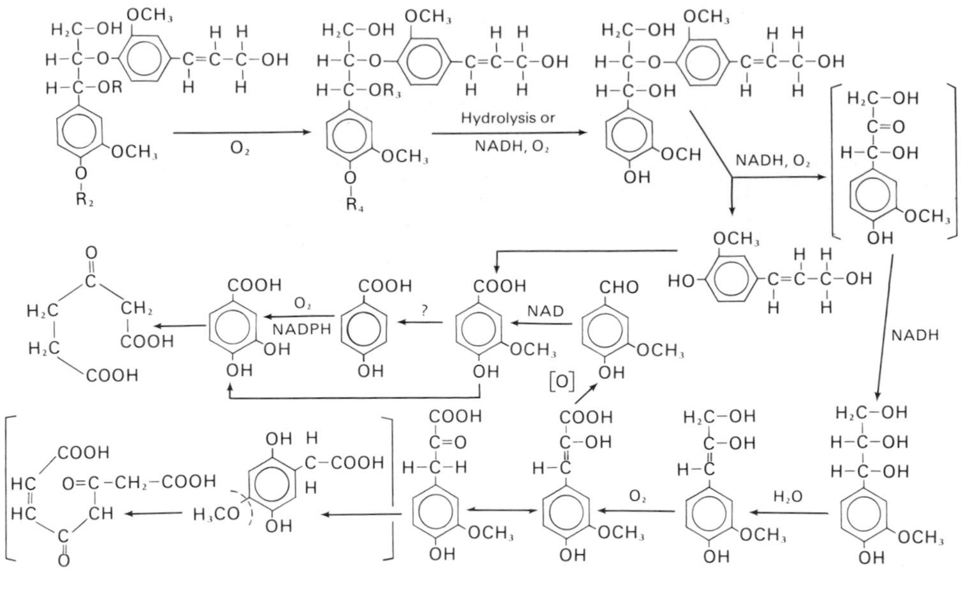

FIGURE 6.   A scheme for the enzymatic degradation of a guaiacylglycerol-β-coniferyl ether moiety: R_1 = hemicellulose, R_2 = guaiacylpropanols, R_3 = hemicellulose, and R_4 = carbonyl compound. (a) Pathway shown in Figures 3 and 4.

products accumulated from cinnamic acid metabolism by *Trametes* sp. However, even if these enzyme systems prevent the polymerization of phenols, the role of such enzyme systems in the biodegradation of lignin remains an important question to be studied further.

NADH may also be required for the formation of guaiacylglycerol from the cleavage of veratrylglycerol-β-ether compounds. The enzyme responsible may be a reducing enzyme which acts on the hypothetical intermediate, guaiacyldihydroxyacetone, shown in Figure 5. However, it still remains a possibility that guaiacylglycerol was directly produced from guaiacylglycerol-β-ether compound by a hydrolase since it was detected in the hydrolysis of lignin.[17] In any case, the enzymatic cleavage of β-ether bonds went slowly. The addition of FMN or of FAD was not effective in accelerating the reaction. Other membrane-bound cofactors may be required. The overall scheme of degradation of the guaiacylglycerol-β-coniferyl ether structure in lignin is illustrated in Figure 6 according to the results mentioned hitherto and the references.[32-36]

II. BACTERIAL DEGRADATION OF LIGNIN-RELATED DIMERS

A. Introduction

Decay and degradation of lignins are caused by wood-rotting fungi and also by soil microorganisms in nature. Studies on these microbial degradations of lignins have been difficult[20] because the structure of lignin is complex and microorganisms capable of metabolizing lignins degrade them comparatively slowly under laboratory culture conditions. However, a large number of studies on bacterial degradation of phenolic compounds have contributed to the elucidation of the cleavage of aromatic rings by oxygenases[21] (Volume I, Chapter 2). Literature on the metabolism of lignin dimers by bacteria, on the other hand, is still sparse.[22-26]

We isolated bacteria from activated sludges for treatment of kraft waste liquor, which grew well in cultures containing a lignin-related dimer. Two of the isolated strains, FK-1 and FK-2, were identified as *Pseudomonas putida*.[27]

We prepared dimer compounds from coniferyl alcohol on a large scale by a specially designed method[28] using peroxidase and H_2O_2 and obtained sufficient amounts and purities of guaiacylglycerol-β-coniferyl ether, dehydrodiconiferyl alcohol, and dl-pinoresinol for studies of bacterial degradation. Artificial lignin-related dimers, such as dehydrodivanillic acid, 2-veratryl-3-guaiacylpropionic acid, and others, were synthesized.

Initial intermediates from each of the above dimers in cultures of *P. putida* were identified, and oxygen uptake of the dimers and intermediates with bacterial cell suspensions was measured using Warburg's apparatus. Based on these results, initial metabolic pathways for each dimer were determined or hypothesized.

B. Isolation and Identification of Bacteria

Samples of bacteria from several activated sludges of pulp mills were cultivated successively in media with dehydrodivanillic acid as the sole source of carbon. Strains were then isolated by dilution plating on agar media of the same sole carbon source. The isolates were then cultivated on a medium of beef extract to confirm the purity of the strain. Finally, the isolates were identified by Dr. Kazuo Komagata of the Institute of Applied Microbiology, University of Tokyo. The two strains with the most rapid growth, FK-1 and FK-2, also happened to be the most effective in defoaming pulping waste liquors (see Volume II, Chapter 11).

These two strains, both *P. putida*, assimilated lignin-related dimers well, but the degree of degradation was somewhat different between the strains. *P. putida* FK-1 degraded dehydrodivanillic acid better, and FK-2 degraded native lignin-related dimers better.

When the bacteria were cultured successively on pure beef extract (Difco®) agar media, they lost the ability to assimilate dehydrodivanillic acid. The laboratory strain of *P. putida* which had the ability to oxidize and assimilate low molecular weight phenolic compounds did not show as strong an ability to degrade lignin-related dimers as the above two strains isolated from activated sludges.

For studies of dimer metabolism, the bacteria were grown on the following medium (to 1ℓ): dimer, 1 g; $MgSO_4 \cdot 7H_2O$, 580 mg; $CaCl_2 \cdot 2H_2O$, 67 mg; $(NH_4)MoO_4 \cdot 4H_2O$, 0.2mg; $(NH_4)_2HPO_4$, 1 g; KH_2PO_4, 3.5 g; K_2HPO_4, 4.5 g; pH 7.0. Incubation was at 30°C, with shaking.

C. Metabolism of Guaiacylglycerol-β-Coniferyl Ether by *P. putida*[27]

P. putida FK-2 rapidly assimilated and degraded this substrate. After 24 hr cultivation, measurement of the UV absorption of the culture filtrate showed the appearance of an absorption at 310 nm (perhaps a conjugated carbonyl group) and a shift and decrease of λ_{max} from 268 to 275 nm (Figure 7). The main initial intermediates were β-hydroxypropiovanillone (yield, 285 mg) and coniferyl alcohol (yield, 6 mg) from a 5 ℓ culture. Minor products included were ferulic and vanillic acids.

β-Hydroxypropiovanillone, the main intermediate, came from the guaiacylglycerol moiety of the substrate and not from coniferyl alcohol since it was not found in the extract of a culture medium containing coniferyl alcohol as the sole source of carbon. Ferulic and vanillic acids were produced from coniferyl alcohol.

The bacteria consumed almost 5 equivalent mol of oxygen for oxidation of guaiacylglycerol-β-coniferyl ether. Measurement of oxygen uptake with each monomer intermediate showed rapid and voluminous oxygen consumption by the bacteria. On the basis of these results, a scheme shown in Figure 8 for the degradation of guaiacylglycerol-β-coniferyl ether by *P. putida* FK-2 was deduced.

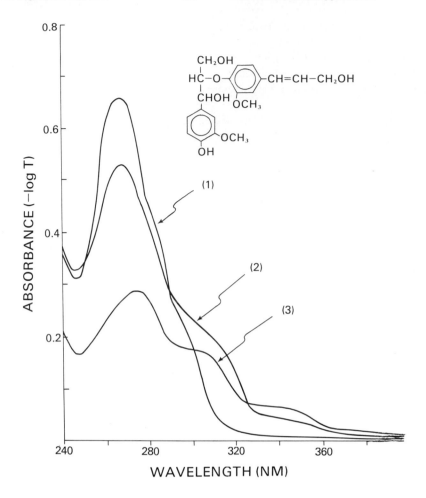

FIGURE 7.   UV absorption spectra of guaiacylglycerol-β-coniferyl ether after various periods of incubation with *Pseudomonas putida* FK-2: (1) control, (2) 18 hr, (3) 24 hr.

D. Metabolism of Dehydrodiconiferyl Alcohol (DHCA) by *P. putida*[29]

P. putida FK-2 showed good growth in the medium containing this substrate as the sole source of carbon. The bacterium consumed the substrate almost completely after 48 hr of cultivation as deduced by UV spectrophotometry. The initial intermediates from 5 *l* of culture after 24 hr cultivation were coniferyl alcohol (85 mg), the γ′--carboxylic compound of the starting material (35 mg), ferulic acid (5 mg), and a small amount of vanillic acid.

P. putida FK-2 which had been cultivated in a medium containing dehydrodivanillic acid could grow in a medium containing dehydrodiisoeugenol as the sole source of carbon; this substrate was degraded as DHCA above. However, the bacterium gradually lost its activity towards dehydrodiisoeugenol, although it retained activity towards DHCA during successive cultivation in media containing dehydrodivanillic acid with a final transfer to DHP.

As shown in Figure 9, cells of the bacterium (FK-2) grown on DHCA medium consumed 5 equivalent mol of oxygen for almost total consumption of the substrate, DHCA. On the other hand, in spite of their similarity in structure, dehydrodiisoeugenol was hardly oxidized by the cells.

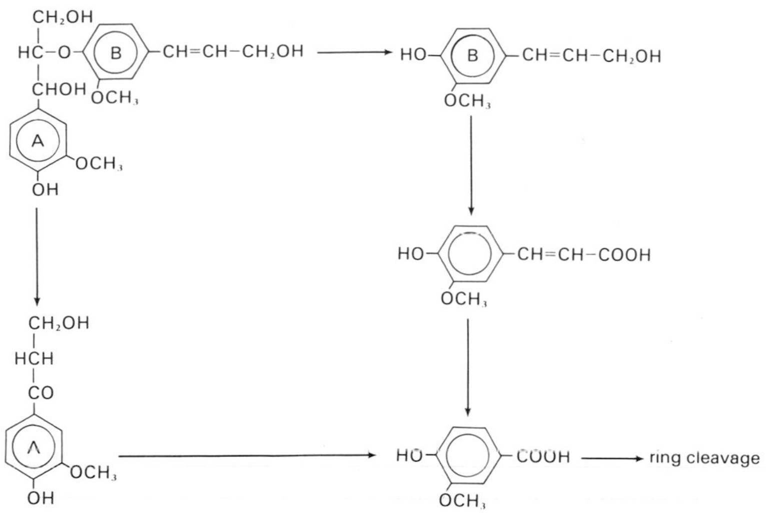

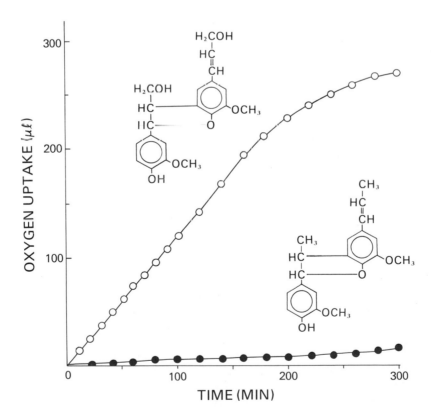

FIGURE 8. Scheme for degradation of guaiacylglycerol-β-coniferyl ether by *Pseudomonas putida* FK-2.

FIGURE 9. Oxygen uptake with substrates dehydrodiconiferyl alcohol and dehydrodiisoeugenol by *Pseudomonas putida* FK-2 grown on dehydrodiconiferyl alcohol-containing medium. Substrate: 2 μM (50 μℓ of oxygen corresponds to 1 equivalent mole uptake).

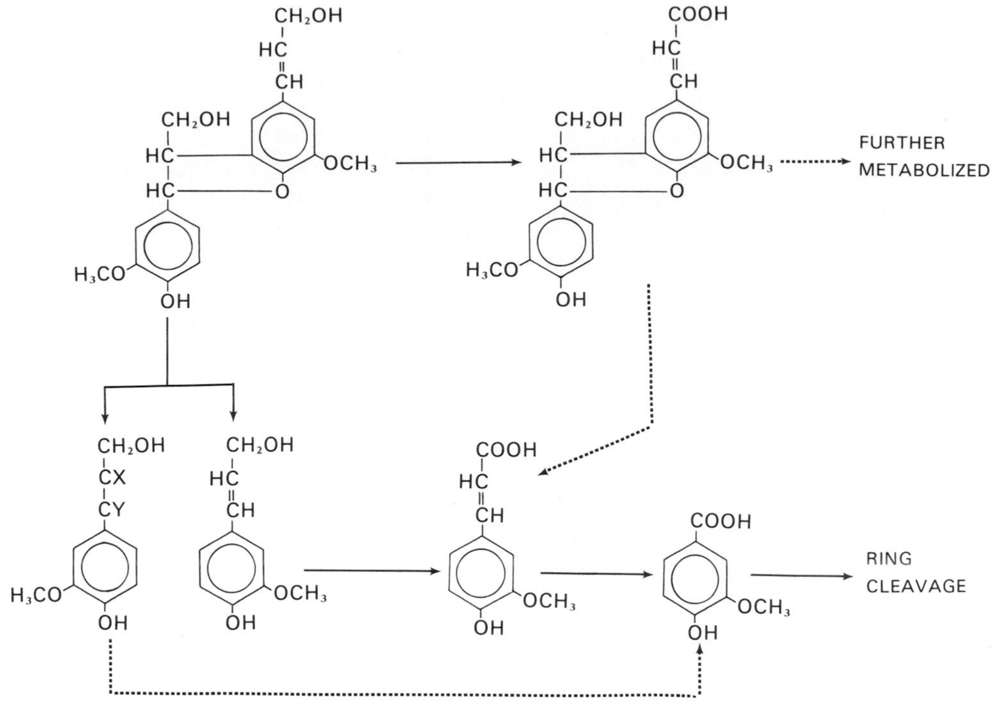

FIGURE 10.   A scheme for degradation of dehydrodiconiferyl alcohol by *Pseudomonas putida* FK-2. X and Y = unknown substituents.

Formation of coniferyl alcohol or ferulic acid from DHCA showed that a cleavage between the β-carbon of the side chain and the benzene ring of the coniferyl alcohol or ferulic acid moiety, which is a stable chemical bond, was brought about by a hitherto unknown enzymatic reaction. An alcohol group on the γ-carbon seems to be necessary for this enzymatic reaction because degradation of dehydrodiisoeugenol hardly occurred. A proposed scheme for the degradation of DHCA by the bacterium is shown in Figure 10.

E. Metabolism of Pinoresinol by *P. putida*

P. putida FK-2 also grew well in the medium with pinoresinol as sole carbon source. Degradative products detected were vanillic acid and two unidentified compounds having ether or alcohol groups. Four equivalent moles of oxygen were consumed per 1 mol of pinoresinol. The bacterium could not assimilate guaiacol and oxidized little α-conidendrin (Figure 11). These results showed that degradation first eliminated the side chain and did not proceed by the mechanism of α-conidendrin degradation by *P. multivorans* as reported by Toms and Wood,[26] in which guaiacol was first liberated.

F. Metabolism of Dehydrodivanillic Acid by *P. putida*

P. putida FK-1 rapidly assimilated dehydrodivanillic acid. One of the degradative products from the substrate was 5-carboxyvanillic acid, identified by comparison of its mass spectrum with that of the authentic compound. The other product was supposed to be a derivative of dehydrodivanillic acid with a diketobutyric acid attached at the meta position (Figure 12).

Oxidation of dehydrodivanillic acid with a cell-free extract of the bacterium in the presence of NADH and GSH showed nearly 1 equivalent mole of oxygen uptake, and

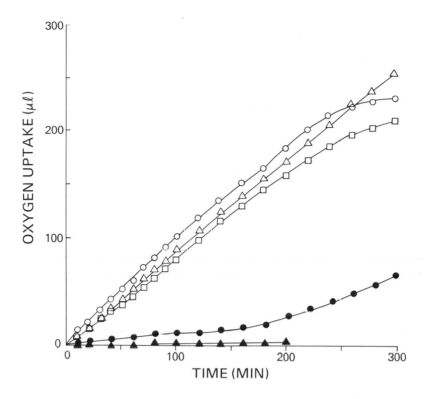

FIGURE 11. Oxygen uptake with various substrates by *Pseudomonas putida* FK-2 grown on a D,L-pinoresinol-containing medium, contaminated with trace amounts of guaiacylglycerol-β-coniferyl ether and dehydrodiconiferyl alcohol: O D,L-pinoresinol, □ dehydrodiferulic acid, • α-conidendrin, Δ 2-veratryl-3-guaiacyl-propionic acid, ▲ guaiacol.

its reaction product was a monodemethylated compound of the substrate. The bacterial cells oxidized dehydrodivanillic acid and dehydrodiferulic acid fairly well, both with an oxygen uptake of 8 equivalent moles per mole of substrate, but oxidized dehydrodi-dihydroferulic acid only slowly (Figure 13).

G. Metabolism of 2-Veratryl-3-Guaiacylpropionic Acid by *P. putida*[30]

P. putida FK-2 rapidly metabolized 2-veratryl-3-guaiacylpropionic acid, a model 1,2-diarylpropane compound structurally related to lignin, producing 2,3-diguaiacyl-propionic acid, veratric acid, and vanillic acid. Measurement of oxygen uptake with 2-veratryl-3-guaiacylpropionic acid showed 10 equivalent moles of oxygen consumption per mole of substrate, and with ferulic and dihydroferulic acids, 5 equivalent moles. 4-Hydroxy-3,3′,4′-trimethoxy-*trans*-stilbene-α-carboxylic acid was oxidized very slowly by the bacterium (Figure 14).

If an oxidative cleavage occurred at the α-β linkage of the side chain, homovanillic acid or vanilloylformic acid should have been produced, but these have not been isolated. Elucidation of this pathway is still in progress.

H. Discussion

The wild type strains of *P. putida* FK-1 and FK-2 showed good growth responses in media containing a lignin-related dimer as the sole carbon source. The bacterium, however, hardly assimilated slightly altered structures of the native dimers, such as α-conidendrin, dehydrodiisoeugenol or dehydrodi-dihydroferulic acid, in spite of broad sub-

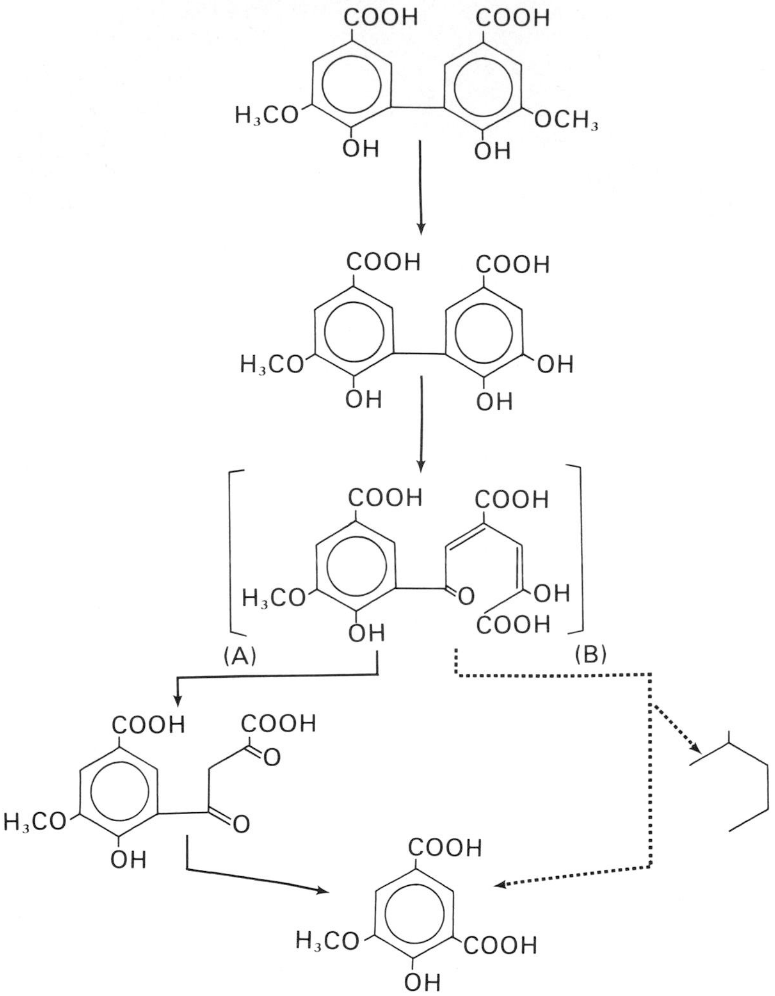

FIGURE 12.   A scheme for degradation of dehydrodivanillic acid by *Pseudomonas putida* FK-1. Structure in brackets is hypothetical.

strate-degrading abilities for native dimers. 2-Veratryl-3-guaiacylpropionic acid was the exception.

When the bacteria were successively cultivated in pure beef extract media, they lost the ability to assimilate lignin-related dimers and could not recover this ability, even if the bacteria were again cultured in media containing the dimers. It is supposed that the genes for the synthesis of the enzyme systems participating in the initial degradation of the dimers are not found in chromosomal DNA, but rather in plasmid DNA in the protoplasm of the bacterium (see Volume II, Chapter 12).

Cleavages of ether bonds or of side chain linkages in the structures of guaiacylglycerol-β-coniferyl ether, DHCA, and 2-veratryl-3-guaiacylpropionic acid studied in this work and initial degradation of dimers are probably due to oxygenation by electron transport systems containing P_{450}[31] since the cleavages occurred with little substrate specificity within the range of lignin-related dimers.

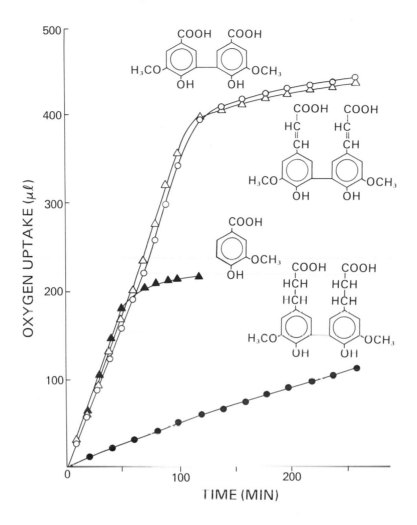

FIGURE 13. Oxygen uptake with biphenyl-type compounds and vanillic acid by *Pseudomonas putida* FK-1 grown on dehydrodivanillic acid-containing medium. Substrate: 2 μM

III. SUMMARY

Early in the course of research on lignin degradation by white-rot fungi, the fungi were found to have laccase, which was capable of polymerizing coniferyl alcohol to lignin. We were one of the first to make a contradictory discovery: laccase-type enzymes also oxidize vanillic acid, vanilloylformic acid, and guaiacylpyruvic acid with CO_2 liberation. We have been very interested in solving this contradiction.

While studying metabolic products from aromatics, we found products from aromatic acid reduction and, more recently, coniferyl alcohol from ferulic acid reduction in the culture. Reduction of trimethoxybenzoic acid to trimethoxybenzyl alcohol by the enzymes of *Trametes* sp. required NADPH and ATP as cofactors, and reverse oxidation was inhibited by ATP.

White-rot fungi degraded almost all lignin-related dimeric compounds studied, although not as rapidly as certain bacteria described below. Degradation of ferulic acid by the enzyme system of *Trametes* sp. produced vanillic acid. This reaction required CoA, ATP and NAD. Enzymatic cleavage of ether bonds of veratrylglycerol-β-ether

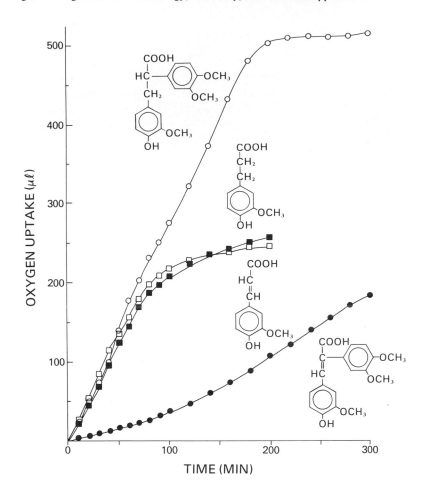

FIGURE 14.   Oxygen uptake with various substrates by *Pseudomonas putida* FK-2 grown on 2-veratryl-3-guaiacylpropionic acid-conaining medium: ○ 2-veratryl-3-guaiacylpropionic acid, ● 4-hydroxy-3,3′,4′-trimethoxy *trans*stilben-α-carboxylic acid, □ ferulic acid, ■ dihydroferulic acid. (Reproduced from Katayami, Y. and Fukuyumi, T., *Mokuzai Gakkaishi*, 25, 67, 1979. With permission.)

compounds was studied, and the reaction was found to require NADH. Cleavage occurred at the β-ether bond and at the *p*-methoxyl ether bond.

Recently, bacteria which grew well in cultures containing a lignin-related compound were isolated from activated sludges of kraft pulp mills. Two of the isolated strains, FK-1 and FK-2, were identified as *P. putida*. In order to study bacterial degradation of dimer compounds, *P. putida* FK-2 was cultivated in media containing one of the following as the sole source of carbon: guaiacylglycerol-β-coniferyl ether, DHCA, pinoresinol, and 2-veratryl-3-guaiaclypropionic acid. *P. putida* FK-1 was cultivated in media containing dehydrodivanillic acid.

The initial intermediates from the dimer compounds were identified, and oxygen uptake of the dimers and intermediates with bacterial cell suspensions was measured using the Warburg apparatus. Based on these results, initial metabolic pathways of each dimer were determined or hypothesized. Guaiacylglycerol-β-coniferyl ether was split to coniferyl alcohol and β-hydroxypropiovanillone, which were further transformed to ferulic and vanillic acids. DHCA was split into coniferyl alcohol and an unknown compound or was oxidized at its terminal alcohol group. Dehydrodivanillic acid was demethylated at the methoxyl group to produce a protocatechuic acid moiety. This ring was probably subsequently cleaved into a diketone side chain and further oxidized to give 5-carboxyvanillic acid. Pinoresinol was degraded to vanillic acid, and 2-veratryl-3-guaiacylpropionic acid was degraded to vanillic acid via 2,3-guaiacyl propionic acid.

REFERENCES

1. **Fukuzumi, T.,** Enzymatic degradation of lignin. I. Paper chromatographical separation of intermediate degradation products of lignin by the wood-rotting fungus *Poria subacida* (Peck) Sacc., *Bull. Agric. Chem. Soc. Jpn*, 24, 728, 1960.

2. **Fukuzumi, T., Uraushihara, S., Ohashi, T., and Shibamoto, T.,** Enzymatic degradation of lignin. III. Oxidation accompanying carbon dioxide liberation from vanillic acid, vanilloylformic acid and guaiacylpyruvic acid by enzymes of *Polystictus sanguineus* and *Poria subacida, Mokuzai Gakkaishi,* 10, 242, 1964.

3. **Fukuzumi, T., Hiyama, T., and Minami, K.,** Metabolic products from aromatic compounds by the wood-rotting fungus *Polystictus sanguineus (Trametes sanguinea). III. Reductive Transformation of veratric acid to veratraldehyde, Mokuzai Gakkaishi,* 11, 175, 1965.

4. **Minami, K., Tsuchiya, M., and Fukuzumi, T.,** Metabolic products from aromatic compounds by the wood-rotting fungus *Polystictus sanguineus (Trametes sanguinea).* IV. Culturing conditions for reduction and demethoxylation of veratric acid, *Mokuzai Gakkaishi,* 11, 179, 1965.

5. **Nishida, A. and Fukuzumi, T.** Formation of coniferyl alcohol from ferolic acid by the white rot fungus *Trametes, Phytochemistry,* 17, 417, 1978.

6. **Nishida, A. and Fukuzumi, T.,** Enzymatic Reduction of Benzoic Acids and Cinnamic Acids and Formation of Vanillic Acid from Ferulic Acid by the Wood-Rotting Fungus, paper presented at the 22nd Symp. Lignin Chemistry, Sapporo, Japan, October 1977, 33.

7. **Nishida, A. and Fukuzumi, T.,** unpublished data, 1978.

8. **Zenk, M. H. and Gross, G. G.,** The enzymic reduction of cinnamic acids, in *Recent Advances in Phytochemistry,* Vol. 4, Runeckles, V. C., Ed., Appleton-Century Crofts, New York, 1972, 87.

9. **Zenk, M. H.,** Biosythese von Vanillin, in *Vanilla planifolia* Andr., *Z. Pflanzenphysiol.,* 53, 404, 1965.

10. **French, C. J., Vance, C. P., and Towers, G. H. N.,** Conversion of *p*-coumaric acid to *p*-hydroxybenzoic acid by cell free extracts of potato tubers and *Polyporus hispidus, Phytochemistry,* 15, 564, 1976.

11. **Van Vliet, W. F.,** The enzymic oxidation of lignin, *Biochim. Biophys. Acta,* 15, 211, 1954.

12. **Fukuzumi, T. and Shibamoto, T.,** Enzymatic degradation of lignin. IV. Splitting of veratrylglycerol-β-guaiacyl ether by enzyme of *Poria subacida, Mokuzai Gakkaishi,* 11, 248, 1965.

13. **Fukuzumi, T., Takatsuka, H., and Minami, K.,** Enzymic degradation of lignin. V. The effect of NADH on the enzymic cleavage of arylalkyl ether bonds in veratrylglycerol-β-guaiacyl ether as lignin model compound, *Arch. Biochem. Biophys.,* 129, 396, 1969.

14. **Matsumoto, H. and Fukuzumi, T.,** unpublished data, 1973.

15. **Matsumoto, H. and Fukuzumi, T.,** unpublished data, 1971.

16. **Shibamoto, T., Fukuzumi, T., and Yanagawa, R.,** Studies on the scheme of decomposition of oxalic acid by some wood-rotting fungi, *Bull. Tokyo Univ. Forest.,* 43, 105, 1952.

17. **Fukuzumi, T.,** Degradation Products from Lignin by Treatment of Wood Meal with Hydrolase, paper presented at the 17th Symp. Lignin Chemistry, Kyoto, Japan, September 1972, 4.

18. **Ishihara, T. and Miyazaki, M.,** Oxidation of milled wood lignin by fungal laccase, *Mokuzai Gakkaishi,* 18, 415, 1972.

19. **Westermark, U. and Eriksson, K. E.,** Cellobiose quinone oxidoreductase, a new wood-degrading enzyme from white-rot fungi, *Acta Chem. Scand. Ser. B,* 28, 209, 1974.

20. **Gottlieb, S. and Pelczar, M. J., Jr.,** Microbiological aspects of lignin degradation, *Bacteriol. Rev.,* 15, 55, 1951.

21. **Hayaishi, O.,** *Oxygenases,* Academic Press, New York, 1962.

22. **Sundman, V. and Haro, K.,** On the mechanism by which cyclolignanolytic agrobacteria might cause humification, *Finska Kemist. Mdd.,* 75, 111, 1966.

23. **Sundman, V.,** A description of some lignanolytic soil bacteria and their ability to oxidize simple phenolic compounds, *J. Gen. Microbiol.,* 36, 171, 1964.

24. **Sundman, V.**, The ability of α-conidendrin-decomposing agrobacterium strains to utilize other lignans and lignin-related compounds, *J. Gen. Microbiol.*, 36, 185, 1964.
25. **Kawakami, K.**, Bacterial degradation of lignin model compounds. I, II, and III, *Mokuzai Gakkaishi*, 21, 93 and 309, and 629, 1975.
26. **Toms, A. and Wood, J. M.**, Early intermediates in the degradation of α-conidendrin by a *Pseudomonas multivorans*, *Biochemistry*, 9, 733, 1970.
27. **Fukuzumi, T. and Katayama, Y.**, Bacterial degradation of dimer relating to structure of lignin. I. β-Hydroxypropiovanillone and coniferyl alcohol as initial degradation products from guaiacylglycerol-β-coniferyl ether by *Pseudomonas putida*, *Mokuzai Gakkaishi*, 23, 214, 1977.
28. **Katayama, Y. and Fukuzumi, T.**, Enzymatic synthesis of three lignin-related dimers by an improved peroxidase-hydrogen peroxide system, *Mokuzai Gakkaishi*, 24, 664, 1978.
29. **Katayama, Y. and Fukuzumi, T.**, Bacterial degradation of dimers structurally related to lignin. II. Initial intermediate products from dehydrodiconiferyl alcohol by *Pseudomonas putida*, *Mokuzai Gakkaishi*, 24, 643, 1978.
30. **Katayama, Y. and Fukuzumi, T.**, Bacterial degradation of dimers structurally related to lignin. III. Metabolism of α-veratryl-β-guaiacylpropionic acid and D,L-pinoresinol by *Pseudomonas putida*, *Mokuzai Gakkaishi*, 25, 67, 1979.
31. **Peterson, J. A.**, Cytochrome content of two Pseudomonads containing mixed-function oxidase systems, *J. Bacteriol.*, 103, 714, 1970.
32. **Arima, K., Morimoto, M., and Yano, K.**, Participations of iron and flavin adenine dinucleotide in the enzymatic hydroxylation of p-hydroxybenzoate, *Agr. Biol. Chem.*, 30, 91, 1966.
33. **Cartwright, N. J. and Smith, A. R.**, Bacterial attack on phenolic ethers: an enzyme system demethylating vanillic acid, *Biochem. J.*, 102, 826, 1967.
34. **Flaig, W. and Haider, K.**, Die Verwertung phenolischer Verbindungen durch Weibfäulepilze, *Arch. Microbiol.*, 40, 212, 1961.
35. **Ishikawa, H., Nord, F. F., and Schubert, W. J.**, Investigations on lignins and lignification. XXX. Enzymic degradation of guaiacylglycerol and related compounds by white rot fungi, *Biochem. Z.*, 338, 153, 1963.
36. **Fukuzumi, T.**, Enzymatic degradation of lignin. II. Oxidation of homogentisic acid and gentisic acid by the enzyme of wood-rotting fungus, *Poria subacida*, *Agr. Biol. Chem.*, 26(7), 447, 1962.

Chapter 7

ISOLATION AND CHARACTERIZATION OF LIGNOCELLULOSE-DECOMPOSING ACTINOMYCETES

Don L. Crawford and John B. Sutherland

TABLE OF CONTENTS

I. INTRODUCTION

A. Microbial Decomposition of Lignocellulose

The ability of certain fungi to decompose the structural components of wood has long been known, and several groups have been distinguished which attack wood in different ways.[1-3] The white-rot fungi are of particular interest because they are the only microorganisms known to oxidize both lignin and cellulose components of wood completely to CO_2 and H_2O.[2,3] The brown-rot and soft-rot fungi, on the other hand, attack primarily the carbohydrate components of wood and alter lignin only slightly[2,3] (see, however, Volume I, Chapter 6). Recognizing that microorganisms other than . fungi may function in lignocellulose decomposition, numerous workers have demonstrated the ability of certain actinomycetes and other bacteria to degrade both lignin and cellulose components.[4-13] The extent of lignin decomposition by bacteria and the changes in the cell-wall structure resulting from attack remain to be elucidated.[1,14]

B. Role of Actinomycetes in Lignocellulose Degradation

Cellulose is decomposed by several actinomycetes,[15] even when it is complexed with lignin. For example, work in our laboratory has shown that *Thermomonospora fusca* decomposes a variety of lignocellulosic plant materials when incubated at 55°C.[5,7,8,10] This thermophilic species depletes primarily the carbohydrate components of lignocellulose[7] and evolves no appreciable $^{14}CO_2$ from lignin when grown on ^{14}C-lignin-labeled lignocellulose.[10] It does, however, convert up to 10% of the lignin to water-soluble products.[16] Other thermophilic actinomycetes are reportedly involved in the degradation of both lignin and cellulosic components in bovine manure.[6]

Waksman and Hutchings suggested in 1937 that actinomycetes may attack lignin in decomposing plant materials,[17] although their evidence was not conclusive. A micromonospora-like actinomycete has been shown by electron microscopy to degrade lignified cell walls of plant tissue,[9] although it is not known which components of the cell wall are attacked. The first conclusive evidence showing that a *Nocardia* species oxidizes ^{14}C-lignin to $^{14}CO_2$ has also recently been published[12] (Volume I, Chapter 3). It seems clear, therefore, that actinomycetes play a role in lignin decomposition, although the extent of their involvement remains unknown.

In our laboratory, we are now examining strains of actinomycetes for lignocellulose-degrading ability. Before we could do this, however, we had to solve some of the methodological problems associated with lignin-biodegradation assays. The long-used Klason lignin assay, which measures lignin as an acid-insoluble residue, has proven too insensitive and prone to interference for use in measuring lignin degradation by microorganisms.[10,18] As an alternative, we have developed a more sensitive assay for lignin degradation which uses substrates labeled with ^{14}C[10,19] (Chapter 3). Natural lignocelluloses, specifically labeled with ^{14}C in their lignin components, are utilized as substrates. Microbial decomposition of ^{14}C-lignin is monitored by measuring both the solubilization of ^{14}C and the evolution of $^{14}CO_2$ in growing cultures. Several ^{14}C techniques for labeling lignin have been developed in other laboratories and are in use in microbiological decomposition studies in those laboratories.[20-23] ^{14}C techniques have also been useful for studying the microbial ecology of lignin degradation in nature.[19,22-24] Recently, we have also prepared ^{14}C-glucan-labeled lignocelluloses for studies of cellulose degradation.[19] With these new techniques at our disposal, we are now able to concentrate on the isolation and characterization of lignocellulose-degrading actinomycetes.

II. ISOLATION AND CHARACTERIZATION PROCEDURES

A. Enrichment Techniques for Isolation

The actinomycete strains used in our laboratory have been isolated primarily by enrichment from a variety of natural substrates containing decomposing plant materials. These include soils collected in forests and gardens, soils adjacent to hot springs, wood-chip piles, composts, and pulp mill waste piles. We have recently described a procedure for isolating lignocellulose-decomposing actinomycetes from these sources.[13] This procedure involves:

1. Serial dilution of soils or plant-material samples followed by a brief treatment with moist heat at 60°C
2. Inoculation of diluted samples into molten agar containing 0.5% newsprint, which is then poured onto a base of nonnutrient agar in petri dishes
3. Incubation of plates until colonies of actinomycetes appear
4. Selection and isolation of any actinomycete colonies which form a clear zone in the newsprint

Although this isolation technique enriches for lignocellulose decomposers, it makes no distinction between strains which degrade lignin and those which degrade only cellulose. Replacing newsprint with other substrates of variable lignin content may increase the probability of isolating strains which preferentially attack either lignin or cellulose.

B. Characterization of Lignin- and Cellulose-Decomposing Abilities

Each actinomycete strain which forms a clear zone in the newsprint is studied to determine its ability to decompose the components of lignocellulose. The basic procedure has been described,[13] although we have recently improved it somewhat. Each isolate is inoculated into tubes of a mineral salts solution containing purified lignocellulose from the inner bark of Douglas fir, *Pseudotsuga menzicsii*, as the primary carbon and energy source, with a 20-amino-acid mixture as a nitrogen source. To determine whether lignin or cellulose is being degraded, either ^{14}C-lignin-labeled[10] or ^{14}C-glucan-labeled[19] lignocellulose is added to the medium.* Evolution of $^{14}CO_2$ and appearance of ^{14}C in the supernatant show that the labeled substrate is being decomposed. Other ^{14}C-lignins can be used in addition to lignocellulose. Each culture is incubated at a constant temperature, usually 30, 37, or 45°C, with continuous aeration. Any $^{14}CO_2$ given off is trapped in 8% NaOH, and its evolution is monitored at regular intervals. After 28 days, residual solids in the tubes are washed, oven-dried, and weighed in order to calculate the weight loss of substrate. Soluble ^{14}C present in culture supernatants at the time of harvest is also determined, and Kjeldahl-nitrogen analyses of residual solids are made to estimate cell mass. Comparing the data with those of uninoculated controls makes it possible to characterize the ability of each culture to attack lignin and cellulose within the lignocellulose complex.

III. CURRENT PROGRESS AND IMPLICATIONS

A. Decomposition of Lignins and Lignocelluloses by Selected Isolates

Using the enrichment techniques described above, numerous lignocellulose-decom-

---

* The source of ^{14}C-lignocelluloses in all experiments reported in this paper was Douglas fir and was prepared by feeding cut twigs either ^{14}C-phenylalanine or ^{14}C-glucose.

TABLE 1

Decomposition of ^{14}C-Lignin- and ^{14}C-Glucan-Labeled Lignocellulose by Selected Strains of *Streptomyces*[a]

Culture	Temperature of incubation (°C)	% ^{14}C recovered from ^{14}C-lignin lignocellulose		% ^{14}C recovered from ^{14}C-glucan lignocellulose	
		as ^{14}CO$_2$[b]	as soluble ^{14}C[c]	as ^{14}CO$_2$[b]	as soluble ^{14}C[c]
28	25	3.45	4.33	31.69	12.20
87A	25	1.38	2.81	21.36	9.33
177	25	1.84	1.66	28.00	5.34
Control	25	—	2.92	—	6.11
201	37	5.81	3.05	28.81	7.20
225	37	7.75	6.32	41.14	9.41
252	37	14.43	5.14	41.20	3.60
526	37	8.65	1.35	28.85	2.40
Control	37	—	4.35	—	9.98

[a]   Cultures were grown in a mineral salts-amino acid medium.[13] Unlabeled lignocellulose prepared from Douglas fir was used as the primary carbon and energy source. ^{14}C-lignin-labeled (75,000 dpm) or ^{14}C-glucan-labeled (50,000 dpm) lignocellulose was added, for a final lignocellulose concentration of 0.5%. All figures represent the average of three replicates.

[b]   dpm recovered as ^{14}CO$_2$ after 28 days (672 hr), expressed as a percentage of the dpm originally present.

[c]   Soluble dpm in the supernatant after 28 days after solids were removed by centrifugation, expressed as a percentage of the dpm originally present.

posing actinomycetes have been isolated, mostly in the genus *Streptomyces.* All strains which cleared newsprint agar readily attacked the cellulosic component of lignocellulose, but only a few attacked the lignin component significantly. Table 1 shows lignin and cellulose decomposition by a selected group of lignin-degrading actinomycetes. In addition to the portion of labeled substrates metabolized to ^{14}CO$_2$, some was found as soluble ^{14}C in the culture supernatant (Table 1). It should be noted that cultures which solubilize lignin or cellulose efficiently, but decompose it immediately to CO$_2$, appear in the table as less efficient at solubilization than the controls.

Figure 1 shows the rate of ^{14}CO$_2$ evolution during the growth of two strains of actinomycetes on lignocellulose labeled in either the lignin or cellulose component. The most rapid substrate decomposition occurs early; with time, the ^{14}CO$_2$ evolution rate begins to decline, but does not reach zero even at 28 days. A significant amount of label is presumably incorporated into cell mass, although it has not yet been measured accurately; ^{14}C incorporated into cell mass and then decomposed to ^{14}CO$_2$ may account for a portion of later ^{14}CO$_2$ evolution.

The data presented here and previously[13] are the first to show that *Streptomyces* can substantially degrade both lignin and cellulose within the lignocellulose complex. For these microorganisms, attack on cellulose may only be possible when the associated lignin is altered or removed. We emphasize that decomposition data reported here, which were obtained during preliminary screening of cultures, do not represent the maximum potential of these streptomycetes under optimum conditions.

Recently, we have found that lignin-degrading strains of *Streptomyces* are also able to attack purified ^{14}C-milled wood lignins (^{14}C-MWLs) and ^{14}C-kraft lignins. ^{14}C-MWL have been prepared from ^{14}C-lignin-labeled lignocelluloses by the Björkman procedure;[25] MWL is considered one of the best purified extractive-free lignins avail-

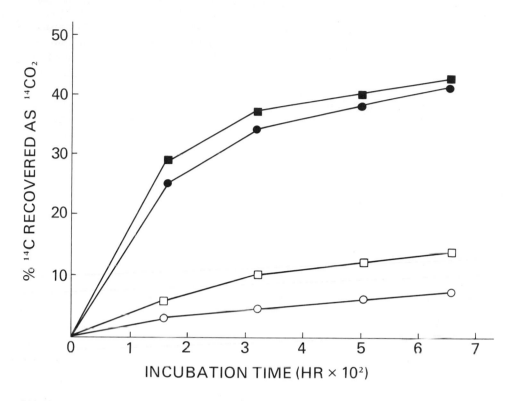

FIGURE 1.   Oxidation of ^{14}C-labeled lignocellulose by *Streptomyces*, strains 225 and 252. Cultures of 10-ml were prepared in triplicate in test tubes containing a mineral salts-amino acid medium[13] to which unlabeled lignocellulose prepared from Douglas fir was added as a primary carbon and energy source. ^{14}C-lignin-labeled (75,000 dpm) or ^{14}C glucan-labeled (50,000 dpm) lignocellulose was added, for a final lignocellulose concentration of 0.5%. After inoculation, tubes were incubated for 28 days (672 hr) at 37°C, and $^{14}CO_2$ evolution was monitored.[13] (o − o strain 225 with ^{14}C lignin-lignocellulose; ● − ● strain 225 with ^{14}C-glucan-lignocellulose; □ − □ strain 252 with ^{14}C-lignin-lignocellulose; ■ − ■ strains 252 with ^{14}C-glucan-lignocellulose).

able for microbiological studies.[2] ^{14}C-kraft lignins, prepared by pulping ^{14}C-lignin-labeled lignocelluloses,[24,26] are somewhat more difficult for microorganisms to decompose than lignocelluloses[24]. Preliminary data from our laboratory[27] show that lignin-degrading streptomycetes vary considerably in their abilities to attack milled wood and kraft lignins. Using ^{14}C-MWL and unlabeled lignocellulose in a 1:9 ratio in the mineral salts-amino acid medium previously mentioned,[13] our most efficient strains convert 15 to 20% of the ^{14}C-MWL to $^{14}CO_2$ in 28 days. Under similar conditions, the same strains convert 2 to 5% of the ^{14}C-kraft lignin to $^{14}CO_2$ in 28 days. It is likely that higher rates of degradation will be achieved with improvement of culture conditions.

B. Potential Applications and Implications for Future Research

Development of processes for bioconversion of lignocellulosic materials to useful products may lead to numerous applications in the future. Microbial conversion of lignocellulose to single-cell protein (SCP) is an obvious possibility. Even more likely is the eventual bioconversion of lignocellulose to organic chemicals. Glucose may be produced from the cellulose portion of lignocellulose, and phenolics and other chemicals may be produced from the lignin portion. Actinomycetes may prove very useful in many of these bioconversions.

By using specifically labeled lignin substrates (labeled in side-chain, ring, or methoxyl groups) that are complexed with unlabeled cellulosic materials, or by pulping these specifically labeled ^{14}C-lignin lignocelluloses to produce labeled waste lignin substrates, it should be possible to study the basic biochemistry of actinomycete attack on lignocellulosics. Industrial lignocellulose bioconversion processes could be developed using selected actinomycetes. For this purpose, the most useful strains may be those which only partially degrade the lignin molecule.

Lignocellulose-degrading actinomycetes also show promise as sources of the cellulase complex. Enzymes able to hydrolyze cellulose which is complexed with lignin, as is the case with most cellulose-containing waste products, would be of particular value for glucose production. Cellulases from selected lignocellulose-degrading actinomycetes, therefore, should be studied, and their efficiencies compared with commercially available cellulases, such as that from *Trichoderma viride*. In particular, cultures should be examined for high exoglucanase activity,[28] for conversion of crystalline cellulose to the amorphous state. It is also important to determine whether purified cellulases from actinomycetes can penetrate the lignin barrier or whether lignin must be altered or partially removed by lignin-specific enzymes before the actinomycete cellulases can act.

Current research in our laboratory has emphasized the isolation of lignocellulose-degrading actinomycete strains and characterization of their abilities to attack lignin and cellulose. Now that promising isolates have been obtained, future work will emphasize the biochemistry of actinomycete attack on lignin and cellulose within the lignocellulose complex. This basic research will hopefully lead to new industrial bioconversion processes for the utilization of this ever more valuable resource, lignocellulose.

IV. SUMMARY

Lignocellulose-decomposing strains of euactinomycetes were isolated by enrichment techniques from a variety of natural habitats rich in decomposing plant materials. All isolates were characterized for their abilities to degrade lignin and cellulosic carbohydrates within the lignocellulose complex. ^{14}C techniques were utilized to differentiate between microbial attack on lignin and cellulose. Several groups of strains were identified, including those which extensively degraded the cellulosic component of lignocellulose to CO_2 while only attacking lignin slightly and those which substantially degraded both components to CO_2. Other strains were able to solubilize considerable amounts of lignin, but did not efficiently degrade these solubilized components to CO_2. Lignin-decomposing actinomycete strains were also shown capable of decomposing ^{14}C-labeled MWLs and kraft lignins to $^{14}CO_2$.

ACKNOWLEDGMENTS

This research was supported, in part, by funding from the National Science Foundation under grants AER75-23401 and AER76-81430 and by the Idaho Agricultural Experiment Station. We thank A. L. Pometto III for his continual and excellent technical assistance and M. B. Phelan and D. L. Sinden for their participation in a portion of this work.

REFERENCES

1. **Liese, W.,** Ultrastructural aspects of woody tissue disintegration, *Annu. Rev. Phytopathol.,* 8, 231, 1970.
2. **Kirk, T. K.,** Effects of microorganisms on lignin, *Annu. Rev. Phytopathol.,* 9, 185, 1971.
3. **Campbell, W. G.,** The biological decomposition of wood, in *Wood Chemistry,* Wise, L. E. and Jahn, E. C., Eds., Reinhold, New York, 1972, 1061.
4. **Sørensen, H.,** Decomposition of lignin by soil bacteria and complex formation between autooxidized lignin and organic nitrogen compounds, *J. Gen. Microbiol.,* 27, 21, 1962.
5. **Crawford, D. L., McCoy, E., Harkin, J. M., and Jones, P.,** Production of microbial protein from waste cellulose by *Thermomonospora fusca,* a thermophilic actinomycete, *Biotechnol. Bioeng.,* 15, 833, 1973.
6. **Bellamy, W. D.,** Single cell protein from cellulosic wastes, *Biotechnol. Bioeng.,* 16, 869, 1974.
7. **Crawford, D. L.,** Growth of *Thermomonospora fusca* on lignocellulosic pulps of varying lignin content, *Can. J. Microbiol.,* 20, 1069, 1974.
8. **Harkin, J. M., Crawford, D. L., and McCoy, E.,** Bacterial proteins from pulps and paper mill sludge, *Tappi,* 57, 131, 1974.
9. **Akin, D. E.,** Ultrastructure of rigid and lignified forage tissue degradation by a filamentous rumen microorganism, *J. Bacteriol.,* 125, 1156, 1976.
10. **Crawford, D. L. and Crawford, R. L.,** Microbial degradation of lignocellulose: the lignin component, *Appl. Environ. Microbiol.,* 31, 714, 1976.
11. **Odier, É. and Monties, B.,** Activité ligninolytique in vitro de bactéries isolées de paille de blé en décomposition, *C. R. Acad. Sci. Ser. D.,* 284, 2175, 1977.
12. **Trojanowski, J., Haider, K., and Sundman, V.,** Decomposition of ^{14}C-labelled lignin and phenols by a *Nocardia* sp., *Arch. Microbiol.,* 114, 149, 1977.
13. **Crawford, D. L.,** Lignocellulose decomposition by selected *Streptomyces* species, *Appl. Environ. Microbiol.,* in press, 1978.
14. **Greaves, H.,** The bacterial factor in wood decay, *Wood Sci. Technol.,* 5, 6, 1971.
15. **Ishizawa, S., Araragi, M., and Suzuki, T.,** Actinomycete flora of Japanese soils. III. Actinomycete flora of paddy soils. (A) On the basis of morphological, cultural and biochemical characters, *Soil Sci. Plant Nutr.,* 15, 104, 1969.
16. **Crawford, D. L.,** unpublished data, 1976.
17. **Waksman, S. A. and Hutchings, I. J.,** Associative and antagonistic effects of microorganisms. III. Associative and antagonistic relationships in the decomposition of plant residues, *Soil Sci.,* 43, 77, 1937.
18. **Crawford, R. L. and Crawford, D. L.,** Radioisotopic methods for the study of lignin biodegradation, *Dev. Ind. Microbiol.,* 19, 35, 1978.
19. **Crawford, D. L., Crawford, R. L., and Pometto, A. L., III,** Preparation of specifically labeled ^{14}C-(lignin) and ^{14}C-(cellulose) lignocelluloses and their decomposition by the microflora of soil, *Appl. Environ. Microbiol.,* 33, 1247, 1977.
20. **Haider, K. and Trojanowski, J.,** Decomposition of specifically labeled phenols and dehydropolymers of coniferyl alcohol as models for lignin degradation by soft and white rot fungi, *Arch. Microbiol.,* 105, 33, 1975.
21. **Kirk, T. K., Connors, W. L., Bleam, R. D., Hackett, W. F., and Zeikus, J. G.,** Preparation and microbial decomposition of synthetic (^{14}C)-lignins, *Proc. Nat. Acad. Sci. U.S.A.,* 72, 2515, 1975.
22. **Hackett, W. F., Connors, W. J., Kirk, T. K., and Zeikus, J. G.,** Microbial decomposition of synthetic ^{14}C-labeled lignins in nature: lignin biodegradation in a variety of natural materials, *Appl. Environ. Microbiol.,* 33, 43, 1977.
23. **Haider, K., Martin, J. P., and Rietz, E.,** Decomposition in soil of ^{14}C-labeled coumaryl alcohols; free and linked into dehydropolymer and plant lignins and model humic acids, *Soil Sci. Soc. Amer. Proc.,* 41, 556, 1977.
24. **Crawford, D. L., Floyd, S., Pometto, A. L., III, and Crawford, R. L.,** Degradation of natural and Kraft lignins by the microflora of soil and water, *Can. J. Microbiol.,* 23, 434, 1977.
25. **Björkman, A.,** Studies of finely divided wood. I. Extraction of lignin with neutral solvents, *Sven. Papperstidn.,* 59, 447, 1956.
26. **Chang, H. M. and Sarkanen, K. V.,** Species variation in lignin: effect of species on the rate of kraft delignification, *Tappi,* 56, 132, 1973.
27. **Crawford, D. L.,** unpublished data, 1978.
28. **Eriksson, K. E.,** Enzyme mechanisms involved in cellulose hydrolysis by the rot fungus *Sporotrichum pulverulentum, Biotechnol. Bioeng.,* 20, 317, 1978.

Chapter 8

DEGRADATION OF LIGNIN-RELATED AROMATICS AND LIGNINS BY SEVERAL PSEUDOMONADS

Hidekuni Kawakami

TABLE OF CONTENTS

I. INTRODUCTION

Lignin is one of the most recalcitrant natural materials in the biosphere. A large quantity is introduced into soil and natural waters in plant tissues every year. Pulping waste water also introduces lignin into natural waters, although the total quantity of lignin from this source is relatively small.[1,2] Such a large quantity of natural organic matter, however, does not accumulate on the earth. A part of the lignin is gradually decomposed and changed to humic or fulvic acids, and these materials may be gradually mineralized by many organisms (see Volume I, Chapter 4). Lignin forms a significant part of the biospheric carbon cycle on the earth. However, the mechanisms of lignin biodegradation have not been elucidated satisfactorily.

There are many microorganisms in forest soils and in sediments of natural waters. In forest soils aerobic and heterotrophic bacteria dominate, especially Gram-negative rods, such as pseudomonads.[3] As is generally known, *Pseudomonas* and *Acromobacter* are ubiquitous in both fresh waters and sea waters. Therefore, it is assumed that these bacteria directly or indirectly may participate in lignin biodegradation. In fact, it has been reported that lignin is degraded by the genera *Pseudomonas*,[4,5] *Flavobacterium*,[4,5] *Mycobacterium*,[5] *Micrococcus*,[5] and *Xanthomonas*.[5] We have investigated in some detail the possibility that lignin and lignin-related aromatics are degraded by several pseudomonads, both stock cultures and newly isolated strains. Those investigations are summarized here.

II. DECOMPOSITION OF LIGNIN-RELATED AROMATICS BY STOCK CULTURE STRAINS

The following studies were carried out: (1) oxidation of a variety of lignin-related aromatics, using Warburg manometry, (2) cleavage mechanism of the guaiacyl nucleus, such as in vanillic acid, (3) decomposition of lignin-related "monomers" (C_6, C_6-C_1, C_6-C_2, and phenylpropanoids), and (4) decomposition of model "dimers" and 5- or 6-position condensed-type compounds having guaiacyl nuclei.

The bacteria used for this study were *P. ovalis* Chester IAM 1002 and *P. fluorescens* Migula IAM 1006 obtained from the Institute of Applied Microbiology, Tokyo University. These strains were found to have a considerably higher activity toward the substrates tested compared to several other stock cultures; they also do not have any phenol-oxidizing enzyme activity, such as laccase or peroxidase.

A. Cleavage of Aromatic Nuclei[6]

In general, when the phenol carboxylic acids are subjected to the action of mono- and dioxygenases of microorganisms, the aromatic rings cleave rapidly through protocatechuic acid(20)* or gentisic acid(40). In our experiments (Table 1), *o*-, *m*-, and *p*-hydroxybenzoic acids(12, 16, 4) were readily decomposed by *P. ovalis* or *P. fluorescens*. Cleavages of the aromatic nuclei were affected remarkably by methylation of phenolic hydroxyl groups. However, the methoxyl group of the *meta* position was easily cleaved by the bacteria. The presence of a carboxyl group or a group easily oxidized to a carboxyl on the aromatic ring, together with the free phenolic hydroxyl group at the *para* position, was required for rapid aromatic ring rupture. Therefore, cleavage of a guaiacyl nucleus in C_6-C_1 compounds, such as vanillin(26) or vanillic acid(27), took place easily. On the contrary, compounds with a methoxyl group at the *para* position, such as anisaldehyde(8), anisic acid(9), isovanillin(22), isovanillic acid(23), veratraldehyde(34), and veratric acid(35), were not cleaved as indicated by

---

*   Numbers in parenthesis refer to structures in Section VII.

TABLE 1

Aromatic Ring Rupture in Lignin-Related Compounds by *Pseudomonas ovalis* and *P. fluorescens* and Protocatechuate Oxygenase-Inducing Ability by Lignin-Related Aromatics in *P. fluorescens*

| Substrate or inducer | Oxygen uptake | | PCA-oxygenase-inducing ability |
	P. fluores-cens	*P. ovalis*	
p-Hydroxy benzaldehyde (3)	A	A	+
p-Hydroxy benzoic acid (4)	A	A	+
p-Coumaric acid (5)	A	A	−
Anisaldehyde (8)	B	A[a]	−
Anisic acid (9)	—	A[a]	−
Catechol (11)	A	A	−
Salicylic acid (12)	—	A	−
m-Hydroxy benzoic acid (16)	—	A	−
Protocatechualdehyde (19)	A	A	+
Protocatechuic acid (20)	A	A	+
Isovanillic acid (23)	—	A[a]	−
Vanillyl alcohol (25)	—	A	−
Vanillin (26)	A	A	+
Vanillic acid(27)	A	A	+
Coniferyl alcohol (29)	—	A[a]	−
Ferulic acid (31)	—	A[a]	−
Guaiacyl glycerol (32)	—	A[a]	−
Veratraldehyde (34)	B	A[a]	−
Veratric acid (35)	—	A[a]	−
Gentisic acid (40)	A	A	−
Gallic acid (48)	—	A	−
Syringaldehyde (52)	B	A[b]	−
Syringic acid (53)	—	A[b]	−

Note: A: More than 1.1 mol of oxygen consumption per mol of substrate, and disappearance of all UV absorption; B : 0.2 to 0.6 mol oxygen consumption, by oxidation of an aldehyde group.

[a] Reaction time, 24 to 48 hr.
[b] Incubation with vanillin or vanillic acid.

Reproduced from Kawakami, H., *Mokuzai Gakkaishi*, 22, 246, 1976. With permission.

Warburg manometry experiments. However, with extended reaction times of up to 2 days, the aromatic rings in these compounds were also cleaved by *P. ovalis*.

The aromatic nuclei of 5- or 6-position condensed guaiacyl compounds (59 to 65) and a dimer(66) were not cleaved by the bacteria. Phenylpropanoids(5, 10, 21, 29 to 32, 37) were also resistant, except for *p*-coumaric acid(5). With extended reaction times, however, coniferyl alcohol(29), ferulic acid(31), and guaiacylglycerol(32) were decomposed completely by *P. ovalis* (Table 1). It is assumed that these compounds were decomposed to C_6-C_1 compounds by β-oxidation of the side chain prior to cleavage of aromatic nuclei.

Phenolalcohols, such as benzyl alcohol and vanillyl alcohol(25), were oxidized only by *P. ovalis*. Methoxylated C_6 compounds, such as guaiacol(13), veratrole(23), anisole(6), pyrogallol-1,3-dimethyl ether(51), pyrogallol trimethyl ether(54), and methoxylated C_6-C_2 compounds, such as homovanillic acid(28) and homoveratric acid(36), were more resistant to ring rupture than methoxylated C_6-C_1 compounds.

The syringyl nucleus, which comprises a major portion of aromatic nuclei in angios-

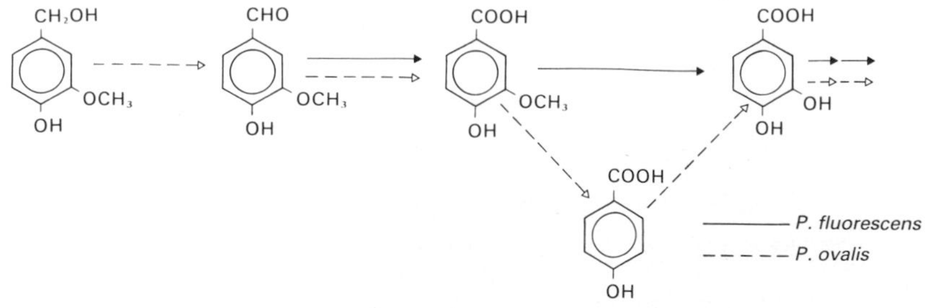

FIGURE 1.   A scheme for protocatechuic acid formation from vanillyl alcohol and vanillin by *P. ovalis* and *P. fluorescens* (Reproduced from Kawakami, H., *Mokuzai Gakkaishi*, 22, 246, 1976. With permission.)

perm lignins (see Volume I, Chapter 1), was not cleaved by the bacteria in the manometry experiments. Only when syringaldehyde(52) and syringic acid(53) were incubated together with vanillin(26) or vanillic acid(27) were they completely oxidized by *P. ovalis*. The ability to cleave the syringyl nucleus in this strain was not induced by syringic acid or syringaldehyde. The utilization degree of syringic acid by the strain was approximately proportional to the concentration of coexistent vanillic acid or vanillin. This synergistic effect is limited to vanillin or vanillic acid; no effect was observed with other compounds, such as benzoic acid, *p*-hydroxybenzoic acid(4), protocatechuic acid(20), veratric acid(35), catechol(11), the acids of the tricarboxylic acid cycle, or C₁ compounds. Protocatechuate oxygenase-inducing ability was examined only in the case of *P. fluorescens*; *P. ovalis* has the enzyme constitutively.

B. Cleavage of the Guaiacyl Nucleus

Aromatic ring cleavage, which is clearly an important step in lignin decomposition by microorganisms, was studied in *P. ovalis* and *P. fluorescens* by using vanillic acid as a sole carbon source.[7] A metabolic pathway of vanillic acid(27) to protocatechuic acid(20) was investigated by the method of sequential induction (sucessive adaptation). Metabolic pathways in relation to the strains are shown in Figure 1 (vanillyl alcohol oxidations to vanillic acid is included). Vanillic acid was metabolized directly via demethylation to protocatechuic acid by *P. fluorescens*; however, in the case of *P. ovalis*, vanillic acid was metabolized via demethoxylation to *p*-hydroxybenzoic acid followed by hydroxylation to protocatechuic acid. Demethoxylation of the guaiacyl nucleus by microorganisms is an unusual phenomenon; it has been reported also in *Polystictus sanguineus* (*Trametes sanguinea*)[8] and in a Sâke yeast.[9]

In earlier studies of a bacterium,[10,11] an imperfect fungus,[12] and a basidiomycete,[13] the manner of ring fission in protocatechuic acid arising from vanillic acid was of the "intradiol type", in which the bond between the carbon atom bearing the hydroxyl groups is cleaved. Our pseudomonads apparently use the "extradiol-type" cleavage. This is suggested on the basis of experiments in which monofluoroacetate was added to the culture medium prior to complete disappearance of vanillic acid in order to inhibit the TCA cycle and accumulate intermediates. *cis*-Aconitic acid, L-malic acid, and lactic acid were isolated from the medium. The strain accumulated *cis*-aconitic acid as a main intermediary product of aromatic ring cleavage. It is thus suggested that the "extradiol-type" cleavage is operative.

C. Decomposition of Lignin Model Monomers in Shaking Culture

In the manometric conditions, many lignin models were not cleaved because the

TABLE 2

Degradation of C_6 and C_6-C_1 Compounds by *P. ovalis*

Substrate	λ_{max}	$D^a_{1/2}$	$D_{1/5}$	$D_{1/10}$	D
Guaiacol (67)	274	3	3	5	7
Veratrole (68)	271	3	4	5	8
Pyrogallol (69)	263	—	—	—	—
Pyrogallol monomethyl ether (70)	266	—	—	—	—
2,6-Dimethoxy phenol (71)	266	11	21	—	—
Isovanillin (72)	251	2	2	2	3
Isovanillic acid (73)	251	1	2	2	2
Veratryl alcohol (74)	275	3	4	4	5
Veratraldehyde (75)	252	2	2	2	3
Veratric acid (76)	283	2	2	2	2
3-*O*-methyl gallic acid (77)	260	2	2	3	6
Syringyl alcohol (78)	265	3	3	3	4
Syringaldehyde (79)	258	3	3	3	4
Syringic acid (80)	262	3	3	3	4
Trimethyl gallic acid (81)	254	4	4	4	5
Homocatechol (82)	280	—	—	—	—
Ethyl vanillin (83)	254	4	21	21	23
Creosol (84)	278	6	6	6	8
p-Methyl anisol (85)	275	7	7	8	9
Anise alcohol (86)	270	4	6	7	9
Anisaldehyde (87)	282	2	2	2	2
Anisic acid (88)	257	2	2	2	2

^{a} $D_{1/2}$, $D_{1/5}$, $D_{1/10}$, and D : days for 50%, 80%, 90%, and complete decomposition of each substrate, respectively.

Reproduced from Kawakami, H., *Mokuzai Gakkaishi*, 21, 309, 1975. With permission.

reaction time was too short. In the natural world, biodegradation of lignin proceeds very slowly. Therefore, incubation time of the recalcitrant models was prolonged for up to 30 days using shake cultures (115 r/min, 30°C).[14] *Pseudomonas ovalis*, which has the strongest ability for the model decomposition among the tested stock culture strains, was employed for this study. About 50 mg (dry weight) of bacteria, precultured on nutrient broth, was used to inoculate 500 mℓ shake flasks containing 100 mℓ of medium (containing 100 mg NH_4NO_3, 100 mg K_2HPO_4, and 50 mg $MgSO_4 \cdot 7H_2O$, pH = 7.0) supplemented with 15 mg of aromatics as the sole carbon source.*

1. C_6 Compounds

In the manometric conditions, C_6 compounds which have methoxyl groups were demethylated or demethoxylated only with difficulty. In the shaking cultures, guaiacol(67) and veratrole(68) were decomposed completely after 7 to 8 days (Table 2). Since cleavage of catechol occurs in a matter of hours, demethylation or demethoxylation and hydroxylation of guaiacol or veratrole were probably rate-limiting, and the numbers of methoxyl groups (one or two) had no apparent influence on the rate of aromatic ring rupture. Pyrogallol(69) and its methyl ethers(70,71) resisted decomposition.

2. C_6-C_1 Compounds

The above-described work indicated that demethoxylation or demethylation of the 3-position of the guaiacyl nucleus was caused by the strains. In the case of *P. ovalis*,

* Degradation was followed by the decrease in UV absorbance at the λ_{max} for the various compounds.

TABLE 3

Degradation of C_6-C_2 Compounds by *P. ovalis*

Substrate	λ_{max}	$D^a_{1/2}$	$D_{1/5}$	$D_{1/10}$	D
P-Hydroxyacetophenone (89)	274	7	8	8	9
Apocynol (90)	276	4	5	8	13
Acetoguaiacone (91)	275	10	11	11	11
Homovanillic acid (92)	278	6	7	7	7
3-Methoxy-4-hydroxy mandelic acid (93)	276	4	6	6	12
Acetoveratrone (94)	273	10	11	11	11
Homoveratric acid (95)	275	10	13	25	30
Acetosyringone (96)	272	—	—	—	—

ᵃ $D_{1/2}$, $D_{1/5}$, $D_{1/10}$, and D : days for 50%, 80%, 90%, and complete decomposition of each substrate, respectively.

Reproduced from Kawakami, H., *Mokuzai Gakkaishi*, 21, 309, 1975. With permission.

demethylation of the *para* position in the anisyl or veratryl nuclei was also indicated. Isovanillin(72), isovanillic acid(73), veratraldehyde(75), veratric acid(76), anisaldehyde(87), and anisic acid(88) were cleaved in 2 to 3 days (Table 2).

Gallic acid methyl eters(77, 80, 81) were decomposed completely by this bacterium in 4 to 5 days, and they were easily decomposed compared to the corresponding C_6 compounds (Table 2).

C_6-C_1 models having a methyl group in the side chain(82, 84) were only slightly affected. (Table 2). Ethyl vanillin(83) was not cleaved by the strain (Table 2).

3. C_6-C_2 Compounds

In the manometric experiments, C_6-C_2 compounds having methoxyl groups were not decomposed, as described above. In an extended incubation period, they were also more resistant than in the corresponding C_6-C_1 compounds (Table 3). This strain induces homoprotocatechuate oxygenase (EC. 1.13.1.7). Although homoprotocatechic acid was cleaved in several hours, homovanillic acid(92) and homoveratric acid(95) required 7 and 30 days, respectively, for complete decomposition. If decomposition of homovanillic acid or homoveratric acid takes place via homoprotocatechuic acid, the results indicate that demethoxylation and hydroxylation or demethylation of each aromatic ring is extremely difficult compared with that of corresponding C_6-C_1 acids. Acetoguaiacone(91) and acetoveratrone(94), which have an α-carbonyl group in the side chain, were decomposed only very slowly, and acetosyringone(96) was not decomposed (Table 3).

In the kraft cooking process, many γ-carbon atoms of the side chains are removed from lignin; thus, leaving C_6-C_2 units. The resistance of C_6-C_2 units described here may contribute stability in kraft lignin toward degradation by bacteria.

4. Phenylpropanoids

It has been reported (Volume II, Chapter 6) that one type of lignin biodegradation by basidiomycetes takes place through phenylpropanoid intermediates. Under our manometric conditions, only *p*-coumaric acid(5) was decomposed by *P. ovalis* among the phenylpropanoids investigated. In an extended incubation period, however, dihydroconiferyl alcohol(106), coniferyl alcohol(107), guaiacylglycerol(108), coniferaldehyde(109), dihydroferulic acid(110), and ferulic acid(111) were easily decomposed dur-

109

TABLE 4

Degradation of Phenylpropanoids by *P. ovalis*

Substrate	λ_{max}	$D^a_{1/2}$	$D_{1/5}$	$D_{1/10}$	D
p-Hydroxy phenyl pyruvic acid (97)	266	2	3	3	5
Anethole (98)	258	7	8	9	11
Caffeic acid (99)	277	2	—	—	—
Coerulignol (100)	279	6	10	11	11
Isoeugenol (101)	258	5	6	10	—
Eugenol (102)	278	6	—	—	—
α-Guaiacyl propanol (103)	275	8	11	13	14
Propioguaiacone (104)	275	9	10	12	12
Guaiacyl acetone (105)	276	2	3	4	11
Dihydroconiferyl alcohol (106)	276	2	3	5	5
Coniferyl alcohol (107)	263	3	3	3	4
Guaiacyl glycerol (108)	260	1	2	2	2
Coniferyl aldehyde (109)	284	2	3	3	4
Dihydroferulic acid (110)	275	1	2	2	2
Ferulic acid (111)	283	2	2	2	2
Guaiacylpyruvic acid (112)	282	—	—	—	—
Isoeugenol methyl ether (113)	259	—	—	—	—
Propioveratrone (114)	273	9	10	11	12
Ferulic acid methyl ether (115)	283	3	4	4	8
Sinapic acid (116)	295	6	—	—	—

[a] $D_{1/2}$, $D_{1/5}$, $D_{1/10}$, and D: days for 50%, 80%, 90%, and complete decomposition of each substrate, respectively.

Reproduced from Kawakami, H., *Mokuzai Gakkaishi*, 21, 309, 1975. With permission.

ing several days by the strain (Table 4). However, guaiacylpyruvic acid(112), which is one of the reported pheylpropanoid intermediates of the decomposition of lignin or lignin model dimers by basidiomycetes,[15] was not decomposed by *P. ovalis*. *p*-Hydroxyphenylpyruvic acid(97) was decomposed with more difficulty than *p*-coumaric acid(5). Therefore, it seems that the pathway after rupture of β-0-4 bonding by this bacterium differs from that of the basidiomycete pathway reported by Ishikawa.[15]

Caffeic acid(99) formed a colored substance by air oxidation and was decomposed incompletely. The decomposition of phenylpropanoids by *P. ovalis*, however, proceeds with either formation of C_6-C_1 compounds by β-oxidation of the side chain before demethylation of the guaiacyl ring or by formation of *p*-coumaric acid by demethoxylation.

Phenylpropanoids having a terminal methyl group on the side chain (98, 100 to 105, 113, 114) were only slowly decomposed (Table 4). This fact shows that the terminal methyl group is difficult to oxidize to the carbinol group by this strain.

Sinapic acid(116), having a syringyl nucleus, was not decomposed (Table 4). Since *P. ovalis* lacks phenoloxidases, phenylpropanoids were decomposed without any condensation or polymerization such as has been observed in basidiomycetes.[16]

D. Decomposition of Model Dimers and 5- or 6-Position Condensed-Type Compounds Having Guaiacyl Nuclei [17]

1. Arylglycerol-β-aryl Ether-Type Compounds
It has been reported that the aryl-alkyl ether bond, an important unit in the lignin structure, is cleaved by certain basidiomycetes,[18,19,20] and by bacteria in the genera *Agrobacterium*[21] and *Pseudomonas*.[22]

TABLE 5

Degradation of Dimeric β-Aryl Ethers by *P. ovalis*

Substrate	λ_{max}	$D^a{}_{1/2}$	$D_{1/5}$	$D_{1/10}$	D
ω-(2-Methoxy phenoxy)-apocynol (117)	274	6	7	7	14
ω-(2-Methoxy phenoxy)-acetoguaiacone (118)	275	4	4	4	20
Guaiacylglycerol-β-guaiacyl ether (119)	271	4	4	5	5
α-(2-Methoxy phenoxy)-β-hydroxypropioguaiacone (120)	272	—	—	—	—
1-Guaiacyl-1-methoxy-2-guaiacoxy ethane (121)	275	26	—	—	—
1-Guaiacyl-1-ethoxy-2-guaiacoxy ethane (122)	275	21	—	—	—
Guaiacoxyacetic acid (123)	271	2	3	3	—
Glycerol-1-guaiacyl ether (124)	271	5	—	—	—

[a] $D_{1/2}$, $D_{1/5}$, $D_{1/10}$, and D: days for 50%, 80%, 90%, and complete decomposition of each substrate, respectively.

Reproduced from Kawakami, H., *Mokuzai Gakkuishi*, 21, 629, 1975. With permission.

In our investigation, guaiacylglycerol-β-guaiacyl ether(119) was also decomposed easily by *P. ovalis* (Table 5). The dimer, however, devoid of the terminal side chain carbon atom(117), was rather poorly decomposed, as was the analogous monomer (C_6-C_2 compound, 90). The dimers with α-carbonyl(118, 120) and α-ether groups(121, 122) were poorly decomposed. Since guaiacylglycerol-β-guaiacyl ether(119) was decomposed easily and guaiacoxyacetic acid(123) and guaiacyl-α-glycerol ether(124) were decomposed only slightly, it may be that the β-aryl ether linkage is cleaved in preference to cleavages within the side chain of the aromatic compounds.

2. 5- or 6-Position Condensed-Type Compounds of Guaiacyl Nucleus

In general, the 5- or 6-position condensed guaiacyl nuclei withstands chemical oxidation. Such compounds were also resistant to the action of *P. ovalis* (Table 6), except for some compounds which have a free phenolic hydroxyl group at the 4-position and a short side chain, such as a formyl(142,143) or a carboxyl(144) group, at the 5-position. Isohemipinic acid(150) and the compounds which have a C_3-condensed moiety at the C-5 position of the aromatic ring(145, 146, 147, 148) were hardly decomposed by the strain. This suggests that the propyl and allyl groups at C-5 are not β-oxidized by the bacterium.

Biphenyl-type compounds, such as dehydrodivanillin(151) and dehydrodivanillic acid(152), and diphenyl methane(153) were the most recalcitrant materials for the strain. 6-Position condensed-type compounds(154,155) were also resistant (Table 6).

3. Miscellaneous Compounds

Although α-benzyl ethers were only poorly metabolized by the bacterium, vanillyl ethers(127, 128) were decomposed easier than veratryl ethers(129, 130) (Table 7). A similar result was obtained with basidiomycetes.[20] The aryl-alkyl ether bonds at both the α- and β-positions of the side chain were cleaved easily by the basidiomycetes.[20] However, that at the β-position was preferentially cleaved by *P. ovalis*. Therefore, the putative β-aryl-ether-splitting enzyme of this bacterium has a relatively high specificity compared with that of the basidiomycetes.

The phenylcoumaran-type structure(137 to 140), which is important in lignin (Volume I, Chapter 1), was not cleaved by this strain (Table 7). This is consistent with the above since the structure contains both condensed 5-unit and α-benzyl ether linkages, both significantly resistant to the bacterial action in other compounds studied.

TABLE 6

Degradation of 5- and 6-Substituted Phenyl Compounds by *P. ovalis*

Substrate	λ_{max}	$D^a_{1/2}$	$D_{1/5}$	$D_{1/10}$	D
o-Vanillin (141)	277	13	15	30	—
5-Formyl vanillin (142)	266	9	11	13	15
5-Formyl vanillic acid (143)	260	3	4	4	5
5-Carboxy vanillic acid (144)	260	9	10	12	13
5-Propyl vanillin (145)	260	—	—	—	—
5-Allyl vanillin (146)	258	—	—	—	—
5-Propyl acetoguaiacone (147)	277	10	—	—	—
5-Allyl acetoguaiacone (148)	280	9	30	—	—
2,3-Dimethoxy benzoic acid (149)	267	7	15	—	—
Isohemipinic acid (150)	250	—	—	—	—
Dehydrodivanillin (151)	263	—	—	—	—
Dehydrodivanillic acid (152)	254	—	—	—	—
2,2'-Dihydroxy-3,3'-dimethoxy-5,5'-dimethyl di-phenylmethane (153)	275	—	—	—	—
m-Hemipinic acid (154)	273	—	—	—	—
4,5-Dimethoxy-o-tolyl methyl ketone (155)	273	—	—	—	—

[a] $D_{1/2}$, $D_{1/5}$, $D_{1/10}$, and D: days for 50%, 80%, 90%, and complete decomposition of each substrate, respectively.

Reproduced from Kawakami, H., *Mokuzai Gakkaishi*, 21, 629, 1975. With permission.

TABLE 7

Degradation of Miscellaneous Compounds by *P. ovalis*

Substrate	λ_{max}	$D^a_{1/2}$	$D_{1/5}$	$D_{1/10}$	D
dl-Pinoresinol (125)	275	—	—	—	—
Dehydrodiferulic acid (126)	280	30	—	—	—
Vanillyl methyl ether (127)	277	12	13	13	13
Vanillyl-α-ethyl ether (128)	277	11	13	13	13
Veratryl methyl ether (129)	277	16	18	19	19
Veratryl guaiacyl ether (130)	278	—	—	—	—
2,3-Di-α-hydroxylvanillyl butane diol 1,4 (131)	272	3	6	7	28
3,4-Diguaiacyl hexane (132)	272	9	12	13	16
Diethylstilbestrol (133)	255	—	—	—	—
Diguaiacyl methane (134)	279	11	24	—	—
3,3'-Dimethoxy-4,4'-dihydroxy chalcone (135)	258	15	25	26	—
Benzyl benzoate (136)	263	6	7	9	9
Dihydrodehydrodiisoeugenol (137)	260	—	—	—	—
Dehydrodiisoeugenol (138)	262	—	—	—	—
Dehydrodiconiferyl alcohol (139)	277	—	—	—	—
Erdtman's acid (140)	263	—	—	—	—

[a] $D_{1/2}$, $D_{1/5}$, $D_{1/10}$, and D; days for 50%, 80%, 90%, and complete decomposition of each substrate, respectively.

Reproduced from Kawakami, H., *Mokuzai Gakkaishi*, 21, 629, 1975. With permission.

The pinoresinol type structure (125, 126) was not decomposed. However, a lignin(131) which has interunit bonding only between the β-carbons of the side chains was completely cleaved, although it required a long period (Table 7). Therefore, the recalcitrant nature of the pinoresinol-type units is due to α-benzyl ether bonding, which forms an oxolane ring at the side chains.

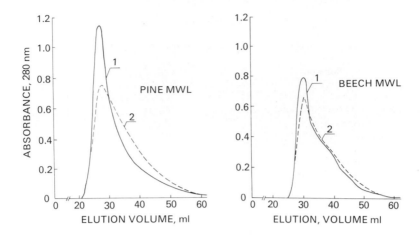

FIGURE 2.   Elution curves obtained on gel filtration of pine and beech MWL before and after degradation by *P. ovalis*. Sephadex® G-50, 80% DMF, 1; control, 2; after incubation for 60 days. (Reproduced from Kawakami, H., *Mokuzai Gakkaishi*, 22, 252, 1976. With permission).

A dimeric compound linked between the α-carbon atoms of the side chains(132) was metabolized. However, a stilbene structure(133) was not decomposed. Diguaiacyl methane(134) and a chalcone(135) were poorly metabolized by this strain. Ester-type linkages(136), which are contained in the lignin of true grasses or poplar (see Volume I, Chapter 1), were easily split (Table 7).

III. DEGRADATION OF LIGNIN BY A SELECTED STOCK CULTURE STRAIN

The above-described series of investigations on the biodegradation of lignin-related aromatics, carried out with a stock culture of *P. ovalis*, allow a prediction of the behavior of the strain in the degradation of lignin structural units. However, the results with the low molecular weight aromatic cannot be assumed to be applicable for prediction of biodegradation of the lignin polymer. The effectiveness of the strain in degrading lignin was therefore investigated by means of analyses of residual lignin isolated after incubation in cultures.

A. Degradation of Milled Wood Lignin (MWL)

Pine and beech MWL were incubated in stationary cultures of *P. ovalis* for 60 days. About 1 g (dry weight) of precultivated bacteria was used to inoculate 3-ℓ Erlenmeyer flasks containing 1 ℓ of medium supplemented with 1 g of lignin as sole carbon source; noninoculated lignin controls were carried through the incubations. After 60 days, the medium was evaporated *in vacuo,* and degraded lignins were extracted with acetone and dioxane from the evaporated residue.[23]

In the ether-soluble fraction of the extract low molecular weight intermediates of lignin biodegradation would be expected. However, only small yields of ether solubles were obtained. Thus, low molecular weight intermediates did not accumulate in the residue or in the culture medium, suggesting a successive breakdown of MWL from terminals rather than extensive depolymerization.

A significant portion of the high molecular weight of MWL was decomposed, as shown by gel filtration of the control and residual lignins. The results also indicated that the pine MWL was more degradable than the beech MWL (Figure 2).

The degraded MWL contained several percent nitrogen. This nitrogen was due to protein and nucleic acid from the bacterium because acid hydrolysis gave a ninhydrin-positive test and guanine was detected in the hydrolysate.

TABLE 8

Yields of Ethanolysis Products from Pine MWL Before and After Degradation by *P. ovalis*

MWL	Guaiacyl acetone (%)	Vanilloyl methyl ketone (%)	2-Ethoxy-1-guaiacyl-1-propanone (%)	1-Ethoxy-1-guaiacyl-2-propanone (%)	Total (%)
Control	1.9	3.6	3.1	1.3	9.9
Degraded	0.7	4.1	1.5	Trace	6.3

Changes in functional groups and in aromatic structure were observed with the progress of bacterial incubation. Degraded MWL showed an increased absorption at 1710 to 1720 cm^{-1} in the IR spectrum. This absorption was decreased by sodium borohydride treatment, but was not decreased by sodium dithionite treatment of the sample, indicating that the absorption was due to the nonconjugated carbonyl groups and not to quinonoid structures.

The β-aryl-ether-type structure in the MWL decresed during incubation. The decrease in major ketones, which was determined by means of Nakano's method,[24] was 43.2% (pine), 20.2% (beech), and 43.3% (straw), respectively, in 60 days. A similar ratio of major ketones was given by ethanolysis of pine MWL as shown in Table 8. In the experiment with β-ether models, the β-aryl ether bond was cleaved easily. However, the linkage cannot be cleaved completely in the MWLs. It seems that the enzyme reaction is restricted because of the three-dimentional structure of the lignin molecule. Also, poor decomposition of some of the β-aryl ether linkages could be due to linkage of the involved units in resistant 5-condensed and phenylcoumaran units. However, it is of importance that the β-aryl ether bond, which is the major interunit linkage in conifer MWL,[25] was cleaved rather easily.

Although they comprise only a few percent of the units in the lignin molecule, the coniferaldehyde-type structure was readily degraded by the strain. This result was similar to that obtained with the low molecular weight compounds.

Residual degraded lignin showed a decrease in units yielding aromatic aldehydes on nitrobenzene oxidation and an increase in isohemipinic, metahemipinic, and dehydrodiveratric acid-yielding groups on permanganate oxidation following methylation (Table 9). These results show that the aromatic aldehyde-yielding groups are preferentially decomposed, but phenylcoumaran-type structures and biphenyl structures are resistant. Methoxyl and phenolic hydroxyl groups in the residual lignin were lower than in the original. In addition, *p*-anisic and 4-methoxyisophthalic acid-yielding groups on permanganate oxidation after metheylation increased with the progress of bacterial incubation; therefore, demethoxylation of MWL was caused by this strain. Hardwood lignin was degraded rather poorly because of the difficulty of metabolism of the syringyl nucleus. This interpretation is based on the information obtained from studies of the metabolism of lignin-related aromatics.

B. Degradation of Pulping Byproduct Waste Lignins

Bacterial degradation of pine and beech pulp waste lignins was also studied using stationary cultures of *P. ovalis*. The samples used in this study were kraft lignin,[26] lignin sulfonate,[27] oxygen-alkali waste lignin,[28] and hydrotropic lignin.[29] Incubation was carried out for up to 60 days, except for the oxygen-alkali waste lignin which after 30 days had been degraded very well by the bacterium. The medium used in studying degradation of waste lignins by *P. ovalis* was similar to that used with the MWLs. The extent of degradation was lignin sulfonate < kraft lignin < hydrotropic lignin < oxygen alkali waste lignin.

TABLE 9

Yields of Aromatic Acids Following Methylation and Permanganate Oxidation of Control and Degraded (*P. ovalis*) MWL[a]

Sample	Veratric	Anisic	Trimethyl gallic	4-Methoxy isophthalic	Isohemipinic	Metahemipinic	Dehydrodiveratric	Veratric acid (% of sample)
Pine								
Control	100.0	8.0	4.2	Trace	18.1	Trace	10.9	5.2
Degraded	100.0	11.9	2.6	2.2	22.4	0.9	19.1	4.2
Beech								
Control	100.0	1.6	26.5	Trace	14.4	Trace	6.2	3.7
Degraded	100.0	4.7	10.9	Trace	17.1	Trace	5.5	1.9

[a] Yields of each acid are calculated as % of veratric acid (100.0).

Reproduced from Kawakami, H., *Mokuzai Gakkaishi*, 22, 252, 1976. With permission.

TABLE 10

Yields of Aromatic Acids Following Methylation and Permanganate Oxidation of Control and Degraded (*P. ovalis*) Kraft Lignin[a]

Sample	Veratric	Anisic	Trimethyl gallic	4-Methoxy isophthalic	Isohemipinic	Metahemipinic	Dehydrodiveratric	Veratric Acid (% of sample)
Pine								
Control	100.0	1.2	12.5	5.8	20.5	Trace	24.5	4.1
Degraded	100.0	4.3	15.9	15.7	17.4	6.6	32.2	2.6
Beech								
Control	100.0	Trace	156.3	4.0	26.9	Trace	9.0	2.1
Degraded	100.0	5.0	213.9	5.3	11.3	8.2	18.6	1.5

[a] Yields of each acid are calculated as % of veratric acid (100.0).

Reproduced from Kawakami, H., *Mokuzai Gakkaishi*, 22, 252, 1976. With permission.

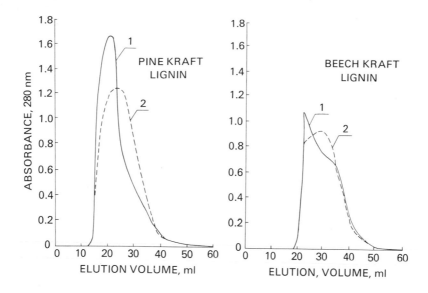

FIGURE 3.. Elution curves obtained on gel filtration of pine and beech kraft lignin before and after degradation by *P. ovalis*. Sephadex G-50, 80% DMF, 1; Control, 2; After incubation for 60 days. (Reproduced from Kawakami, H., *Tappi (Japan)*, 29, 309, 1975.)

A portion of the biodegraded lignins was adsorbed to bacterial cells. There was a tendency for this adsorption to increase with the extent of biodegradation.[29]

1. Kraft Lignins

After incubation, pine kraft lignin was much more decomposed than beech kraft lignin by *P. ovalis*. A portion of the high molecular weight part of the pine kraft lignin was appreciably affected, as shown in Figure 3, and carbonyl groups were found to have increased remarkably. Vanillin-yielding groups, on nitrobenzene oxidation of the pine lignin, decreased extremely. However, neither vanillin- nor syringaldehyde-yielding groups were degraded significantly in beech lignin. Biphenyl-, 5-position condensed guaiacyl-, and syringyl-type structures had all increased in the residual lignin after decay. Methoxyl groups had decreased with the progress of biodegradation. However, phenolic hydroxyl and catechol groups in the residual lignin also had decreased. Among the products formed on permanganate oxidation after methylation of the residual lignin, *o*-anisic acid and 4-methoxyisophthalic acid had increased in comparison to the original lignin, (Table 10). Therefore, it is likely that demethoxylation was also caused in the kraft lignins. (Small amounts of elemental sulfur and inorganic sulfur compounds in the kraft lignin were shown not to affect the activities of dioxygenase or the decomposition of lignin-related "monomers" by this strain.[30])

2. Lignins Sulfonates

Lignin sulfonates were only slightly decomposed by this bacterium compared with other waste lignins. Biodegraded residual lignin sulfonates contained only small amounts of nitrogen, which derived from protein and nucleic acid of the bacterial cells. In the residual lignin, however, a decrease in aromatic aldehyde-yielding groups, on nitrobenzene oxidation, was observed. *p*-Anisic acid- and 4-methoxyisophthalic acid-yielding groups (on permanganate oxidation after methylation) of pine lignin sulfonate increased with biodegradation. Methoxyl groups and phenolic hydroxyl groups decreased. These facts suggest that demethoxylation was effected during a relatively long

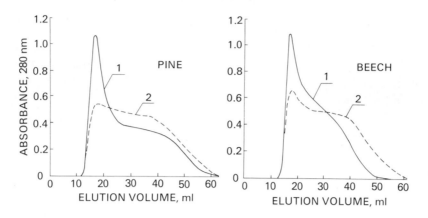

FIGURE 4.. Elution curves obtained on gel filtration of pine and beech oxygen-alkali waste lignin before and after degradation by *P. ovalis.* Sephadex G-100, 80% DMF, 1; Control, 2; After incubation for 60 days. (Reproduced from Kawakami, H. and Kanda, T., *Tappi (Japan)*, 30, 165, 1976. With permission.)

incubation period by this strain. Consequently, although it was only to a very small extent, biodegradation of lignin sulfonates was observed to proceed in a manner similar to that of the other lignins. Since the sulfonic acid group is very stable, its content in the samples increased during incubation.

3. Oxygen-Alkali Waste Lignin

Since oxygen-alkali waste lignin has been appreciably oxidized and has an increased hydrophilic property, it was more easily affected by the *P. ovalis* than the other waste lignins. After incubation for 30 days, a great portion of the high molecular weight part of pine and beech oxygen-alkali waste lignin was appreciably decomposed, as shown in Figure 4. The nitrogen content of the residual lignin increased with incubation time. The amino acid composition of the hydrolysate of the biodegraded lignin differed markedly from that of the bacterium cells. In the hydrolysate, valine, leucine, serine, and tyrosine were absent or nearly so; glycine, alanine, isoleucine, proline, and aspartic acid were found. The nitrogen content of the degraded samples was almost unaffected by treatment with a proteolytic enzyme (Pronase®). Small amounts of guanine and adenine arising from bacterial nuleic acids were detected; however, there were no detectable quantities of nucleic acids.

Vanillin- and syringaldehyde-yielding groups (on nitrobenzene oxidation) were decreased extremely by incubation with the bacterium. As a result, 5-position condensed guaiacyl-type and biphenyl-type structures were increased. Methoxyl groups and phenolic hydroxyl groups decreased with the progress of bacterial decomposition, but units yielding *p*-anisic acid and 4-methoxyisophthalic acid increased, as in the case of kraft lignins. Therefore, demethoxylation of this lignin was also affected.

4. Hydrotropic Lignin

Hydrotropic lignin was also affected by *P. ovalis.* In the lignin molecule, the following changes were observed with the progress of the bacterial incubation: the lignin molecule was appreciably decomposed, and the carbonyl groups and nitrogen content increased; Aromatic aldehyde-yielding groups (nitrobenzene oxidation) decreased; 5-position condensed guaiacyl-type structures increased; methoxyl groups decreased; and phenolic hydroxyl groups increased. It is suggested that demethoxylation was caused by the strain, as in the other lignins studied, because anisic acid- and 4-methoxyisophthalic acid-yielding structures increased on incubation.

IV. DEGRADATION OF KRAFT LIGNIN BY WILD TYPE PSEUDOMONADS

As described above, the lignin biphenyl linkage and the biphenyl-type lignin model compounds were hardly degraded by the stock culture strain of *P. ovalis*. A screening test was therefore undertaken with the intention of isolating bacteria which have the ability to utilize the biphenyl-type compounds. A number of promising strains were isolated from natural waters in enrichment shaking cultures using dehydrodivanillin as a sole carbon source. Dehydrodivanillin was decomposed completely by these strains in several days. All of the strains were identified as pseudomonads on the basis of physiological and biological criteria.

A. Degradation of Kraft Lignin by Pseudomonads Isolated from Fresh Water

From fresh water samples of rivers and ponds, three promising strains were isolated by the enrichment technique.[31] Demethylation of vanillic acid was caused by the three strains, but demethoxylation, as seen with *P. ovalis*, was not. The strains also easily decomposed phenylcoumaran and pinoresinol-type models,[32] which were not decomposed by *P. ovalis*. Physiological tests indicated that these strains belonged to *P. fluorescens* (PA-101 and RH-102) and *P. putida* (RJ-108). They have a much stronger ability to degrade lignin than the *P. ovalis* stock culture.[32] After incubation with these strains for 1 to 2 weeks, kraft lignin, as the sole carbon source, rose to the surface of the medium with the evolution of a large quantity of gas, as shown in Figure 5. Only small amounts of degraded lignin were recovered from the medium. The high molecular weight portion of the lignin was decomposed appreciably, and fine lignin was more degraded than beech lignin. Degraded lignin exhibited an increased nitrogen content, increased contents of phenolic hydroxyl, catechol, and carboxyl groups, and a decreased methoxyl content compared to the control lignin. However, these differences were not large.

On permanganate oxidation of the lignin, following methylation, the ratio of dehydrodivanillin to veratric acid was scarcely changed by biodegradation. It seems that the lignin was not subjected to selective degradation of particular parts of the molecule by these strains; the lignin was degraded by the strains quite uniformly, and a successive breakdown of the lignin from terminals is suggested.

B. Degradation of Kraft Lignin by Pseudomonads Isolated from Sea Water

Discharged pulp waste effluents are carried away to the sea in a relatively short period. Therefore, waste lignin decomposition would be carried out predominantly in the brackish water region at the mouths of rivers and along the sea coast. Such waters should be a good source of lignin-degrading bacteria.

From about 150 kinds of sea water samples, three strains of dehydrodivanillin-utilizing bacteria were isolated.[33] These strains were also identified as pseudomonads. Demethylation of guaiacyl lignin models was caused by the strains, but demethoxylation was not. Two strains are most likely fresh water bacteria and degraded kraft lignin to some extent. The higher molecular weight portion of the lignin was decomposed somewhat. The degraded lignin had an increased carboxyl and a decreased methoxyl content. However, the lignin was decomposed by these bacteria rather uniformly. Nevertheless, the ability of these strains to degrade lignin was lower than that of the strains isolated from fresh water samples.

One strain is perhaps a marine bacterium. The lignin was degraded to a much smaller extent by this strain than by the other two strains.

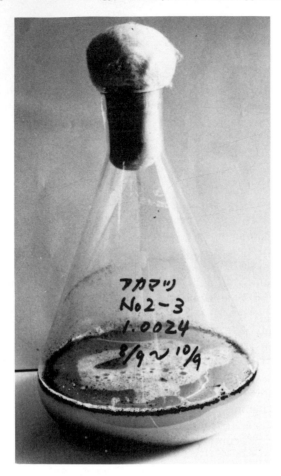

FIGURE 5.   Incubation of kraft lignin with RH-102,
a wild type pseudomonad. Gas formation after 7-day
incubation is shown.

V. CONCLUSION

In general, the carbon-carbon bonds, such as in the biphenyl-type linkage and in 5-
or 6-position condensed-type structures of the guaiacyl nucleus, in lignin are stable
both chemically and microbiologically. This fact is supported by some investigations
on chemical oxidation of humic acids.[34,35] Although *P. ovalis* Chester IAM 1002 has
a strong degradative ability for lignin-related aromatics and lignins compared to other
stock culture strains, it also led to increased amounts of biphenyl-, and 5- or 6-position
condensed-type structures in the residual lignin, and compounds containing such structures were generally poorly metabolized.

However, several strains of pseudomonads which have an ability to metabolize biphenyl structures were isolated from natural waters. These strains have a much
stronger ability for kraft lignin degradation than the stock culture strain. Kraft lignin
was decomposed by the wild type pseudomonads quite uniformly, and little differences
in the functional groups between the degraded and control lignins were oserved.

Syringyl nucleus-assimilating bacteria have now been isolated from fresh waters.[32]
Therefore, it is suggested that hardwood lignin may also be decomposed by bacteria.
The studies reported here suggest strongly that bacteria participate in lignin decomposition in nature and especially in the decomposition of waste lignins in streams and
other waters.

VI. SUMMARY

Pseudomonads occur widely in forest soil and natural waters. Biodegradation of lignin and lignin models were studied by *P. ovalis*, which was found to have the highest activity among several stock cultures. About 120 lignin models, MWLs, and waste lignins were used as samples. Several strains of pseudomonads which have much strong ability for lignin degradation than *P. ovalis* were isolated from fresh water and sea water and used for studying the biodegradation of kraft lignin.

In general, phenylpropanoids and C_6-C_1 compounds were decomposed more easily than corresponding C_6-C_2 and C_6 compounds. bAryl-ether- and ester-type compounds were decomposed rather easily. 5-Position condensed guaiacyl-phenylcoumaran- and pinoresinol-type compounds were resistant to the action of *P. ovalis*, but easily decomposed by wild type pseudomonads isolated by the use of culture media containing biphenyl-type compounds as a sole source of carbon.

Pine and beech MWL were degraded by stationary cultures of *P. ovalis* for 60 days; the former was considerably more decomposed than the latter. Residual degraded lignin showed a decrease of aromatic aldehyde-yielding groups on nitrobenzene oxidation, the baryl-ether-type structures increased relatively, and demethoxylation of MWL was evident. Low molecular weight intermediates did not accumulate in the culture media, and a successive breakdown of MWL from terminals was suggested.

The extent of degradation of waste lignins was in the following order: Lignin sulfonate < kraft lignin < hydrotropic lignin < oxygen-alkali lignin. Kraft lignin was decomposed by the wild type pseudomonads quite uniformly, and little difference in the functional groups between the degraded and control lignins was observed. The extent of biodegradation by the strains from fresh water was much greater than by the strains from sea water.

VII. MODEL COMPOUNDS

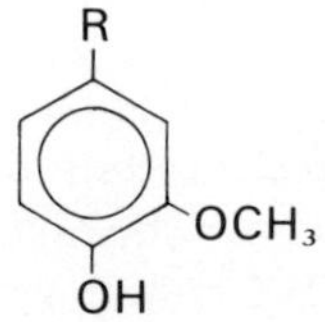

R

24 R = CH$_3$
25 R = CH$_2$OH
26 R = CHO
27 R = COOH
28 R = CH$_2 \cdot$COOH

29 R = CH: CH$\cdot$CH$_2$OH
30 R = CH: CH$\cdot$CHO
31 R = CH: CH$\cdot$COOH
32 R = CHOH$\cdot$CHOH$\cdot$CH$_2$OH

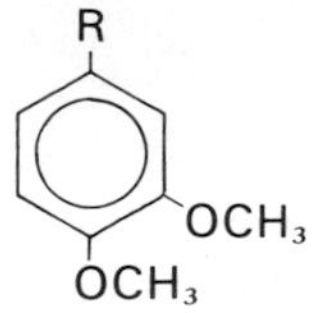

OCH$_3$
OH

R

33 R = H
34 R = CHO
35 R = COOH
36 R = CH$_2 \cdot$COOH
37 R = CH:CH$\cdot$COOH

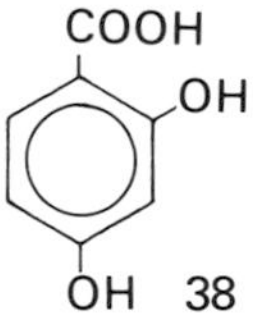

OCH$_3$
OCH$_3$

COOH
OH
OH 38

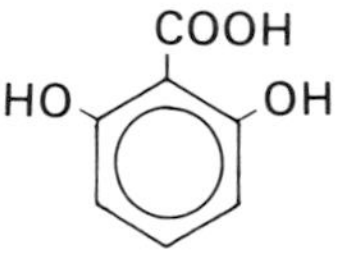

COOH
HO   OH
39

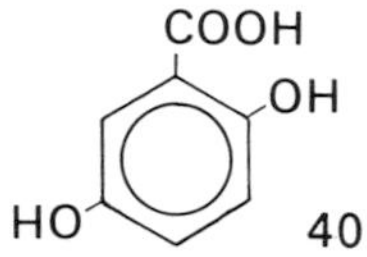

COOH
OH
HO   40

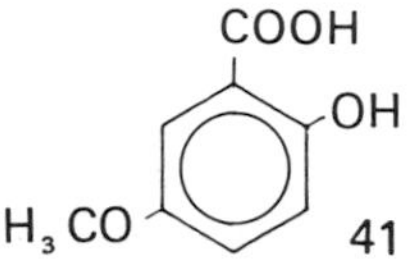

COOH
OH
H$_3$CO   41

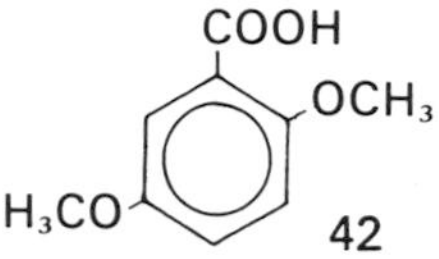

COOH
OCH$_3$
H$_3$CO   42

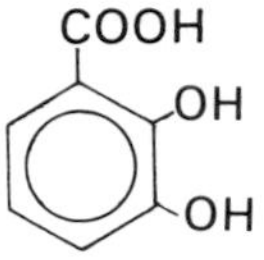

COOH
OH
OH   43

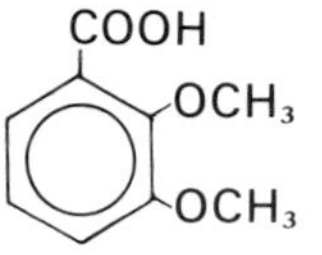

COOH
OCH$_3$
OCH$_3$ 44

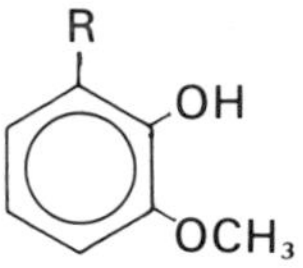

R
OH
OCH$_3$
45 R = CHO
46 R = COOH

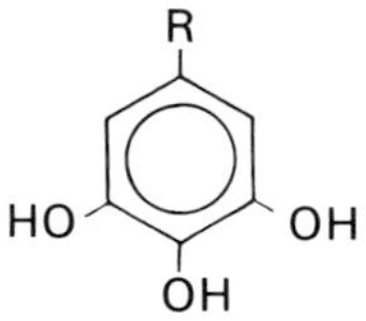

R
HO   OH
OH
47 R = H
48 R = COOH

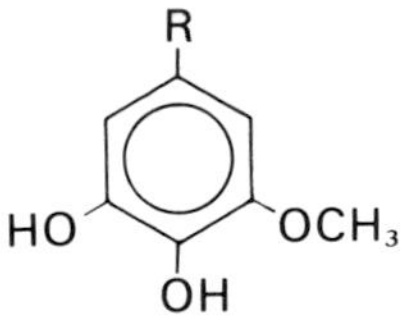

R
HO   OCH$_3$
OH
49 R = H
50 R = COOH

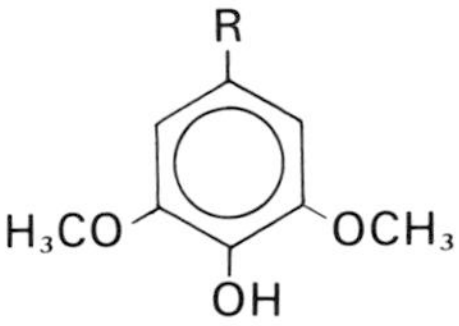

R
H$_3$CO   OCH$_3$
OH
51 R = H
52 R = CHO
53 R = COOH

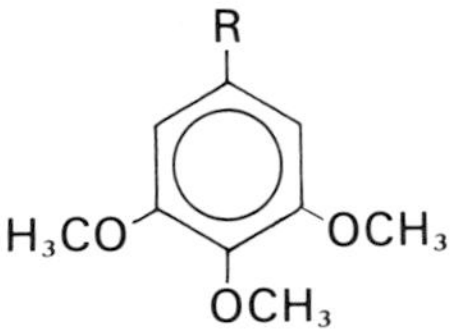

R
H$_3$CO   OCH$_3$
OCH$_3$
54 R = H
55 R = COOH

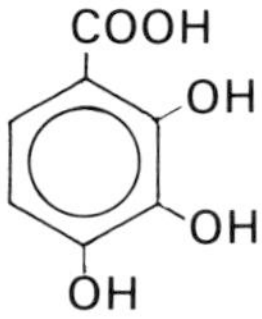

COOH
OH
OH
OH   56

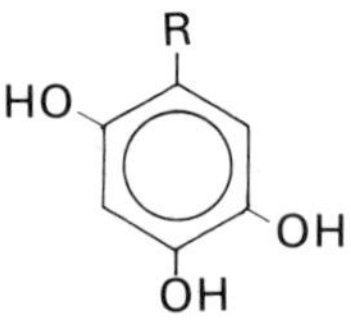

R
HO
OH
OH
57 R = H
58 R = COOH

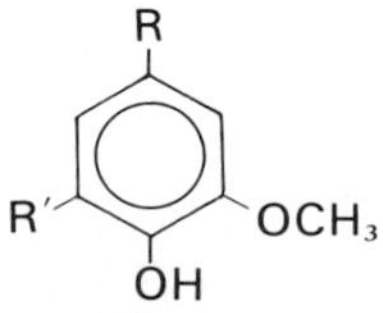

R
R′   OCH$_3$
OH

59 R = CHO;   R′ = CHO
60 R = COOH;   R′ = CHO
61 R = COOH;   R′ = COOH
62 R = CHO;   R′ = CH$_2 \cdot$ CH: CH$_2$
63 R = CHO;   R′ = CH$_2 \cdot$CH$_2 \cdot$CH$_3$

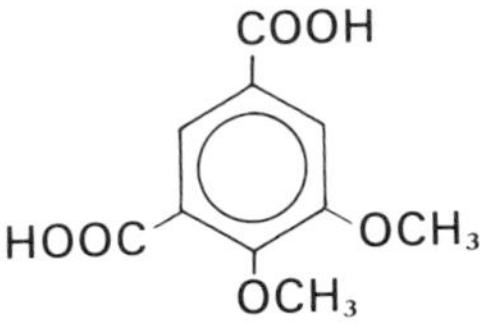

COOH

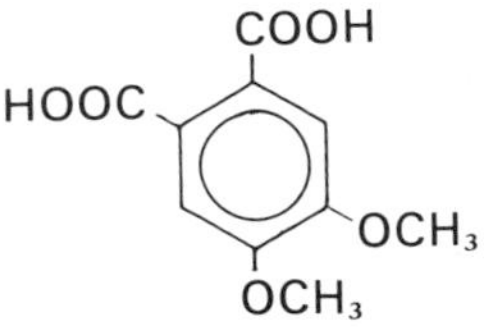

HOOC OCH₃
OCH₃   64

COOH

HOOC OCH₃
OCH₃   65

CH₂OH
HC–O
CHOH   OCH₃

OH OCH₃   66

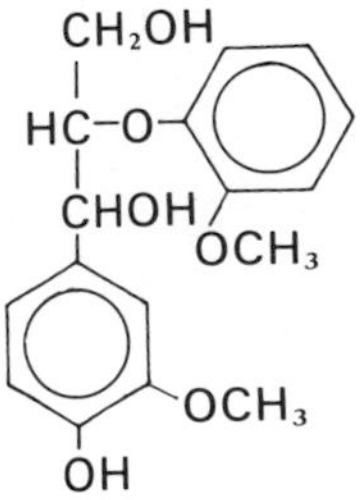

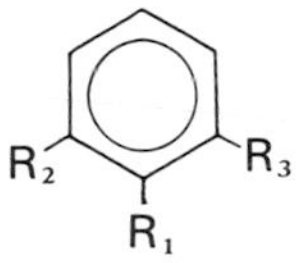

67 R₁ = OH;   R₂ = OCH₃;   R₃ = H
68 R₁ = OCH₃;   R₂ = OCH₃;   R₃ = H
69 R₁ = OH;   R₂ = OH;   R₃ = OH
70 R₁ = OH;   R₂ = OH;   R₃ = OCH₃
71 R₁ = OH;   R₂ = OCH₃;   R₃ = OCH₃

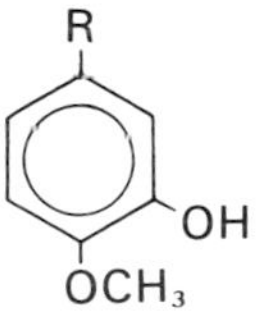

R₂ R₃
R₁

R
OH
OCH₃

72 R = CHO
73 R = COOH

R
OCH₃
OCH₃

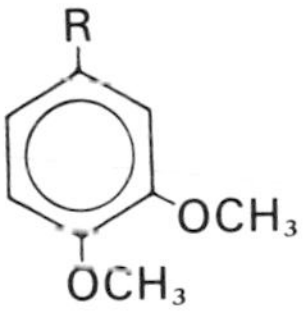

74 R = CH₂OH
75 R = CHO
76 R = COOH

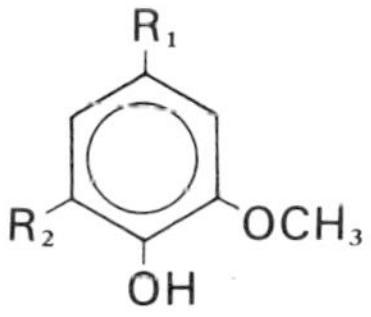

R₁
R₂ OCH₃
OH

77 R₁ = COOH,   R₂ = OH
78 R₁ = CH₂OH,   R₂ = OCH₃
79 R₁ = CHO,   R₂ = OCH₃
80 R₁ = COOH,   R₂ = OCH₃

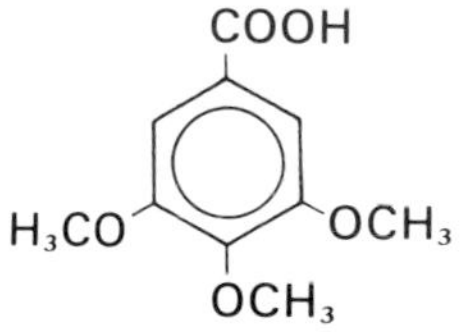

COOH
H₃CO OCH₃
OCH₃   81

CH₃
OH
OH   82

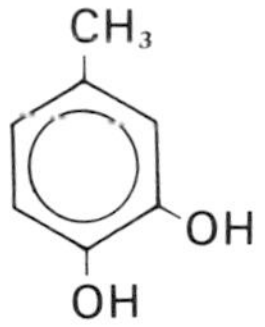

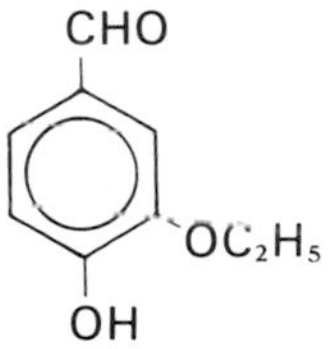

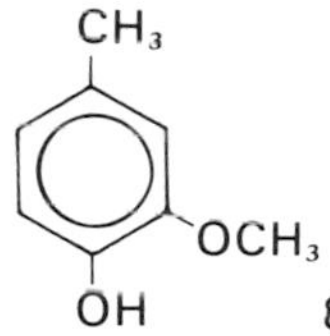

CHO
OC₂H₅
OH   83

CH₃
OCH₃
OH   84

R
OCH₃

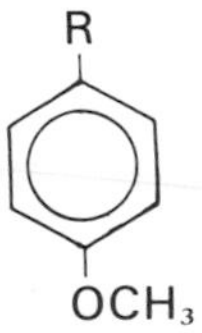

85 R = CH₃
86 R = CH₂OH
87 R = CHO
88 R = COOH

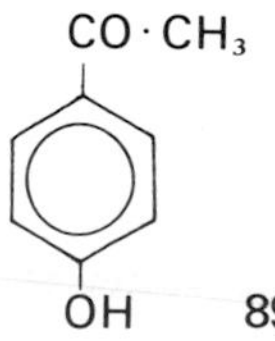

CO·CH₃
OH   89

R
OCH₃
OH

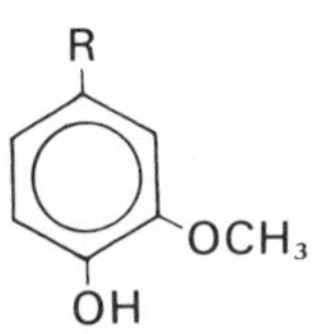

90 R = CHOH·CH₃
91 R = CO·CH₃
92 R = CH₂·COOH
93 R = CHOH·COOH

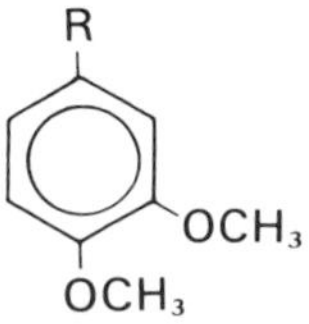

R
OCH₃
OCH₃

94 R = CO·CH₃
95 R = CH₂·COOH

CO·CH₃ ... (structure) ... H₃CO, OCH₃, OH — **96**

CH₂·CO·COOH ... (structure) ... OH — **97**

CH:CH·CH₃ ... (structure) ... OCH₃ — **98**

CH:CH·COOH ... (structure) ... OH, OH — **99**

100	R = CH₂·CH₂·CH₃
101	R = CH:CH·CH₃
102	R = CH₂·CH:CH₂
103	R = CHOH·CH₂·CH₃
104	R = CO·CH₂·CH₃
105	R = CH₂·CO·CH₃
106	R = CH₂·CH₂·CH₂OH
107	R = CH:CH·CH₂OH
108	R = CHOH·CHOH·CH₂OH
109	R = CH:CH·CHO
110	R = CH₂·CH₂·COOH
111	R = CH:CH·COOH
112	R = CH₂·CO·COOH

(structure with R, OCH₃, OH)

113	R = CH:CH·CH₃
114	R = CO·CH₂·CH₃
115	R = CH:CH·COOH

(structure with R, OCH₃, OCH₃)

CH:CH·COOH ... (structure) ... H₃CO, OCH₃, OH — **116**

(structure with R, HC—O, CX, OCH₃, OCH₃, OH)

117	R = H;	X = <H/OH
118	R = H;	X = O
119	R = CH₂OH;	X = <H/OH
120	R = CH₂OH;	X = O
121	R = H;	X = OCH₃
122	R = H;	X = OC₂H₅

(structure with R, H₂C—O, OCH₃)

123	R = COOH
124	R = CHOH·CH₂OH

(structure OH, OCH₃, XC—O—CH, HC——CH, HC—O—CX, OCH₃, OH)

125	X = H
126	X = O

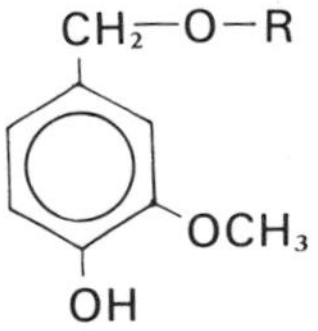

127 R = CH₃

128 R = C₂H₅

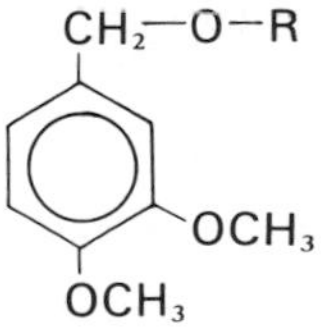

129 R = CH₃

130 R =

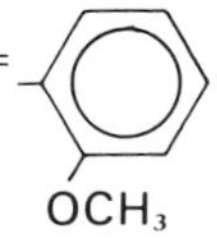

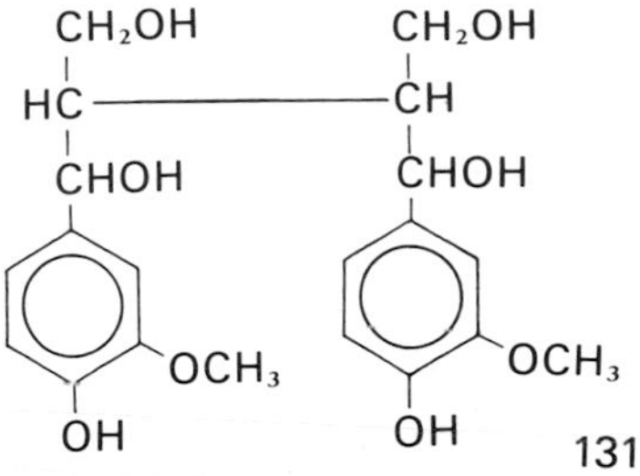

131

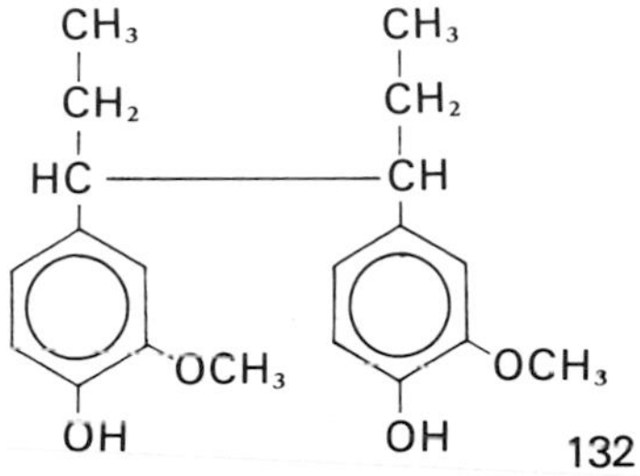

132

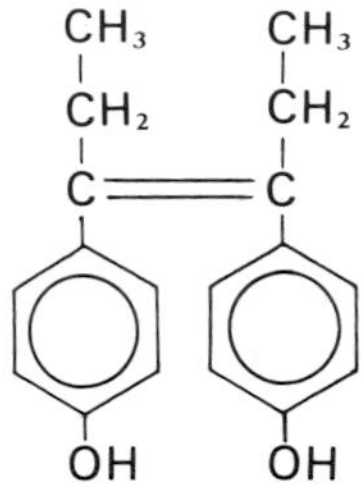

133

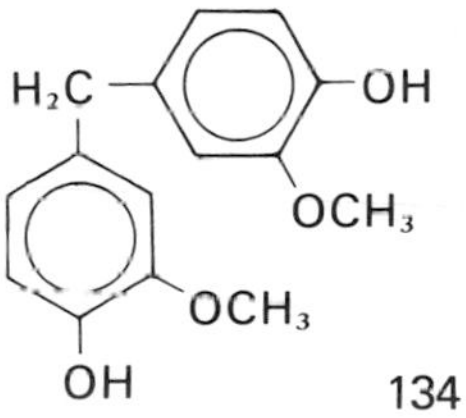

134

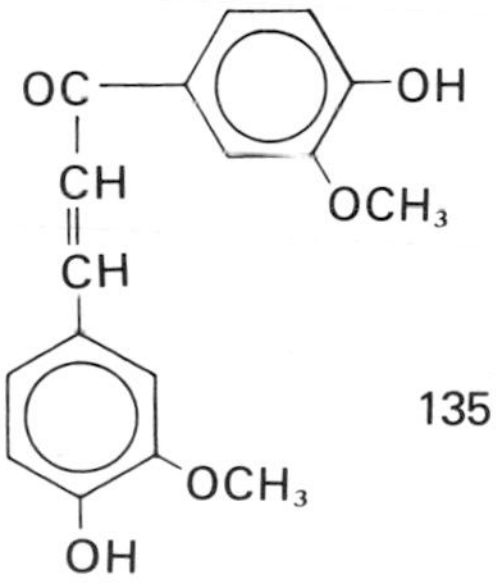

135

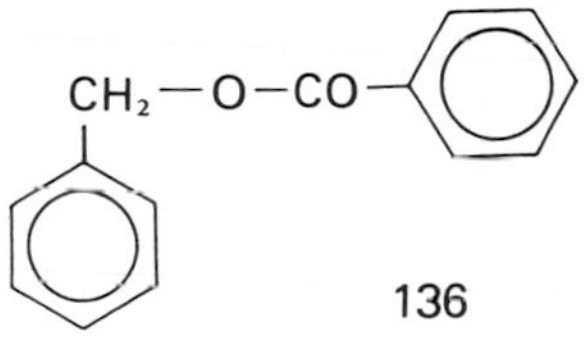

136

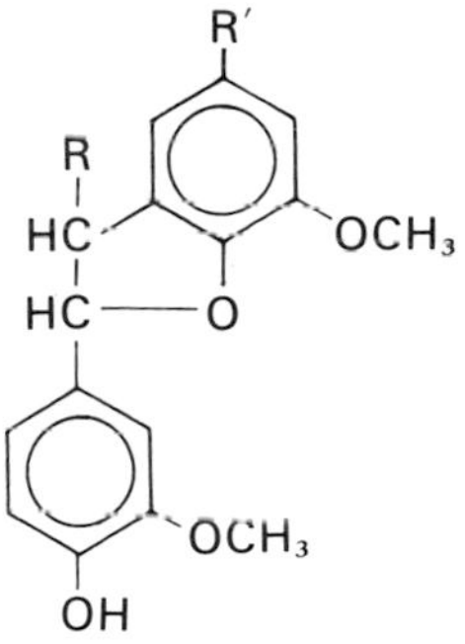

137 R = CH₃;   R' = CH₂·CH₂·CH₃

138 R = CH₃;   R' = CH:CH·CH₃

139 R = CH₂OH; R' = CH:CH·CH₂OH

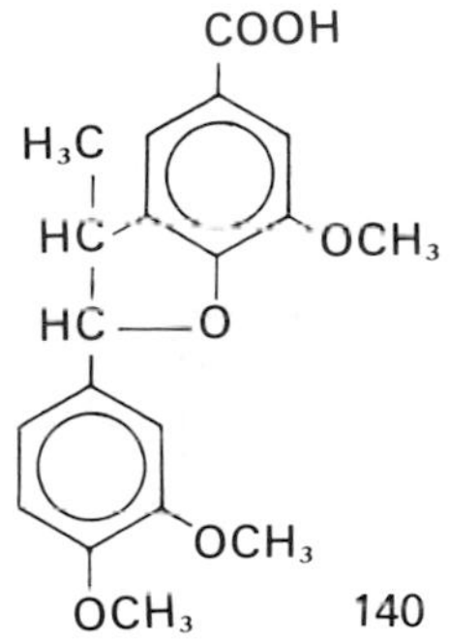

140

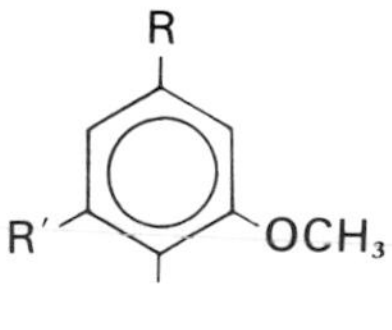

141 R = H;   R' = CHO   145 R = CHO; R' = CH₂·CH₂·CH₃

142 R = CHO;   R' = CHO   146 R = CHO; R' = CH₂·CH:CH₂

143 R = COOH; R' = CHO   147 R = CO·CH₃; R' = CH₂·CH₂·CH₃

144 R = COOH; R' = COOH   148 R = CO·CH₃; R' = CH₂ CH:CH₂

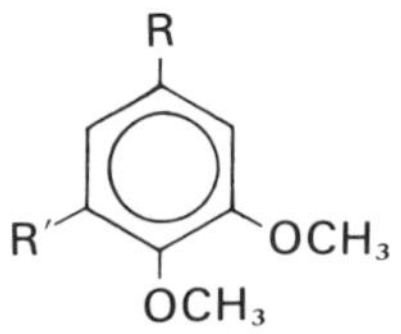

149 R = H; R' = COOH   151 R = CHO

150 R = COOH; R' = COOH   152 R = COOH

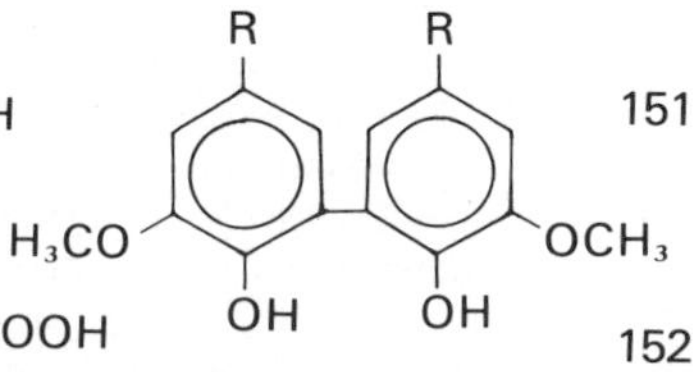

154 R = COOH; R' = COOH

153

155 R = CO·CH₃; R' = CH₃

(structures: first guaiacyl-type ring with CH₃ at top, H₃CO and OH; second ring with CH₃, CH₂, OCH₃, OH labeled 153; third ring with R, R', OCH₃, OCH₃ with labels 154 and 155)

REFERENCES

1. **Pocklington, K. and MacGregor, C. D.**, The determination of lignin in marine sediments and particulate form in seawater, *J. Environ. Anal. Chem.*, 3, 81, 1973.
2. **Gardener, W. S. and Menzel, D. W.**, Phenolic aldehydes as indicators of terrestrially derived organic matter in the sea, *Geochim. Cosmochim. Acta*, 38, 813, 1974.
3. **Nioh, I.**, Characteristics of bacteria in the forest soils under natural vegetation, *Soil Sci. Plant Nutri. (Tokyo)*, 23, 523, 1977.
4. **Sørensen, H.**, Decomposition of lignin by soil bacteria and complex formation between autoxidized lignin and organic nitrogen compounds, *J. Gen. Microbiol.*, 27, 21, 1962.
5. **Kirk, T. K.**, Effects of microorganisms on lignin, *Annu. Rev. Phytopathol.*, 9, 203, 1971.
6. **Kawakami, H.**, Bacterial degradation of lignin model compounds. I. On the cleavage of aromatic nuclei, *Mokuzai Gakkaishi*, 21, 93, 1975.
7. **Kawakami, H.**, Bacterial degradation of lignin model compounds. IV. On the aromatic ring cleavage of vanillic acid, *Mokuzai Gakkaishi*, 22, 246, 1976.
8. **Minami, K., Tsuchiya, M., and Fukuzumi, T.**, Metabolic products from aromatic compounds by the wood-rotting fungus *Polystictus sanguineus* (*Trametes sanguinea*). IV. Culturing condition for reduction and demethoxylation of veratric acid, *Mokuzai Gakkaishi*, 11, 179, 1965.
9. **Omori, T., Yamamoto, A., and Yasui, H.**, Studies on the flavors of Sâke. X. Confirmation of demethoxylation of vanillin by yeasts, *Agric. Biol. Chem.*, 32, 539, 1968.
10. **Tabak, H. H., Chambers, C. W., and Kabler, P. W.**, Bacterial utilization of lignins. I. Metabolism of α-conidendrin, *J. Bacteriol.*, 78, 469, 1959.
11. **Cartwright, N. J. and Smith, A. R. W.**, Bacterial attack on phenolic ethers an enzyme system demethylating vanillic acid, *Biochem. J.*, 102, 826, 1967.
12. **Henderson, M. E. J.**, The metabolism of aromatic compounds related to lignin by some hyphomycetes and yeast-like fungi of soil, *J. Gen. Microbiol.*, 26, 155, 1961.
13. **Flaig, W. and Haider, K.**, The utilization of phenolic compounds by white rot fungi, *Arch. Microbiol.*, 40, 212, 1961.
14. **Kawakami, H.**, Bacterial degradation of lignin model compounds. II. On the degradation of model monomers by shaking culture, *Mokuzai Gakkaishi*, 21, 309, 1975.
15. **Ishikawa, H., Schubert, W. J., and Nord, F. F.**, Investigations of lignin and lignification. XXVIII. The degradation by *Polyporus versicolor* and *Fomes formentarius* of aromatic compounds structurally related to softwood lignin, *Arch. Biochem. Biophys.*, 100, 140, 1963.
16. **Isikawa, H. and Oki, T.**, The oxidation decomposition of lignin. I. The enzyme degradation of softwood lignin and related aromatic compounds by peroxidase, *Mokuzai Gakkaishi*, 10, 207, 1964.
17. **Kawakami, H.**, Bacterial degradation of lignin model compounds. III. On the degradation of model dimers and 5-position condensed type compounds of guaiacyl nuclei, *Mokuzai Gakkaishi*, 21, 629, 1975.
18. **Fukuzumi, T. and Shibamoto, T.**, Enzymatic degradation of lignin. IV. Splitting of veratrylglycerol-β-guaiacylether by enzyme of *Poria subacida*, *Mokuzai Gakkaishi*, 11, 248, 1965.
19. **Fukuzumi, T., Takatuka, H., and Minami, K.**, Enzymic degradation of lignin. V. The effect of NADH on the enzymic cleavage of arylalkylether bond in veratrylglycerol-β-guaiacylether as lignin model compound, *Arch. Biochem. Biophys.*, 129, 396, 1969.
20. **Ishikawa, H. and Oki, T.**, The oxidative degradation of lignin. IV. The enzymic hydrolysis of ether linkages in lignin, *Mokuzai Gakkaishi*, 12, 101, 1966.
21. **Trojanowski, J., Wasilewska, M. W., and Wolska, B. J.**, The decomposition of veratrylglycerol-β-coniferyl ether by *Agrobacterium* sp., *Acta Microbiol. Pol. Ser. B*, 2, 13, 1970.
22. **Crawford, R. L., Kirk, T. K., Harkin, J. M., and MaCoy, E.**, Dissimilation of the lignin model compound veratrylglycerol-β-(o-methoxyphenyl) ether by *Pseudomonas acidovorans*: initial transformations, *Appl. Microbiol.*, 25, 322, 1973.
23. **Kawakami, H.**, Bacterial degradation of lignin. I. Degradation of MWL by *Pseudomonas ovalis*, *Mokuzai Gakkaishi*, 22, 252, 1976.

24. **Nakano, J., Ka, S., Kato, K., and Migita, N.**, Studies on arylglycerol-β-arylether structure in hard wood lignin, *Mokuzai Gakkaishi*, 13, 108, 1967.

25. **Lai, Y. Z. and Sarkanen, K. V.**, Isolation and structural studies, in *Lignins Occurrence, Formation, Structure and Reactions*, Sarkanen, K. V. and Ludwig, C. H., Eds., Interscience, New York, 1971, chap. 5.

26. **Kawakami, H., Sugiura, M., and Kanda, T.**, Biodegradation of components of pulp waste effluents by bacteria. I. On the degradation of kraft lignin, *Tappi (Japan)*, 29, 309, 1975.

27. **Kawakami, H., Mori, N. and Kanda, T.**, Biodegradation of components of pulp waste effluents by bacteria. II. On the degradation of lignin sulfonates, *Tappi (Japan)*, 29, 596, 1975.

28. **Kawakami, H. and Kanda, T.**, Biodegradation of components of pulp waste effluents by bacteria III. On the degradation of lignin preparation from waste liquor of oxygen-alkali pulping, *Tappi (Japan)*, 30, 165, 1976.

29. **Kawakami, H. and Kanda, T9**, Biodegradation of components of pulp waste effluents by bacteria IV. On the degradation of hydrotropic lignin, *Tappi*, 30, 215, 1976.

30. **Kawakami, H. and Takata, M.**, Inorganic sulfur in kraft lignin and their influence on bacterial decomposition of lignin models, *Tappi (Japan)*, 31, 173, 1977.

31. **Kawakami, H.**, Isolation and identification of the aquatic bacteria, utilizing biphenyl type lignin models. *Mokuzai Gakkaishi*, 22, 537, 1976.

32. **Kawakami, H. and Ohyama, T.**, unpublished data.

33. **Kawakaoi, H. and Ohyama, T.**, odegradation of components of pulp waste effluents by bacteria, V. On the degradation of kraft lignin by bacteria isolated from sea water, *Tappi (Japan)*, 32, 359, 1978.

34. **Jackson, M. P., Swift, R. S., Posner, A. M., and Knox, J. R.**, Phenolic degradation of humic acid, *Soil Sci.*, 114, 75, 1972.

35. **Matsuda, K. and Schnitzer, M.**, The permanganate oxidation of humic acids extracted from acid soils, *Soil Sci.*, 114, 185, 1972.

Chapter 9

METABOLISM OF LIGNIN-RELATED COMPOUNDS BY BACTERIA

Masaaki Kuwahara

TABLE OF CONTENTS

I. INTRODUCTION

The over-all degradation of lignin in natural environments is achieved by the combined and sequential actions of a variety of organisms in a process which advances very slowly (Volume I, Chapters 4 and 5). Lignin is thought to be initially attacked by fungi, especially by the white-rot fungi. Conceivably, degradation of lignin is caused by their extracellular enzyme systems. This results in a lowering in methoxyl content, in the cleavage of certain ether linkages, and in the fission of certain ring structures of the lignin. During these partial and sequential degradation processes, lignin yields soluble and lower molecular weight compounds. Aromatic compounds which have been isolated from rotted wood and the culture broth of fungi grown with isolated lignins are: vanillic, ferulic, syringic, *p*-hydroxybenzoic, *p*-hydroxycinnamic, and 3-methoxy-4-hydroxyphenyl pyruvic acids; vanillin, coniferaldehyde, syringaldehyde, *p*-hydroxycinnamaldehyde, guaiacylglycerol, its *β*-coniferyl ether, and others.[1-4] Soil microorganisms including bacteria, yeasts and fungi are thought to readily assimilate these substances; thus, they participate in the degradation of lignins at various stages of decomposition.

During fungal degradation, lignin undergoes demethylation. In experiments with isolated lignin, 23 to 41% of the methoxyl content of the lignin was lost after 6 months of aerobic decay.[5] Another study with milled wood lignin from rye straw showed that *Pholiota mutabilis* decreased the methoxyl content of the lignin to about 80% of the original amount.[6] In addition to fungi, bacteria and other soil organisms may be involved in the demethylation of lignin. The direct degradation of pine and beech milled wood lignin (MWL) by a large inoculum of *Pseudomonas ovalis* was examined.[7] Analytical results showed that the methoxyl and hydroxyl groups of the residual lignin decreased after 60 days of stationary culture.

Several enzymes have been reported to be responsible for the demethylation of lignin and its model compounds. *O*-Demethylase catalyzes the demethylation of lignin models to liberate formaldehyde,[8,9] whereas, laccases from wood-rotting fungi have been shown to eliminate methanol.[10] The fate of the methyl units formed from lignin is thought to be related to the metabolism of C_1-compounds by microorganisms in natural environments. Methyl units from lignin and its degradation products are metabolized by lignin- and aromatic compound-utilizing organisms themselves, as well as by a group of organisms which utilize C_1-compounds preferentially as the substrate for growth. Methanol is oxidized to formaldehyde, followed by successive oxidation to formic acid and to carbon dioxide.[11] Through this process, energy is recovered and stored in the form of ATP. Alternatively, formaldehyde or methanol is incorporated into cellular materials via the C_1-assimilating pathways. These metabolic pathways will be reviewed later.

This chapter refers to the microbial degradation of methoxylated aromatic compounds related to the lignin structure, and to the possible metabolic routes for methyl units which might be liberated from lignin and methoxylated compounds. An attempt biologically to convert aromatic compounds to more chemically reactive substances is also mentioned in the last part of this chapter.

II. METABOLISM OF METHOXYLATED AROMATIC COMPOUNDS*

To determine the mechanism of the biodegradation of lignin, various aromatic compounds have been used as model compounds of lignin. Carboxylic acids, alcohols, and aldehydes of methoxylated or hydroxylated aromatic compounds and their dimeric

---

*   See Volume I, Chapter 2.

compounds have been synthesized and used as substrates for the growth of various microorganisms. Similar structures within the lignin polymer may be attacked by microorganisms which can grow on the model compounds.

Henderson[12] showed that protocatechuic acid was an intermediate in the metabolism of vanillin and ferulic acid by the soil fungi *Pullularia pullulans, Margarinomyces heteromorpha,* and *M. mutabilis.* Vanillic acid was also accumulated in the medium supplemented with veratric acid, ferulic acid, and vanillin. These results indicate that methoxylated compounds are metabolized via the β-ketoadipate pathway. Recently, Black and Dix[13] isolated 21 species of microfungi from litter and soil and showed that the majority readily utilized ferulic acid. Kurosawa[14] investigated the anisic acid metabolism of several Aspergilli which have been used for the traditional production of alcoholic beverages in Japan. He reported that *p*-anisic acid was converted to *p*-hydroxybenzoic acid and to hydroquinone-monomethylether by *A. flavus* and *A. oryzae,* respectively. In a recent report on the degradation of aromatic acids by *A. niger,* Thatcher and Cain[15] described the importance of fungi in the degradation of lignin.

The mechanism of methoxylated aromatic acid degradation has been further investigated in detail with bacteria. Several experiments showed that degradation of the phenylpropane-type structure may occur when the side chain is shortened by a two-carbon fragment before ring fission. Ferulic acid is oxidized by a strain of *Pseudomonas fluorescens* to vanillic acid, which in turn is demethylated before further oxidation via protocatechuic acid to products involved in normal intermediary metabolism.[8] Toms and Wood[16] have proposed a scheme of ferulic acid degradation using *P. acidovorans,* in which ferulic acid is metabolized to vanillic acid via 3-methoxyl-4-hydroxyphenyl-β-hydroxypropionic acid and vanillin as the intermediates. Other experiments have shown that phenylpropanoids are degraded by ring fission, but the side chain is left intact. *Achromobacter* sp.[17] and *P. fluorescens*[18] metabolize β-phenylpropionic acid and caffeic acid, respectively, in this fashion.

Degradation of methoxylated compounds with the basic structure of C_6-C_1 has also been investigated. *P. ovalis* and *P. fluorescens* cleave the guaiacyl nucleus in vanillin or vanillic acid. However, the syringyl nucleus is degraded only when it coexists with vanillin and vanillic acid[19] (Volume II, Chapter 8). *Nocardia corallina,* isolated by Crawford et al.[20] from a lignin-rich environment, metabolizes *p*-anisic acid to *p*-hydroxybenzoic acid and isovanillic acid. This actinomycete also demethylates veratric acid to a mixture of vanillic and isovanillic acids, both of which are demethylated to protocatechuic acid, which undergoes ring fission to give β-carboxy-*cis,cis*-muconic acid.

A. Veratric Acid and Other Aromatic Acids

Various organisms were checked for their ability to assimilate aromatic compounds[21] (Table 1). Soil pseudomonads isolated on a medium enriched with veratric acid utilized veratric and vanillic acids for growth. One strain, BVE 1201, assimilated ferulic acid as well. An unidentified bacterial strain (LS 59) obtained from lignin-sulfonate enriched culture utilized vanillic acid. *Corynebacterium glutamicum,* used by us to produce catechol (see III in this chapter), also assimilated a variety of aromatics. *Fusarium solani,*[22] which could grow on the DHP of coniferyl alcohol, degraded vanillic, veratric, and syringic acids. Unidentified fungi isolated from soil showed assimilation patterns similar to that of *Fusarium.* A group of yeasts belonging to *Torula* and *Rhodotorula,* which possess oxidative metabolic properties, utilized vanillic acid readily. Since these organisms assimilated protocatechuic acid rapidly, they might metabolize methoxylated compounds via protocatechuic acid.

The metabolism of veratric acid by pseudomonads isolated from soil samples by an enrichment technique were investigated in detail. Four strains, BVE 3, 1161, 1201, and

TABLE 1

Growth of Microorganisms on Benzoic Acid Derivatives

Substrate for growth	Bacteria[a]				Yeasts[b]		Fungi	
	BVE 1161	BVE 1201	LS 59	C. glu.	4730	4804	F6[c]	F. solani[d]
	A_{600nm}				A_{600nm}		mg/6 mℓ	
Veratric	0.89	1.06	0.35	0.09	0.26	0.35	8.3	8.7
Vanillic	1.06	0.89	1.48	0.80	1.50	1.02	8.3	6.1
Syringic	0.28	0.29	0.09	0.04	0.38	0.15	7.6	4.3
o-Anisic	0.25	0.22	0.27	0.16			1.6	ng
m-Anisic	0.21	0.27	0.64	0.01	0.21	0.05	1.8	0.5
p-Anisic	0.33	0.80	0.31	0.01	0.05	0.05	1.4	ng[e]
Ferulic	0.43	0.99	0.48	0.14	0.18	0.10	1.2	7.9
Gentisic	0.39	0.32	0.36	1.40	0.48	0.97	6.3	5.4
Protocatechuic	1.50	1.32	1.39	1.45	1.39	1.66	7.3	4.6
m-Hydroxybenzoic	0.41	0.42	0.97	1.32	1.30	1.70	12.4	6.8
p-Hydroxybenzoic	1.10	0.79	1.70	1.26	1.70	1.55	6.9	8.0
Benzoic	0.39	0.88	0.32	1.07	1.32	0.05	2.9	5.9
Phthalic	0.39	0.53	0.35		0.24	0.21	2.1	ng
o-Cumaric	0.21	0.22	0.22	0.03	0.11	0.09	1.5	
m-Cumaric	0.33	0.28	0.14	0.02	1.48	1.60	13.4	8.1
p-Cumaric	0.31	0.86	0.25	0.02	0.07	0.06	1.4	8.4
t-Cinnamic	0.35	0.26	0.18	0.02	0.07	0.04	15.5	ng
β-Phenylpropionic	0.30	0.24	0.28		0.09	0.03	1.4	6.1
Phenylacetic	0.38	1.09	1.28	0.46	0.28	0.05	14.9	
Control	0.39	0.32	0.32	0.02	0.22	0.29	2.2	ng

[a]   2-day culture, *C. glu.*: *Corynebacterium glutamicum.*
[b]   3.5-day culture, 4730: *Torula rubra*, 4804: *Rhodotorula minuta.*
[c]   5-day culture.
[d]   3-day culture.
[e]   No growth.

From Kuwahara, M., Kashima, K., Namikawa, W., and Iwahara, S., *Hakkokogaku*, 55, 248, 1977. With permission.

1231, were used. When these strains were grown in a medium containing veratric acid as the sole carbon source, three dissimilation products (compounds A, B, and C) were formed (Figure 1). Compounds B and C were identified as vanillic and isovanillic acids, respectively, by IR, PMR, and UV spectrometry. Compound A was not accumulated sufficiently in the medium to be identified. It was considered to be protocatechuic acid based on its Rf values for paper and thin layer chromatography. These results indicated that veratric acid was metabolized to two isomeric products by demethylation of either of the two O-methyl groups after which it was metabolized to protocatechuic acid (Figure 2). The time courses of veratric acid degradation by the BVE strains are shown in Figure 3. These strains began to grow immediately after inoculation into the medium, except for the short time lag observed for BVE 1161. Strains BVE 3, 1201, and 1231 accumulated isovanillic acid as the main product, whereas BVE 1161 preferentially formed vanillic acid. The accumulation of intermediates in the medium reached a maximum 12 to 24 hr after inoculation. These intermediates disappeared on successive incubation. Table 2 shows that strains BVE 3 and 1201 catabolized vanillic acid more rapidly than they did isovanillic acid, whereas the isovanillic acid-forming strain BVE 1161 grew well on isovanillic acid. These phenomena are probably due to the specificity of the demethylating systems. In addition to this pathway via protocatechuic

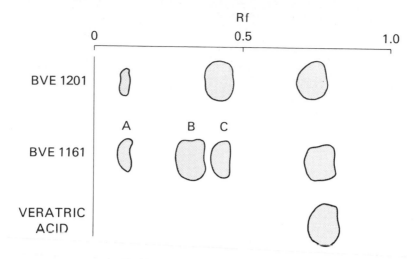

FIGURE 1.   Separation of metabolites by paper chromatogrpahy. Each strain was cultured in 6 m*l* of medium (pH 7.0) containing 0.5% veratric acid, 0.3% NH₄Cl, 0.2% K₂HPO₄, 0.05% MgSO₄·7H₂O, and 0.05% yeast extract for 36 hr at 28°C. An n-propanol-ammonia (7:3, v/v) system was used to develop the chromatogram.

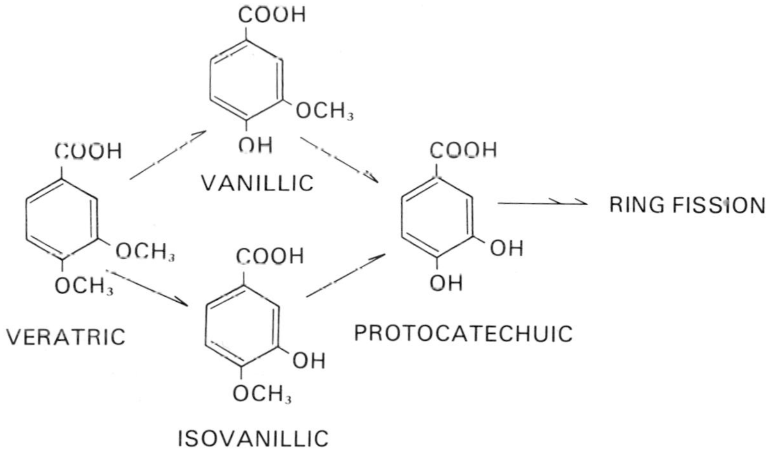

FIGURE 2.   Metabolism of veratric acid by BVE strains. (From Kuwahara, M., Fushima, K., Namikawa, W., and Iwahara, S., *Hakkokogaku*, 55, 248, 1977. With permission.)

acid, BVE 3 seems to have another degradation pathway via gentisic acid, because this acid and *m*-hydroxybenzoic acid supported growth.

Ordinarily, the first step in the degradation of methoxylated aromatic compounds is the liberation of methoxyl units from the aromatic ring. The enzyme which participates in this reaction is an *O*-demethylase. Cartwright and Smith[8] obtained an enzyme preparation from *Pseudomonas fluorescens* which demethylated aromatic methyl ethers. They suggested that vanillate *O*-demethylase consumed only one atom of oxygen per mole of vanillic acid with the concomitant formation of protocatechuic acid and formaldehyde. However, Ribbons[9,23] proposed a different mechanism for *O*-demethylation, catalyzed by monooxygenase-type enzymes from *Pseudomonas aeruginosa* and *P. testosteroni*. It was suggested that the demethylation of 1 mol of 3-methoxybenzoic acid required 1 mol each of oxygen and reduced nicotinamide adenine dinucleotide (NADH).

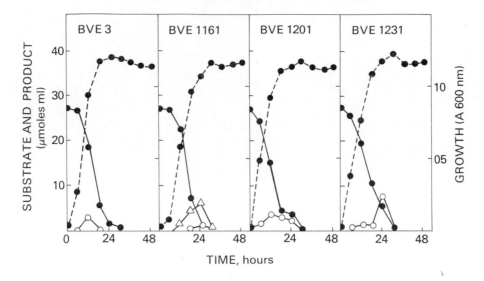

FIGURE 3.   Degradation of veratric acid by BVE strains. — ● — ,Veratric acid remaining; — O — , Isovanillic acid accumulated;   — Δ— ,Vanillic acid accumulated; - - ● - - ,Growth. (From Kuwahara, M., Kashima, K., Namikawa, W., and Iwahara, S., *Hokkokogaku*, 55, 248, 1977. With permission.)

$$\underset{R}{\overset{\overset{\displaystyle COOH}{\big|}}{\bigcirc}}\!\!\!-OCH_3 + NADH_2 + O_2 \longrightarrow \underset{R}{\overset{\overset{\displaystyle COOH}{\big|}}{\bigcirc}}\!\!\!-OH + NAD^+ + HCHO + H_2O$$

3-METHOXYBENZOATE:R = H; VANILLATE:R = OH; VERATRATE:R = OCH$_3$

Recently, Bernhart et al.[24] purified a 4-methoxybenzoate *O*-demethylase from *Pseudomonas putida* and separated in into two components, and NADH-dependent reductase and an iron-containing and acid-labile-sulfur-containing monooxygenase. This investigation is believed to give a clue to the mechanism of demethylation of lignin-related compounds.

B. Distribution and Occurrence of Aromatic Acid- and Methanol-Utilizing Organisms

Low-molecular-weight aromatic compounds liberated by fungi are believed to be finally oxidized to CO_2 and H_2O by soil organisms. Methyl units eliminated from methoxylated compounds may be dissimilated or assimilated both by the aromatic acid-utilizers themselves, and by the C_1-compound-utilizers.

Thus, the existence and distribution of methoxylated aromatic acid-assimilating organisms and methanol-assimilating organisms are thought to influence lignin degradation in specific circumstances. As stated previously, varieties of organisms capable of growing on methoxylated compounds, i.e., veratric, vanillic, and *p*-anisic acids, were isolated from soil by an enrichment technique. However, this technique does not produce quantitative information on the organisms which possess specific properties in a given amount of sample. The direct counting method and agar dilution method are widely used to estimate the number of organisms in a given sample. In addition, the most-probable-number (MPN) technique is one way of determining the potential activity of a microbial population. This technique has been applied to soil organisms to quantify nitrifying organisms.[25] Recently, it has been used to estimate the numbers

TABLE 2

Degradation of Veratric, Vanillic, and Isovanillic Acids by BVE Strains

		Substrate[a]						
		Veratric				Vanillic		Isovanillic
Strain	Growth[b]	Veratric remaning (10^{-3} M)	Vanillic formed (10^{-3} M)	Isovanillic formed (10^{-3} M)	Growth[b]	Vanillic remaining (10^{-3} M)	Growth[b]	Isovanillic remaining (10^{-3} M)
3	1.09	1.0	0	0	1.08	0.8	0.29	18.7
1161	0.99	2.9	9.3	0.5	0.12	27.1	0.52	213
1201	1.09	5.8	0	4.2	1.21	Trace	0.98	3.9
1231	0.31	21.7	0	0.6	0.29	24.2	0.93	Trace

Note: Cultures were incubated for 24 hr at 28°C.
[a] Initially 3×10^{-2} M
[b] Absorbance at 600 nm.

TABLE 3

Most-Probable-Number (MPN) of Methoxylated Aromatic Compound- and Methanol-Utilizing Organisms in Various Material Sample

	Substrate (MPN × 10^3/g of sample)				
Sample	Veratric	Vanillic	p-Anisic	Methanol	Bouillon
Forest land A	1.0	31.7	4.0	0.3	6000
Forest land B	0.9	390	1.6	ND[a]	3900
Cultivated field	0.2	274	4.0	2.1	21300
Reservoir water	ND	0.1	0.1	0.1	17
Soil of cattle shed	130	1300	9.5	79	790000
Rumen content A	ND	490	ND	7.0	2400
Rumen content B	0.5	4.9	ND	0.3	4900
Activated sludge	0.3	7.0	—	0.5	7000

Note: The basal medium was prepared according to the method of Dworkin and Foster.[28] Aromatic acid and methanol were added at concentrations of 0.3 and 1.0%, respectively.
[a] Not detected.

of petroleum-degrading microorganisms in soils,[26] of methanol utilizing bacteria[27] and of lignin-degrading microorganisms in various environments (Volume I, Chapter 3).

In our study, the distribution and occurrence of organisms assimilating veratric, vanillic, and p-anisic acids and methanol as well were examined with the 5-tube MPN technique. Nutrient broth was used as the index to show the number of organisms growing in the medium supplemented with one of the aromatic acids at a concentration of 0.3%, or methanol at 1.0% as the sole carbon source. Results (Table 3) showed that forest land and cultivated fields contained high levels of vanillic acid-utilizers. Soil from cow-shed, enriched with cattle dung, was also a good habitat for aromatic acid-utilizers as well as for methanol-utilizers. Although the percentage of methanol-utilizers in the total number of organisms was small, it was of considerable size when compared to samples from irrigated fields and from sewage treatment plants, in which

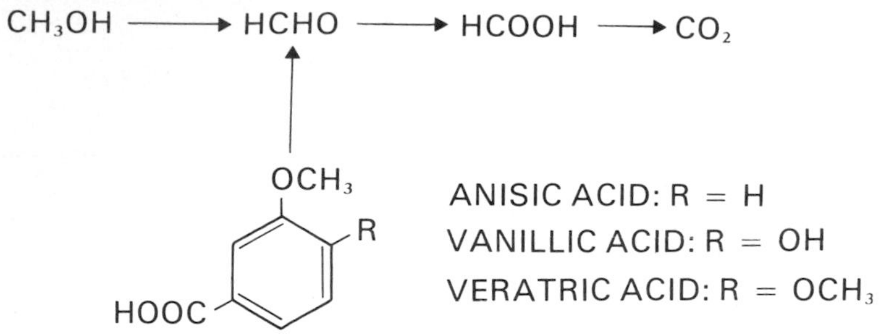

$$CH_3OH \longrightarrow HCHO \longrightarrow HCOOH \longrightarrow CO_2$$

ANISIC ACID: R = H
VANILLIC ACID: R = OH
VERATRIC ACID: R = OCH_3

FIGURE 4.   Oxidative pathway for the methyl unit and methanol.

the number of methanol-utilizers was reported to be at most 4.9×10^4/g of the sample.[27] Since cattle are fed lignin-rich fodder, the rumen contents are good material for the examination of organisms. As expected, a high frequency of vanillic acid-utilizing organisms was detected in one sample. The number of methanol-utilizers was also significantly large.

These results indicate that great numbers and varieties of organisms which possess potent activity to degrade methoxylated aromatic compounds are present in nature. This assumption is conceivable because microorganisms have ample opportunity to come in contact with such aromatic compounds, mostly derived from lignin.

Bacteria were isolated from tubes showing turbidity caused by the growth of organisms. Those isolated from tubes inoculated with the most dilute samples are considered the dominant organisms in a given sample. Isolated strains were mostly Gram-negative and showed an oxidative characteristic in the oxidation-fermentation (O-F) test.

A growth study showed that organisms which were isolated from a methoxylated compound medium could not grow in a methanol medium and vice versa. However, when a culture grown in vanillic-acid medium was cultured on an agar plate of the same substrate, small pink colonies sometimes appeared on the plate, as did large colonies of vanillic acid-utilizing bacteria. The organism isolated from a pink colony was an obligate methylotroph, i.e., one which utilizes only C_1-compounds as growth substrate. In a culture grown on veratric acid, methylotrophic bacteria were also detected. The coexistence of these two different kinds of organisms in the same culture seems to show that methylotrophs may utilize C_1-units released from vanillic acid by vanillic acid degrading bacteria.

III. METABOLISM OF THE C_1-UNIT

As described in IIB, several experimental results suggested a relationship between the metabolism of methoxylated aromatic acids and that of methanol. Liberation of the methyl unit as formaldehyde is thought to be the first step in the degradation of methoxylated compounds. The pathways proposed for the oxidation and assimilation of such C_1-compounds as methanol, methane, and others are thought to be involved in the metabolism of the liberated formaldehyde.

A. Oxidation of the C_1-Unit

Some experiments indicated that free formaldehyde is liberated from vanillic acid and from other methoxylated acids as the reaction product of demethylation catalyzed

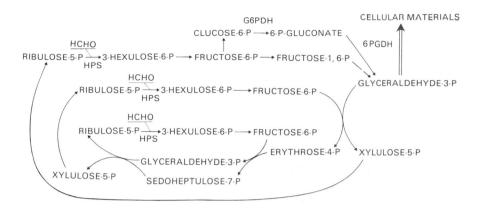

FIGURE 5. Incorporation of formaldehyde by the ribulose monophosphate pathway; *HPS*: Hexulose phosphate synthase, *6PGDH*: 6-Phosphogluconate dehydrogenase, *G6PDH*: Glucose-6-phosphate dehydrogenase.

by *O*-demethylase. One possible route for the metabolism of formaldehyde is the aerobic methanol (or methane) oxidation pathway (Figure 4).

This pathway was first proposed on the basis of experiments with methane-grown *Pseudomonas methanica*.[28] However, the pathway is common to organisms which utilize C_1-compounds as a carbon source.[11]

The enzymes listed below are involved in methanol oxidation. Primary alcohol dehydrogenase: Phenazine methosulfate (PMS) is an apparent electron acceptor for this enzyme, whereas the puteridine derivative is believed to be a real coenzyme. The presence of the NH_4^+ ion is necessary for activity.

1. Alcohol dehydrogenase — NAD^+ is required for the oxidation of methanol.
2. Alcohol oxidase — molecular oxygen is a final electron acceptor and flavine adenine dinucleotide (FAD) is required as the coenzyme. This enzyme has been reported mostly in yeasts, and in one Pseudomonad. The oxidation of methanol by yeasts depends on this enzyme.[29]
3. Catalase — this enzyme has peroxidative properties and oxidizes methanol in the presence of the hydrogen peroxide formed by alcohol oxidase.[30]

The second step, oxidation of formaldehyde to formic acid, is catalyzed by formaldehyde dehydrogenase, which requires NAD^+ and reduced glutathione as cofactors. Nonspecific aldehyde dehydrogenase also participates in the oxidation of formaldehyde in several organisms. 2,6-Dichlorophenolindophenol (DCPIP) is the only electron acceptor.

Cell-free extracts of methoxylated aromatic acid-grown bacteria were examined for enzymic activity in methanol oxidation. Results are summarized in Table 4. No primary alcohol dehydrogenase and NAD^+-linked methanol dehydrogenase activities were detected. Crude extracts commonly showed formaldehyde dehydrogenase activity in the presence of NAD^+ and glutathione. No formate dehydrogenase activity, which is induced in methylotrophs,[11] was detected. This enzyme is thought to be unstable and to easily lose activity, even in the presence of reagents protecting the sulfhydryl (SH) groups of the enzyme. However, the absence of activity may mean that the specific activity of formate dehydrogenase, if the enzyme preparation contains this enzyme, is

TABLE 4

Enzyme Activities Involved in Methanol Oxidation

Strain		Substrate for growth	Specific activity			
			Primary alcohol dehydrogenase	Methanol dehydrogenase[a]	Formaldehyde dehydrogenase[a]	Formate dehydrogenase[a]
BVE	3	Veratric	ND[b]	ND	8.64	ND
	18		ND	ND	7.26	ND
	1161		ND	ND	9.96	ND
	1191		ND	ND	4.20	ND
	1201		Trace	ND	9.40	ND
	1253		ND	ND	7.68	ND
BVA	1	Vanillic	ND	ND	1.38	ND
	7		ND	ND	7.74	ND
	17		ND	ND	0.30	ND
BPA	5	*p-Anisic*	ND	ND	1.44	ND
	6		ND	ND	9.02	ND

[a]   NADH$_2$ formed: μmoles/mg protein/hr.
[b]   Not detected.

low in comparison with that of the formaldehyde dehydrogenase, or that the NADH produced by this enzyme is canceled by the presence of NADH oxidase in the same preparation.

Formaldehyde dehydrogenase in the two strains tested was induced by methoxylated compounds. Veratric, vanillic, ferulic, *p*-hydroxybenzoic, and other aromatic acids were assimilated as carbon sources, and methoxylated compounds functioned as inducers (Table 5). It is not certain whether enzyme activity is induced directly by methoxylated aromatic compounds or by the formaldehyde liberated from the methoxyl groups of these substrates.

B. Assimilation of the C$_1$-Unit.

Two independent pathways for assimilation of the C$_1$-unit have been proposed. They involve either the direct incorporation of formaldehyde into a phosphoric ester of hexulose (the ribulose monophosphate pathway), or a condensation of formaldehyde with glycine to form serine (the serine or *icl*-serine pathway).

The ribulose monophosphate pathway (Figure 5) operates in obligate methylotrophs such as *Pseudomonas methanica*[31] and *Methylococcus capsulatus*,[32] as well as in facultative methylotrophs like *Pseudomonas* C.[33] The key reaction of this pathway is the addition of formaldehyde to ribulose-5-phosphate to give 3-hexulose-6-phosphate. This reaction is catalyzed by 3-hexulose phosphate synthase. The 3-hexulose-6-phosphate formed is first converted to fructose-6-phosphate, then it is phosphorylated to fructose-1,6-diphosphate. This sugar ester is degraded to glyceraldehyde and hydroxyacetone phosphate. Fructose-6-phosphate is alternatively oxidized to 6-phosphogluconate which liberates 3-phosphoglycerate through the pentose phosphate cycle. Glyceraldehyde-3-phosphate and dihydroxyacetone-3-phosphate are metabolized in the glycolytic pathway and are assimilated into cellular materials.

TABLE 5

Effect of the Growth Substrate on Enzyme Formation

Strain	Enzyme	Specific activity[a] Substrate for growth						
		Veratric	Vanillic	Ferulic	Benzoic	p-Hydroxy-benzoic	Protocatecuic	Succinic
BVE 18	Formaldehyde dehydrogenase	3.53	2.12	2.20	—[b]	0.39	0.38	0.07
	Hydroxypyruvate reductase							
	+ Substrate	3.02	2.88	1.55	—	4.45	4.45	1.25
	− Substrate	0.86	1.28	0.41	—	1.50	1.50	0.25
BVE 1201	Formaldehyde dehydrogenase	8.82	8.25	6.42	1.04	1.20	1.55	0.71
	Hydroxypyruvate reductase							
	+ Substrate	3.16	3.48	2.61	2.62	2.91	2.08	1.33
	− Substrate	0.64	0.77	0.61	0.66	0.71	0.77	0.37

[a] Change of $[NADH_2]$: μmoles/mg protein/hr.
[b] This strain could not grow on benzoic acid.

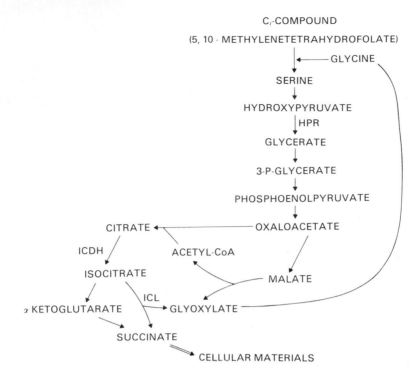

FIGURE 6.   Serine pathway for C_1-compound incorporation; *HPS*: Hydroxypyruvate reductase, *ICDH*: Isocitrate dehydrogenase, ICL: Isocitrate lyase.

The other route for C_1-unit assimilation, the serine pathway (Figure 6), exists in *Methylococcus capsulatus*,[34] an obligate methylotroph, and in *Pseudomonas* AM 1,[35] and other facultative methylotrophs. The first step in this pathway is the condensation of glycine and 5,10-methylenetetrahydrofolate, in which the formaldehyde moiety is bound, to form serine. The key enzyme in this pathway is hydroxypyruvate reductase which catalyses the reduction of hydroxypyruvate to glycerate in the presence of NADH. Glycerate is transformed to oxaloacetate, which is assimilated to cellular components via the tricarboxylic acid (TCA) cycle. The serine pathway is closely related to the glyoxylate cycle, which provides glyoxylate, the precursor of glycine.

If organisms which utilize methoxylated aromatic acid are capable of assimilating methyl units into their cell material, then the enzyme prepararation of the organisms will show enzymic activities involved in the key reactions of the C_1-assimilating pathways. Enzymic activities in C_1-unit assimilation and related pathways were assayed using crude extracts of bacteria grown on veratric, vanillic, or *p*-anisic acid. Results with BVE 1201 (*Pseudomonas* sp.) and other strains are summarized in Table 6.

BVE 1201 and the other strains possessed hydroxypyruvate reductase, whose activity was determined by the decrease in absorption at 340 nm caused by the oxidation of NADH added as the cofactor. Oxidation of NADH in the absence of the substrate indicated the presence of NADH oxidase. However, the initial velocity of NADH oxidation was much lower than in the presence of the substrate. The intensity of hydroxypyruvate reductase activity did not depend on the kind of carbon source on which this organism grew (Table 5). This indicates that the enzyme is constitutive rather than inducible in contrast to formaldhyde dehydrogenase. Isocitrate lyase activity suggested the presence of the gloxylate cycle, which indicates that these organisms possess a route to provide glycine and finally serine. Hexulose phosphate synthase was assayed using

TABLE 6

Enzyme Activities Involved in C₁-Assimilation

Specific activity
Enzyme

Strain	Substrate for growth	Hydroxy-pyruvate reductase (substrate)		Isocitrate dehydrogenase (coenzyme)		Isocitrate lyase	Glucose-6-P dehydrogenase (coenzyme)		6-P-Gluconase dehydrogenase (coenzyme)		Hexulose-P synthase
		+	–	NADP	NAD		NADP	NAD	NADP	NAD	
BVE 3	Veratric	3.3	0.8	14.2	1.9	0.4	0.9	0.6	1.4	0.4	Trace
18		1.2	1.0	32.4	1.3	2.3	0.4	ND[a]	0.2	0.7	Trace
1201		1.0	0.2	54.9	1.2	0.2	1.0	ND	Trace	1.0	Trace
BVA 7	Vanillic	2.1	1.0	40.1	4.3	0.5	0.8	ND	ND	1.3	—
BPA 6	p-Anisic	2.3	0.4	38.6	4.6	1.0	1.2	ND	Trace	Trace	—

[a] Not detected.

ribose-5-phosphate and formaldehyde as the substrates. Although these organisms showed pentose (or pentose phosphate) isomerase activity, no synthase activity was clearly demonstrated. High specific activities were obtained for isocitrate (NADP-dependent) and glucose-6-phosphate (NADP-dependent) dehydrogenases, but the activity of 6-phosphogluconate dehydrogenase was low.

The enzymic studies described above suggest that these organisms which utilize methoxylated compounds possess some enzymes involved in C_1-assimilation. However, it is not certain whether these organisms actually utilize the assimilation pathways to construct cell materials from the methyl units of aromatic compounds. The specific activity of hydroxypyruvate reductase, for example, is lower than that of the methylotrophs,[33] and is considered to be constitutive. Studies of a methylotrophic bacterium, *Pseudomonas* C,[33] indicated that when cells were grown on formaldehyde or formate, the specific activity of the enzyme was extremely high, but that growth on methonal or glucose gave low activities. Experiments with the fungi *Paecilomyces* and *Gliocladium* sp. produced similar results.[36]

The oxidation pathway, in contrast, may play an important role in the metabolism of methyl units in utilizers of methoxylated aromatics. Oxidation of formaldehyde is believed to be linked to the oxidative phosphorylation system to give ATP. In contrast, formaldehyde is utilized by other groups of organisms which preferentially utilize C_1-compounds. The formate produced from formaldehyde also serves as a proper substrate for other organisms.

IV. MICROBIAL CONVERSION OF AROMATIC COMPOUNDS

Lignin, which has an aromatic structure, is one of the most abundant organic materials. Aromatic compounds obtained from coal and petroleum are used as starting materials in the production of various useful chemical compounds. In addition to these fossil fuels, lignin is a prospective source of aromatic compounds. Vanillin and its derivatives are now produced from lignin in several countries. Catechol is another useful chemical; it is produced through the hydrogenolysis of lignin.

Biological conversions of aromatic compounds to organic acids and other compounds have been examined in the field of petroleum fermentation. *Pseudomonas aeruginosa* is reported to produce cumic acid from *p*-cymene.[37] A *Pseudomonas* strain growing on *p*-xylene converts *m*-xylene and 1,2,4-tri-methylbenzene to *m*-toluic acid and 3-methylsalicylic acid, respectively.[38] Bicyclic aromatic hydrocarbons have have also been subjected to microbial conversion. A strain of *Nocardia* gave a yield of 87% salicylic acid from naphthalene.[39] In another experiment with *Pseudomonas* sp., a 91% yield of salicylic acid was reached.[40]

Many aromatic compounds are metabolized via catechol and *cis,cis*-muconic acid. These intermediates are sometimes found in the culture medium, and are believed to be useful starting materials for chemical syntheses. The microbial production of catechol from phenol was examined using *Rhodotorula glutinis* and *Sporobolomyces parasens.*[41] In this experiment, phenol was repeatedly fed to the culture medium and 2.3 g of catechol was formed in a liter of medium after 8 hr of incubation.

Several attempts were made to convert benzoic acid to catechol and *cis,cis*-muconic acid using *Corynebacterium glutamicum*[42,43] which has been used in the industrial production of glutamic acid. Preliminary experiments[21] showed that this bacterium could degrade various aromatic compounds (Table 1) and that these compounds were supposedly metabolized via the β-ketoadipate pathway. This strain was also susceptible to the mutagenesis caused by nitrosoguanidine, a potent mutagen widely used because it induces a high frequency of mutation at doses which result in little killing.

Catechol was first produced from benzoic acid using deletion mutants, whose ben-

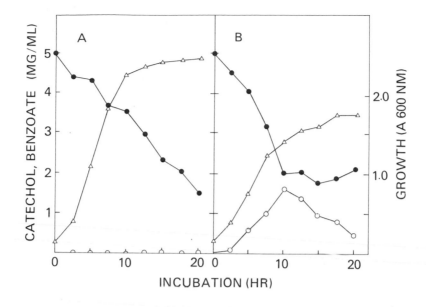

FIGURE 7. Growth and catechol accumulation of the parent strain and bM 9; A: parent, B: bM 9, —O—: catechol, —●—: benzoic acid, —△—: growth. Each strain was cultured in medium (pH 7.0) containing 0.5% benzoic acid, 0.34% yeast extract, 0.3% K₂HPO₄, and 0.05% MgSO₄·7H₂O. (From Kuwahara, M. and Tsuji, M., *J. Ferment. Technol.*, 52, 28, 1974. With permission.)

zoic acid-metabolizing sequence was blocked at the step of catechol oxygenase. A suspension of *C. glutamicum* harvested at the logarithmic growth phase was treated with a solution of nitrosoguanidine. Mutants were concentrated by the penicillin-screening method and treated cells were grown on an agar plate in the benzoic acid medium. The surrounding area of colonies with catechol-producing cells turned dark green when sprayed with an alkaline phloroglucinol solution. Figure 7 shows growth and catechol accumulation of a parent strain and a mutant bM 9 as a function of incubation time. The accumulation of catechol in the medium reached a maximum 10 to 15 hr after inoculation. This accumulated catechol was degraded on successive incubation. This suggested that the mutant was a leaky one whose metabolic pathway after catechol was not completely blocked. To increase the yield of catechol, benzoic acid was fed during cultivation. Figure 8 shows that an addition of the substrate at intervals of 2.5 hr was favorable for catechol accumulation. This procedure as a whole gave a 46% yield based on the amount of benzoic acid added. Catechol was also produced from resting cells in a considerable yield.

We next produced *cis,cis*-muconic acid, which is formed by oxidative ring fission of catechol. Mutants producing *cis,cis*-muconic acid were also induced by nitrosoguanidine mutagenesis. These were isolated from colonies which grew slowly on a benzoic acid agar plate. The maximum accumulation of this acid occurred about 30 hr after inoculation. The yield of the product was about 15% of the amount of benzoic acid added to the medium (Figure 9). Resting cells of the parent strain lacked the ability to degrade exogenous *cis,cis*-muconic acid, but cells of the mutant bMm 86 could consume the substrate rapidly. This indicated that bMm 86 was a permeability mutant capable of incorporating and excreting *cis,cis*-muconic acid. Thus, the *cis,cis*-muconic acid formed by the cleavage of catechol is excreted from cells and accumulated in the medium.

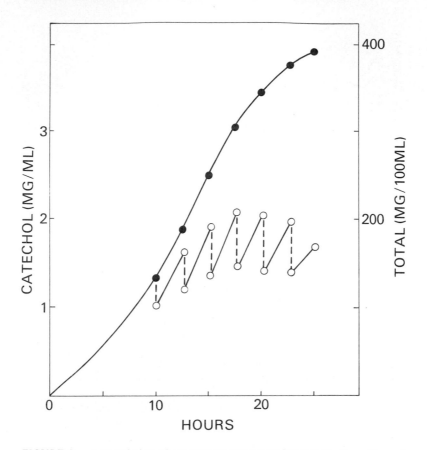

FIGURE 8.   Accumulation of catechol by bM 9 with interval feeding of benzoic acid; —○—: catechol in the culture, —●—: total catechol production. bM 9 was cultured on 100mℓ of medium. After a 10-hr incubation, 20 mℓ samples of the cultured broth were taken at intervals of 2.5 hr and 20 mℓ portions of medium were fed to the culture at the same intervals. (From Kuwahara, M. and Tsuji, M., *J. Ferment. Technol.*, 54, 782, 1976. With permission.)

V. CONCLUSION

There are a great number of natural organisms which degrade methoxyl aromatic compounds. The number of vanillic acid degrading organisms, for example, was 4 to 13×10^5/g of soil and of other samples. Veratric or *p*-anisic acid utilizers were less abundant. Methylotrophic bacteria were frequently found because they were added to the culture with methoxylated compounds as carbon sources. This indicates a close relationship between aromatic compound-utilizing organisms and methylotrophs in the natural degradation of methoxylated compounds.

In the degradation of methoxylated aromatic acids, methyl units are liberated prior to ring fission. *Pseudomonas* sp. isolated from soil samples degraded veratric acid via two isomeric compounds, vanillic and isovanillic acids. Further demethylation formed protocatechuic acid.

The methyl units liberated are metabolized via the methanol oxidizing pathway. Although no enzymic activity responsible for the oxidation of methanol was detected, intense formaldehyde dehydrogenase activity was found in the crude extracts of bacteria grown on methoxylated compounds. This activity was induced by the methoxylated compounds in the culture medium, not by nonmethoxylated compounds. This

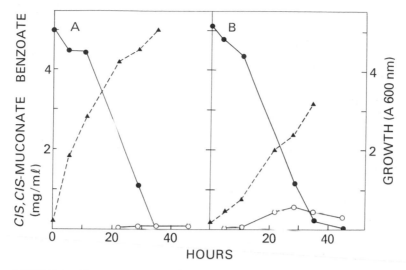

FIGURE 9. Time course of *cis,cis*-muconic acid accumulation by mutant bMm 86; A: parent, B:bMm 86, —○—: *cis,cis* -muconic acid, —●—: benzoic acid, --▲--: growth. The composition of the medium was the same as given in Figure 8. (From Tsuji, M. and Kuwahara, M., *Hakkokogaku*, 55, 95, 1977. With permission.

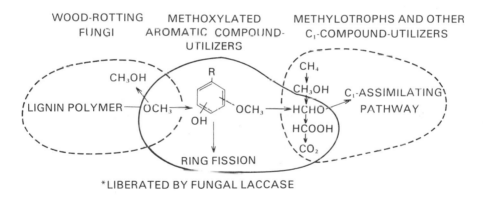

FIGURE 10. Degradation of lignin-related compounds in natural environment.

indicates the participation of the methanol oxidation pathway in the metabolism of methoxylated compounds. The possible involvement of organisms in the over-all degradation of lignin are summarized in Figure 10.

There have been many attempts to exploit lignin. However, microbial use of lignin has not progressed far because of the stability of lignin against biological attack. A basic characteristic of lignin is its composition of aromatic compounds. These components are thought to be good material for the microbial production of useful materials. The availability of lignin for this purpose depends on developing procedures which are economically profitable and which gave advantages over other competitive processes. Chemical procedures, such as hydrogenolysis, which degrade lignin, but leave the aromatic structures intact, provide phenylpropanes and other aromatic components, which might be subjected to further chemical and biological conversion.

We have tried to convert aromatic acids to more chemically reactive substances.

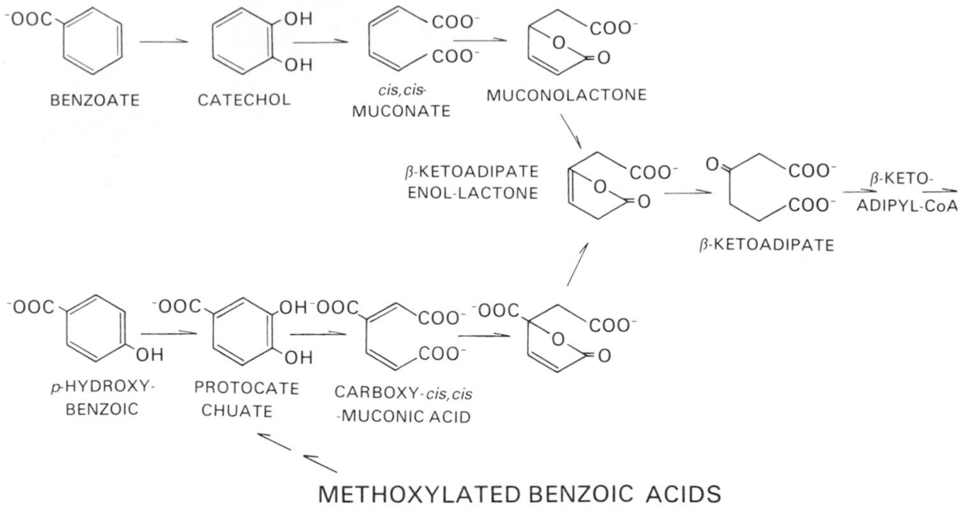

FIGURE 11.   Degradation of aromatic compounds via β-ketoadipate pathway.

Benzoic acid was selected as the simplest model for aromatic compounds. Catechol and *cis,cis*-muconic acid were produced in good yields with mutant strains of *Corynebacterium glutamicum*. If aromatic compounds can be produced from lignin by economical procedures, it should be possible to produce catechol, *cis,cis*-muconic acid, β-carboxy-*cis,cis*-muconic acid and other useful materials through the metabolic pathway shown in Figure 11.

VI. SUMMARY

The metabolism of lignin-related aromatic compounds by bacteria was investigated. Bacteria which assimilated methoxylated aromatic compounds were isolated by enrichment on veratric, vanillic, or *p*-anisic acid as sole sources of carbon. Several soil pseudomonads accumulated isovanillic acid as a major intermediate of veratric acid dissimilation, whereas only a few preferentially formed vanillic acid. These intermediates were subject to further demethylation to afford protocatechuic acid.

The fate of methyl units liberated from methoxylated compounds is discussed in relation to the pathways of methanol oxidation and assimilation. Cell-free extracts of organisms grown on methoxylated aromatic acids oxidized formaldehyde. However, primary alcohol, methanol, and formate dehydrogenase activities were not detected. Activity of hydroxypyruvate reductase, a key enzyme in the serine pathway for the assimilation of C_1-compounds, was found in the extracts. Numbers of organisms which assimilate methoxylated compounds and methanol in soil and other environments were estimated using the most probable number techniques.

Lignin-related compounds will be prospective resources for the production of useful materials. Benzoic acid was selected as a model compound for the microbial conversion of aromatic compounds. Mutant strains of *Corynebacterium glutamicum* accumulated catechol or *cis,cis*-muconic acid in the medium containing benzoic acid as a substrate.

REFERENCES

1. Higuchi, T., Kawamura, I., and Kawamura, H., Properties of the lignin in decayed wood, *Nippon Ringaku Kaishi,* 37, 298, 1955.
2. Fukuzumi, T., Enzymatic degradation of lignin, *Bull. Agric. Chem. Soc. Jpn.,* 24, 728, 1960.
3. Ishikawa, H., Schubert, W. J., and Nord, F. F., Investigations on lignins and lignification. XXVII. The enzymic degradation of softwood lignin by white-rot fungi, *Arch. Biochem. Biophys.,* 100, 131, 1963.
4. Henderson, M. E. K., Release of aromatic compounds from birch and spruce sawdusts during decomposition by white-rot fungi, *Nature (London),* 1975, 634, 1955.
5. Bartlett, J. B. and Norman, G. G., Changes in the lignin of some plant materials as a result of decomposition, *Soil Sci. Soc. Am. Proc.,* 3, 210, 1938.
6. Higuchi, T., Formation and biological degradation of lignins, in *Advances in Enzymology,* Vol. 34, Nord, F. F., Ed., John Wiley & Sons, New York, 1971, 207.
7. Kawakami, H., Bacterial degradation of lignin 1, Degradation of MWL by *Pseudomonas ovalis,* *Mokuzai Gakkaishi,* 22, 252, 1976.
8. Cartwright, N. J. and Smith, A. R. W., Bacterial attack on phenolic ethers, *Biochem. J.,* 102, 826, 1967.
9. Ribbons, D. W., Stoicheiometry of O-demethylase activity in *Pseudomonas aeruginosa,* *FEBS Lett.,* 8, 101, 1971.
10. Ishihara, T. and Miyazaki, M., Demethylation of lignin and lignin models by fungal laccase, *Mokuzai Gakkaishi,* 20, 39, 1974.
11. Cooney, C. L. and Levine, D. W., Microbial utilization of methanol, in *Advances in Applied Microbiology,* Vol. 15, Perlman, D., Ed., Academic Press, New York, 1971, 337.
12. Henderson, M. E. K., The metabolism of aromatic compounds related to lignin by some hyphomycetes and yeast-like fungi in soil, *J. Gen. Microbiol.,* 26, 155, 1961.
13. Black, R. L. B. and Dix, N. J., Utilization of ferulic acid by microfungi from litter and soil, *Trans. Br. Mycol. Soc.,* 66, 313, 1976.
14. Kurosawa, Y., On the discoloration phenomena of conidia of Aspergilli, Metabolic products of o-, m-methoxybenzoic acid, and p-ethoxybenzoic acid by *A. oryzae* and *A. flavus,* *Nippon Nogei Kagaku Kaishi,* 32, 419, 1958.
15. Thatcher, D. R. and Cain, R. B., Metabolism of aromatic compounds by fungi, 1. Purification and physical properties of 3-carboxy-*cis,cis*-muconate cyclase from *Aspergillus niger,* *Eur. J. Biochem.,* 48, 549, 1974.
16. Toms, A. and Wood, J. M., The degradation of trans-ferulic acid by *Pseudomonas acidovorans,* *Biochemistry,* 9, 337, 1970.
17. Dagley, S., Chapman, P. J., and Gibson, D. T., The metabolism of β-phenylpropionic acid by an *Achromobacter,* *Biochem. J.,* 97, 643, 1965.
18. Seidman, M. M., Toms, A., and Wood, J. M., Influence of side-chain substituents on the position of cleavage of the benzene ring by *Pseudomonas fluorescens,* *J. Bacteriol.,* 97, 1192, 1969.
19. Kawakami, H., Bacterial degradation of lignin model compounds, *Mokuzai Gakkaishi,* 21, 93, 1975.
20. Crawford, R. L., McCoy, E., Harkin, J. M., Kirk, T. K., and Obst, J. R., Degradation of methoxylated benzoic acids by a *Nocardia* from a lignin-rich environment: significance to lignin degradation and effect of chloro substituents, *Appl. Microbiol.,* 26, 176, 1973.
21. Kuwahara, M., Kashima, K., Namikawa, W., and Iwahara, S., Metabolism of veratric acid by *Pseudomonas* sp., *Hakkokogaku,* 55, 248, 1977.
22. Iwahara, S., Kuwahara, M., and Higuchi, T., Microbial degradation of the dehydrogenation polymer of coniferyl alcohol (DHP), *Hakkokogaku,* 55, 325, 1977.
23. Ribbons, D. W., Requirement of two protein fractions, for O-demethylase activity in *Pseudomonas testosteroni,* *FEBS Lett.,* 12, 161, 1971.
24. Bernhardt, F-H., Pachowsky, H., and Staudinger, H., A 4-methoxybenzoate O-demethylase from *Pseudomonas putida,* *Eur. J. Biochem.,* 57, 241, 1975.
25. Alexander, M., Most probable number for microbial populations, in *Methods of Soil Analysis,* Part 2, American Society Agronomy, ed., Madison, 1965, 1467.
26. Walker, J. D. and Colwell, R. R., Enumeration of petroleum-degrading microorganisms, *Appl. Environ. Microbiol.,* 31, 198, 1976.
27. Kouno, K. and Ozaki, A., Distribution and identification of methanol-utilizing bacteria, in *Microbial Growth on C₁-Compounds,* Society Fermentation Technology, Osaka, Japan, 1975, 11.
28. Dworkin, M. and Foster, J. W., Studies on *Pseudomonas methanica* (Söhngen) Nov. Comb., *J. Bacteriol.,* 72, 646, 1956.
29. Tani, Y., Miya, T., Nishikawa, H., and Ogata, K., The microbial metabolism of methanol, Part I. Formation and crystallization of methanol oxidizing enzyme in a methanol-utilizing yeast, *Kloeckera* sp. No. 2201, *Agric. Biol. Chem.,* 36, 68, 1972.

30. **Fujii, T. and Tonomura, K.,** Oxidation of methanol and formaldehyde by a system containing alcohol oxidase and catalase purified from *Candida* sp., *Agric. Biol. Chem.,* 39, 2325, 1975.

31. **Kemp, M. B. and Quayle, J. R.,** Microbial growth on C_1 compounds, *Biochem. J.,* 99, 41, 1966.

32. **Lawrence, A. J., Kemp, M. B., and Quayle, J. R.,** Synthesis of cell constituents by methane-grown *Methylococcus capsulatus* and *Methanomonas methanooxidans, Biochem. J.,* 116, 631, 1970.

33. **Goldberg, I. and Mateles, R. I.,** Growth of *Pseudomonas* C on C_1 compounds: activities in extracts of *Pseudomonas* C cells grown on methanol, formaldehyde and formate as sole carbon sources, *J. Bacteriol.,* 122, 47, 1975.

34. **Whittenbury, R., Dalton, H., Eccleston, M., and Reed, H. L.,** The different types of methane oxidizing bacteria and some of their more unusual properties, in *Microbial Growth on C_1-Compounds,* Society Fermentation Technology, Osaka, Japan, 1975, 1.

35. **Large, P. J., Peel, D., and Quayle, J. R.,** Microbial growth on C_1 compounds. 2. Synthesis of cell constituents by methanol-and formate-grown *Pseudomonas* AM-1, and Methanol-grown *Hyphomicrobium vulgare, Biochem. J.,* 81, 470, 1961.

36. **Sakaguchi, K., Kurane, R., and Murata, M.,** Assimilation of formaldehyde and other C_1-compounds by *Gliocladium deliquescens* and *Paecilomyces variotti,* in *Microbial Growth on C_1-Compounds,* Society Fermentation Technology, Osaka, Japan, 1975, 163.

37. **Horiguchi, S. and Yamada, K.,** Studies on the utilization of hydrocarbons by microorganisms. Part XI, *Agric. Biol. Chem.,* 32, 555, 1968.

38. **Omori, T., Horiguchi, S., and Arima, K.,** Studies on the utilization of hydrocarbons by microorganisms, Part X, *Agric. Biol. Chem.,* 31, 1337, 1967.

39. **Klausmeier, R. E. and Strawinski, R. J.,** Microbial oxidation of naphthalene, *J. Bacteriol.,* 73, 461, 1957.

40. **Ishikura, T., Nishida, H., Tanno, K., Miyachi, N., and Ozaki, A.,** Microbial production of salicylic acid from naphthalene, *Agric. Biol. Chem.,* 32, 12, 1968.

41. **Nei, N., Tanaka, Y., and Tanaka, N.,** Production of catechol from phenol by a yeast, *J. Ferment. Technol.,* 52, 28, 1974.

42. **Kuwahara, M. and Tsuji, M.,** Accumulation of catechol from benzoic acid by a mutant induced from *Corynebacterium glutamicum, J. Ferment. Technol.,* 54, 782, 1976.

43. **Tsuji, M. and Kuwahara, M.,** Accumulation of *cis,cis*-muconic acid from benzoic acid by a mutant induced from *Corynebacterium glutamicum, Hakkokogaku,* 55, 95, 1977.

Chapter 10

BIODEGRADATION OF LIGNIN-RELATED POLYSTYRENES

Takafusa Haraguchi and Hyoe Hatakeyama

TABLE OF CONTENTS

I. INTRODUCTION

Petrochemical industries produce many types of high polymers, among which polyethylene, polypropyrene, polystyrene, and polyvinylchloride are most widely made and used. These and other synthetic polymers are very useful as packaging and construction materials because of their physical and chemical properties. However, the disposal of these polymers has become a serious public problem because they are not easily degraded in nature.

Biodegradation of water-soluble polymers is thought to be easier than that of water-insoluble ones. Poly(ethylene glycol)[1,2] and poly(vinyl alcohol),[3] for example, are known to be attacked easily by microorganisms. Oligomers of water-insoluble polymers such as polyethylene,[4] polyurethane,[5] aliphatic polyester,[6] and nylon-6[7] are also known to be degraded by microorganisms, although they are not really suitable for biodegradation. On the other hand, polystyrene, which is an aromatic polymer, is thought to be highly resistant to microorganisms and other biodeteriogens.

Lignin is a natural polymer which also is aromatic, but which is degraded in nature. With this in mind, the authors have tried to find a model biodegradable synthetic polymer;* poly(p-hydroxystyrene) (PHS), poly(3-methoxy-4-hydroxystyrene) (PMHS), and poly(3,5-dimethoxy-4-hydroxystyrene) (PDMHS) were prepared as model polymers. These styrene derivatives have simpler structures than lignin and their biodegradation was expected to be achieved via the p-hydroxyphenyl, guaiacyl-, and syringyl-groups, which are the main constituents of lignin. This paper is concerned with the biodegradation of these lignin-related polystyrenes.

---

* Structurally related to lignin.

$$\left[\begin{array}{c} -\mathrm{CHCH_2}- \\ \bigcirc \end{array} \right]_n$$

POLYSTYRENE

$$\left[\begin{array}{c} -\mathrm{CHCH_2}- \\ \mathrm{OH} \end{array} \right]_n \quad \left[\begin{array}{c} -\mathrm{CHCH_2}- \\ \mathrm{OCH_3} \\ \mathrm{OH} \end{array} \right]_n \quad \left[\begin{array}{c} -\mathrm{CHCH_2}- \\ \mathrm{H_3CO} \quad \mathrm{OCH_3} \\ \mathrm{OH} \end{array} \right]_n$$

PHS   PMHS   PDMHS

II. SYNTHESIS OF LIGNIN-RELATED POLYSTYRENE DERIVATIVES

By a procedure similar to the method reported previously *p*-hydroxystyrene was prepared from *p*-hydroxycinnamic acid.[8,9] The yield was about 70%.

PHS was obtained by bulk polymerization using 1% of 2,2'-azo-*bis*-isobutyronitrile as an initiator. In a glass tube under 10^{-4} mm Hg, *p*-hydroxystyrene was sealed with the initiator and the tube was treated at 90°C for 4 hr. The polymerization mixture was dissolved in methanol, then added to water. The white precipitate was collected by filtration. The number-average molecular weight (Mn) obtained from the gel permeation chromatographic measurement of PHS was 1800.

PMHS and PDMHS were prepared from ferulic acid and sinapic acid, respectively, by procedures similar to that mentioned above. The $\overline{\mathrm{Mn}}$ of these polymers were 1200 and 1100, respectively.

III. BIODEGRADATION OF THE POLYSTYRENE DERIVATIVES

A. Polystyrene Derivative-Degrading Microorganisms

An enrichment technique was used to isolate microorganisms which can degrade the lignin-related polystyrenes. Soil extract was used as the source of microorganisms. Fresh garden soil (25 g) collected from the campus of Tokyo University of Agriculture and Technology was added to 1ℓ of water. The mixture was stirred for 1 hr and filtered through Toyo® filter paper No. 4. Then the following mineral salts, and each polymer as carbon source, were added to the filtrate (g/ℓ): NH₄NO₃ (20.0), K₂HPO₄ (1.5), MgSO₄·7H₂O (0.5), polymer (0.5). The pH was adjusted to 7.0. The solution was placed in a respirometer (Figure 1) and incubated at 30°C. Oxygen uptake was measured during the incubation in order that appearance of the polymer-degrading microbes might be observed. Endogenous respiration (no polystyrene) was subtracted. After 2 month's incubation, 5 mℓ of the supernatant of the culture solution was pipetted into another respirometer containing new culture medium. After growth, 5 mℓ of the supernatant was transferred into new culture medium at intervals of 1 week. This procedure was repeated through 5 transfers, and finally one loopful of the supernatant of the culture solution was spread on agar medium which contained yeast extract (4 g/ℓ), malt extract (10 g/ℓ), glucose (4 g/ℓ), and agar (20 g/ℓ). Isolates obtained were examined microscopically and biochemically.

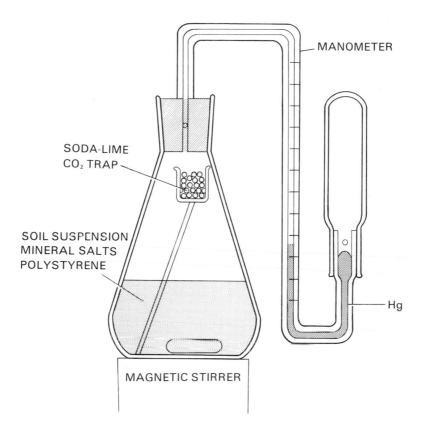

MANOMETER

SODA-LIME
CO₂ TRAP

SOIL SUSPENSION
MINERAL SALTS
POLYSTYRENE

Hg

MAGNETIC STIRRER

FIGURE 1.   Respirometer used to study biodegradation of polystyrenes. Prior to incubation, the system was equilibrated with the atmosphere. O₂ uptake was followed during incubation at 30°C.

Oxygen uptake during the incubation in the respirometer demonstrated the appearance of microorganisms which were effective in the degradation of lignin-related polystyrenes (Figure 2). The pH of the culture solution was lowered during incubation and this also suggested that microbial attack on the polymers was taking place. After the enrichment culture, bacteria, actinomycetes, and filamentous organisms were found. Judging from the morphological and biochemical properties, the bacterium chosen for study is thought to be in the genus *Moraxella* (Table 1) and one of the studied actinomycetes to be a *Micromonospora* sp. The isolated fungus that was studied appeared to be a *Penicillium* sp.[10] The bacterium was the most effective isolate in degrading the polystyrenes.

B. Degradation of Poly(4-Hydroxystyrene), Poly(3-Methoxy-4-Hydroxystyrene), and Poly(3,5-Dimethoxy-4-Hydroxystyrene) by Mixed Populations of Soil Microorganisms

The soil extract was prepared from fresh soil (50 g) and distilled water (1 ℓ). After addition of the mineral salts as described above, a solution (100 mℓ) was dispensed into shaking flasks. The carbon source PHS, PMHS, and PDMHS samples (each 0.05 g), which were sterilized by steam, were added. The flasks were shaken reciprocally at 30°C. When large quantities of degradation products were required for chemical and physical determination, incubation was carried out on a large scale. Absence of autoxidation of the polystyrenes was verified in blank tests carried out with the same media containing no microorganisms.

Incubation of PHS, PMHS, and PDMHS was stopped at different stages of sub-

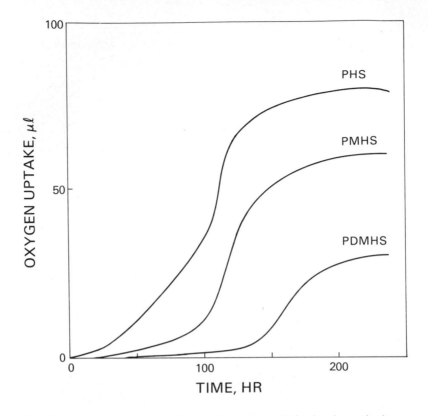

FIGURE 2.   Time course of oxidation of polystyrene derivatives by a mixed population of soil microorganisms (5th transfer; see text) in a respirometer.

TABLE 1

**Criteria used to tentatively identify as
a *Moraxella* a bacterium isolated after
enrichment on lignin-related polystyrenes**

Microscopic and bio- chemical tests	Test results
Gram stain	−
Shape	Rod
Motility	−
Growth in air	+
Growth anaerobically	−
Catalase	+
Oxidase	+
Glucose (acid)	−
Carbohydrates (F/O)	−

strate utilization, and each degraded polymer was separated from the culture by filtration with Toyo® filter paper No. 4. Each sample for UV spectroscopy was prepared
by dissolution in pure ethanol (0.1%). The absorption peak of the phenyl groups of
PHS appearing at 280 nm decreased with incubation time as shown in Figure 3. Results
with PMHS and PDMHS were similar to those with PHS. The change in the UV
spectra showed that these polymers were degraded by the microorganisms. In particu-

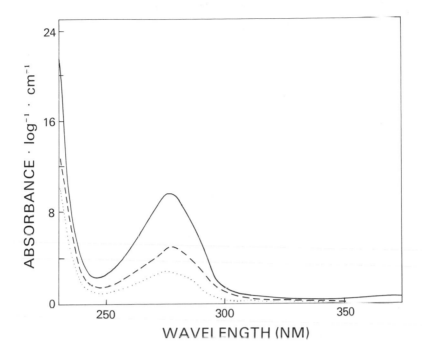

FIGURE 3. Change in the UV spectrum of PHS on incubation with mixed soil microorganisms: ———, original, ——, 10 days, ..., 20 days. Samples were dissolved in 100% ethanol.

lar, the decreased intensity of the absorption peak at 280 nm is considered to be the result of cleavage of phenyl groups in the polymers.

The IR spectrum of PMHS was measured in KBr pellets using a sample recovered by the same method used for the UV-spectrum measurement. The change in the IR spectrum at 1500 to 2000 cm⁻¹ of PMHS incubated with microorganisms showed a new absorption band at 1675 to 1725 cm⁻¹ which can be attributed to C=O stretching vibrations (Figure 4).[11] The intensity of the carbonyl bands relative to the 2900 cm⁻¹ band was estimated by the base-line method.[12] The relative intensity is not significantly affected by concentration in the pellet or variation in scattered light.

The relative intensity of the carbonyl stretching band increased markedly with incubation time, as shown in Figure 5. This change in band intensity suggests that the degradation of PMHS, PHS, and PDMHS increased exponentially with incubation time.

The gel permeation chromatograms of the ether-soluble fraction of the polymers incubated for 10 days are shown in Figures 6 and 7, in which changes in the pattern of molecular weight distribution were observed. Analysis of the gel chromatograms of the biodegraded PDMHS, including comparisons with known standards, indicated that the tetrameric fraction decreased and that the dimeric and trimeric fractions increased on 10 days' attack (Table 2).

Biodegradation products of PMHS were fractionated by the method shown in Figure 8. After incubation at 30°C for 40 days, cells (and other isolubles were collected by centrifugation, washed, and then extracted with ethyl ether. The ether-soluble part was then separated into acid and phenol fractions by washing with 5% NaHCO₃ and 5% NaOH solutions. From the acid fractions, cyrstalline compounds A and B were obtained. Crystalline compound C was obtained after methylation (CH₃I) of the solution. The methyl esters of the acid products were analyzed by gas chromatography.

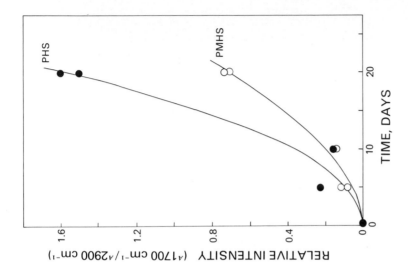

FIGURE 5. Change in relative intensity of the carbonyl stretching band (1700 cm⁻¹) in the IR spectrum of PHS and PMHS during incubation with mixed soil microflora.

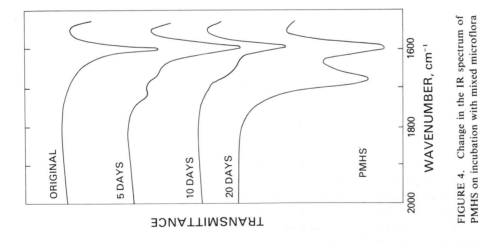

FIGURE 4. Change in the IR spectrum of PMHS on incubation with mixed microflora of soil (KBr pellets).

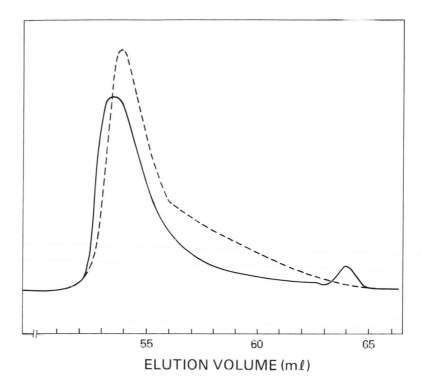

FIGURE 6. Change in the gel permeation chromatogram of PHS with incubation-time: ——— , original, ——, incubated for 10 days with a mixed population of microorganisms. (Column: Toyo® Soda styrogel 2000H × 2 and 4000H × 2; solvent: THF; flow rate 1.2 mℓ/min; T = 40°.)

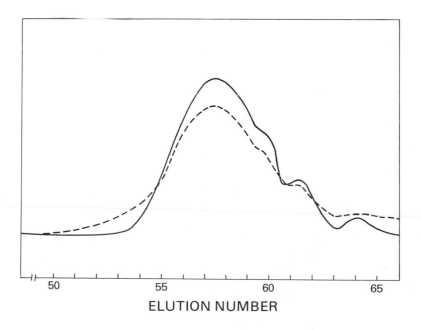

FIGURE 7. Change in the gel permeation chromatogram of PDMHS with incubation time: ——— , original, ——, incubated for 10 days with a mixed population of microorganisms. (Chromatography as described in Figure 6).

TABLE 2

Change of Molecular Weight Distribution of PDMHS After 10 Days' Exposure to Soil Microorganisms

	Original %	10 days %
Tetramer	2.3	0.9
Trimer	1.7	2.2
Dimer	2.5	12.4

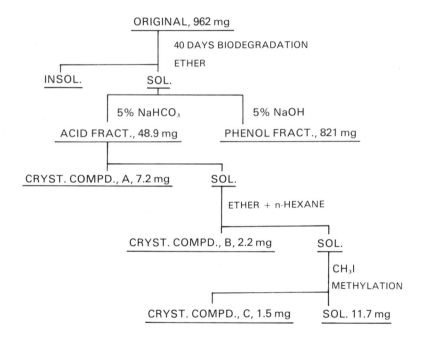

FIGURE 8.   Fractionation of PMHS biodegraded by mixed microflora of soil.

Compound A was obtained as colorless needles (melting point 101 to 102°C). Physical and chemical investigation indicated that compound A was a dibasic acid. Its molecular weight was 126, in agreement with the value for empirical formula $C_2H_2O_4 \cdot 2H_2O$. Elemental analysis for the anhydride showed: calculated, C 26.68; H 2.24; found, C 26.70, H 2.20. Mass spectrometry of compound A showed fragments identical with oxalic acid (m/e, 44, 45, 46 corresponding to COO, COOH, and HCOOH, respectively). The IR specrum of compound A was identical with oxalic acid as shown in Figure 9.

Compound B melted at 143 to 144°C. The molecular weight was 200. Elemental analysis showed C 48.06, H 4.01; calculated for $C_8H_8O_6$: C 48.01, H 4.03. The IR spectrum of compound B was identical with the monomethyl ester of β-carboxymuconic acid, as shown in Figure 10, which was compared to the authentic compound synthesized according to the procedure of Husband et al.[13]

Compound C, crystallized as colorless needles from the ether solution of the acid fraction methylated by CH_3I, melted at 59 to 60°C. Its molecular weight was 196. The IR spectrum of compound C corresponded to methyl veratrate, as shown in Figure 11.

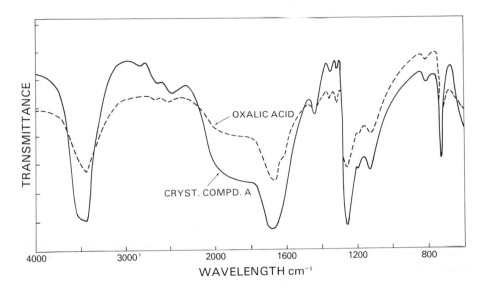

FIGURE 9.   IR spectra of crystalline compound A and oxalic acid.

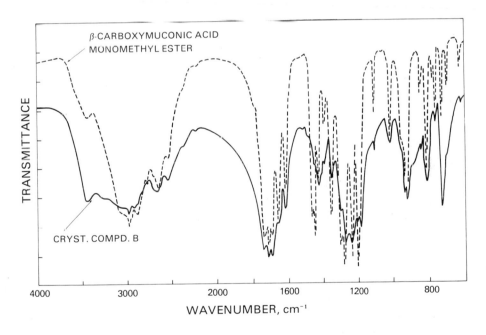

FIGURE 10.   IR spectra of crystalline compound B and monomethyl ester of β-carboxymuconic acid.

Control incubations (without microbes) showed no degradation. Thus, *Moraxella*, *Penicillium*, and other species of microorganisms in the soil were essential for the degradation of lignin-related polystyrenes. The PMHS sample used here is considered to be degraded in a manner analogous to that of the degradation of lignin and its model compounds by microorganisms. Figure 12 is tentatively put forward as the most likely representation of the events. Intermediate III is hypothetical. PMHS is degraded to β-carboxymuconic acid derivative (IV), presumably through cleavage of the benzene ring between C_3 and C_4 via either or both of the following two pathways. One possible

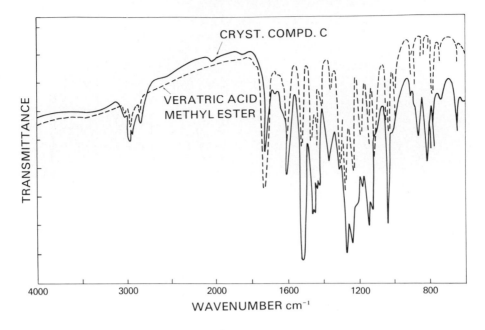

FIGURE 11.   IR spectra of crystalline compound C and methyl veratrate.

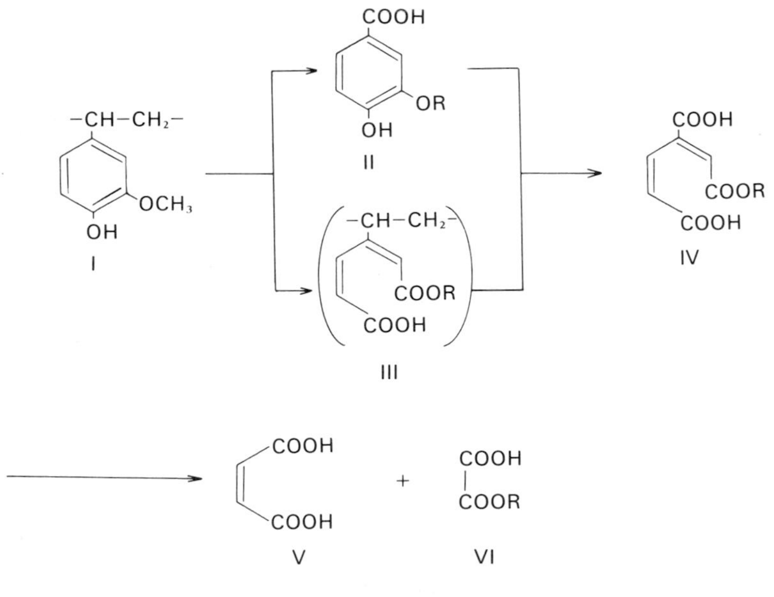

FIGURE 12.   Tentative scheme for degradation of PMHS by mixed microorganisms from soil.

path is through vanillic acid (II). The other possible path is through the intermediate (III). The β-carboxymuconic acid derivative (IV) presumably is degraded to maleic (V) and oxalic acid (VI).

In the above, the presence of the intermediate II was deduced from identification of methyl veratrate (compound C), obtained as crystals, after methylation of the extract. Accordingly, it is difficult to determine whether either vanillic (R=CH₃ in II) or protocatechuic (R=H in II) acids, or both, are essential in the degradation from compound I to compound IV. It is known that protocatechuic acid is degraded by many microorganisms via β-carboxy-*cis,cis*-muconic acid.[14] Kawakami[15] reported that vanillic acid is degraded to protocatechuic acid by *Pseudomonas ovalis* and *Pseudomonas fluorescens* (see Volume II, Chapter 8) This report seems to suggest that the demethylation of vanillic acid to protocatechuic acid takes place before the cleavage of the aromatic ring. On the contrary, Zabinski et al.[16] showed that the methyl ester of 3-carboxy-5-hydroxy-*cis,cis*-muconic acid monomethyl ester was formed when protocatechuic acid-5-oxygenase cleaves 3-0-methyl gallic acid. As described previously, the monomethyl ester of β-carboxymuconic acid was obtained and identified in the authors' experiment. These results appear to show that ring fission takes place without prior demethylation. Therefore, it may be assumed that the degradation of PMHS via vanillic acid followed by aromatic ring cleavage to the monomethyl ester of β-carboxymuconic acid is the most likely process. Maleic acid was detected by gas chromatography; oxalic acid was obtained and identified as crystals. Thus, β-carboxymuconate (IV) may be degraded to maleate (V) and oxalate (VI) via a pathway similar to that reported by Ornston et al.[17]

It is difficult at present to gauge how the catabolic pathway of PHS, PMHS, and PDMHS operates in nature, although the hypothetical process of dissimilation may be assumed by comparing it with the better known process of dissimilation of lignin and its model compounds.

C. Degradation of Poly(3-Methoxy-4-Hydroxystyrene) by an Isolated Bacterium

The question of whether the lignin-related polystyrenes can be degraded by pure cultures of microorganisms isolated from soil was investigated. As pointed out in section III. A. (above), several microorganisms were isolated by enrichment procedures. Further study revealed that a bacterium tentatively identified as a *Moraxella,* was the most active. After PMHS was exposed to a pure culture of the *Moraxella,* the polymer was analyzed by gel permeation chromatography (Figure 13). Changes in the molecular weight distribution were noted, especially in the low molecular weight region; this suggests that the culture of *Moraxella* attacks the polymer. Although not yet studied thoroughly, the isolated degradation products seem similar to those produced by the mixture culture.

IV. CONCLUSIONS

When the lignin-related polystyrene derivatives (PHS, PMHS, and PDMHS) were incubated with a mixed population of soil microorganisms, the following phenomena were observed on 3 month's incubation: oxygen absorption during incubation, lowering of pH in incubation media, changes in UV and IR spectra, and changes in pattern of gel permeation chromatograms in the residual polystyrenes. These results suggest that the polystyrene derivatives were attacked by microorganisms in soil. The fact that veratrate (vanillate ?) and the monomethyl ester of β-carboxymuconate were isolated from culture solution containing PMHS suggests that the degradation pathway of the polymer resembles somewhat that of lignin.

It can therefore be concluded that one of the putatively most recalcitrant polymers, polystyrene, is changed into a biodegradable polymer by introducing some functional groups, which are characteristic of lignins, into the benzene ring.[5]

V. SUMMARY

Water-soluble polymers, polyethyleneglycol and polyvinyl alcohol, for example, are

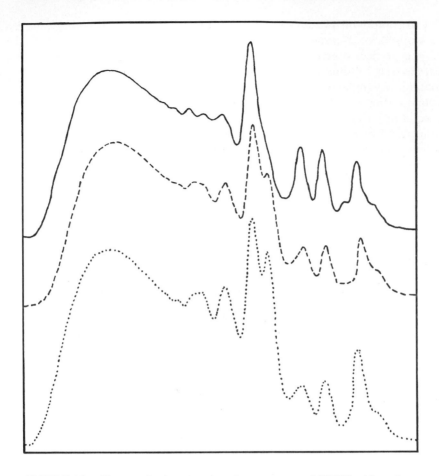

FIGURE 13.   Change of gel permeation chromatogram of PMHS with various
incubation times *(Moraxella)* ———— , original,————10 days,. . . 22 days.

An investigation was undertaken to find a good model for a biodegradable synthetic
polymer among natural polymers all of which apparently are biodegradable, with special attention to lignin. Poly(4-hydroxy styrene), poly(3-methoxy-4-hydroxy styrene),
and poly(3,5-dimethoxy-4-hydroxy styrene) were consequently prepared as model polymers. These polymers were subjected to microorganisms in soil, and several organisms which were effective for the degradation of the lignin-related polystyrenes were
found.

Four degradation products were separated from culture medium in the course of
degradation of poly(3-methoxy-4-hydroxy styrene) by a mixed soil microflora. Among
the products, vanillic acid seems to be the first degradation product in the metabolism
of the polymer. β-carboxymuconic, maleic, and oxalic acids were also identified, and
are thought to be derived from the vanillic acid. This result suggests that the degradation of the above polystyrenes may be somewhat similar to the degradation of lignin
by microorganisms.

REFERENCES

1. **Fincher, E. L. and Payne, W. J.**, Bacterial utilization of ether glycols, *Appl. Microbiol.*, 10, 542, 1962.
2. **Payne, W. J.**, Pure culture studies of the degradation of detergent compounds, *Biotechnol. Bioeng.*, 5, 355, 1963.
3. **Suzuki, T.**, Degradation of polyvinyl alcohol by microorganisms, *Kogyo Gijutsu*, 15(3), 48, 1974.
4. **Liu, J.-H. and Schwartz, A.**, Action of bacterial mixtures on polyethylenes of various molecular weights, 51, 315, 1961.
5. **Darby, R. T. and Kaplan, A. M.**, Fungal susceptibility of polyurethanes, *Appl. Microbiol.*, 16(6), 900, 1968.
6. **Berk, S. E. and Teitell, L.**, Utilization of plasticizers and related organic compounds by fungi, *Ind. Eng. Chem.* 49, 1115, 1957.
7. **Fukumura, T.**, Bacterial breakdown of ε-caprolactam and its cyclic oligomers, *Plant Cell Physiol.*, 7(1), 93, 1966.
8. **Hatakeyama, H., Hayashi, E., and Haraguchi, T.**, Biodegradation of poly(3-methoxy-4-hydroxystyrene), *Polymer*, 18, 759, 1977.
9. **Sovish, R. C.**, Preparation and polymerization of p-vinyl-phenol, *J. Org. Chem.*, 24, 1345, 1959.
10. **Hayashi, E., Hatakeyama, H., Haraguchi, T.**, Isolation of styrene derivative degrading microorganisms, *Polym. Prepr. Jpn.*, 25, 230, 1976.
11. **Kato, M.**, Synthesis of Functional Polymers Obtained from Hydroxystyrene Derivatives, Ph.D. thesis, Tokyo Metropolitan University, Tokyo, 1969.
12. **Heigl, J. J., Bell, M. F., and White, J. V.**, Application of infrared spectroscopy to the analysis of liquid hydrocarbon, *Anal. Chem.*, 19, 293, 1947.
13. **Husband, R. M., Logan, C. D., and Purves, C. V.**, The polyhydroxyphenol series. VII. The oxidation of vanillin with sodium chloride and chlorine dioxide, *Can. J. Chem.*, 33, 68, 1955.
14. **Schlegel, H. G.**, *Allgemeine Mikrobiologie*, Georg Thieme Verlag, Stuttgart, 1972, 397.
15. **Kawakami, H.**, Bacterial degradation of lignin model compounds. IV. On the aromatic ring cleavage of vanillic acid, *J. Jpn. Wood Soc.*, 22, 246, 1976.
16. **Zabinski, R., Munck, E., Champion, P. M., and Wood, J. M.**, Kinetics and mossbauer studies on the mechanism of protocatechuic acid 4,5-oxygenase, *Biochemistry*, 11, 3212, 1972.
17. **Ornston, L. N. and Stanier, R. Y.**, The conversion of catechol and protocatechuate to β-ketoadipate by *Pseudomonas putida*, *J. Biol. Chem.*, 241, 3776, 1966.

Chapter 11

MICROBIAL DECOLORIZATION AND DEFOAMING OF PULPING WASTE LIQUORS

Toshio Fukuzumi

TABLE OF CONTENTS

I. INTRODUCTION

In the treatment of pulp-industry effluents, aggregated precipitations and in some cases activated sludges are used in Japan. These present methods hardly control color or foam since these phenomena occur even at low concentrations of waste liquor. In this study, we investigated the removal of these materials by treatment with selected microorganisms.

Lignins and their derivatives are abundant in pulping waste liquors and are important causes of their blackish-brown color. Since white-rot fungi are capable of degrading lignin in wood and metabolizing it along with carbohydrates, it was considered that their ability might be applicable to the clarification of waste liquors. We have already shown that white-rot fungi can degrade thiolignin (a kraft lignin) to some degree.[1] Sundman and Näse[2] indicated that several white-rot fungi degraded industrial lignins. Furthermore, Tono et al.[3] reported that a mould belonging to the genus *Aspergillus* could clarify a medium containing thiolignin by adsorption, and Marton et al.[4] showed that decolorization of kraft black liquor by *Polyporus (Coriolus) versicolor* also was due mainly to surface adsorption by the mycelia. Here, we attempted the decolorization by white-rot fungi of the first alkaline extract after chlorination in the kraft pulping process, hereafter referred to as AE waste liquor. This extract is the richest in organic materials.

Detergents cause much foam in rivers and research on the defoaming of detergent-containing effluents has already been carried out elsewhere, but pulp industry waste liquors do not have as much foaming power as effluents with detergents, so that ordinary measurements of foaming power designed for detergent solutions cannot be applied. Therefore, we designed a new semiautomatic apparatus specifically for waste liquor, and measured foam characteristics before and after treatment with various microorganisms.

II. DECOLORIZATION OF KRAFT WASTE LIQUOR BY WHITE-ROT FUNGI[5]

A. Screening of the Fungi

Bavendamm's (see Volume II, Chapter 2) and Sundman's[2] tests were carried out on 25 tropical species (isolated from Papua and New Guinea by Dr. K. Aoshima) and 10 Japanese species of wood-rotting fungi. The results of both tests are shown in Table 1. Tests of decolorization of culture solutions containing chlorine-oxylignin from AE waste liquor are also summarized in the same table.

On the basis of the above tests, 10 strains, Numbers 15, 22, 26, 28, 214, 356, 374, 398, *Collybia velutipes* and *Tyromyces* sp. in the table were selected. These fungi grew rapidly in media containing AE waste liquor. Changes in color, in UV absorption (at 260 and 280 nm, which represents chlorine-oxylignin content in the medium), and in pH of the culture filtrate were recorded for each fungus. Six species bleached the brown color of the medium to a faint yellow: Numbers 15, 22, 28, 398, *Collybia velutipes,* and *Tyromyces* sp. *Collybia velutipes* seemed to decrease the chlorine-oxylignin content to a lesser extent than the others by measurement of UV absorption at 260 nm, and so this fungus was eliminated. The remaining five species of fungi: four tropical isolates, *Schizophyllum* sp. (No. 15), *Trametes* sp. (No. 22), *Tinctoporia borbonica* (No. 28), *Ganoderma* sp. (No. 398), and one Japanese isolate, *Tyromyces* sp., were therefore selected for study.

B. Decolorization of Waste Liquor by the Fungi

The five fungi were inoculated into three kinds of media: 1) basal medium with 75

TABLE 1

Activity of Various White-rot Fungi in Sundman's Lignin Degradation Color Test[2], the Bavendamm Test for Phenol-Oxidizing Enzymes, and Decolorization of Kraft Pulp Bleaching Waste Liquor[a]

Test No.	Fungus species	Sundman Test[a,b]			Bavendamm Test[b]	Decolorization[b]
		T.L.	O.L.	L.S.A.		
2	Coriolus sp.	+	+	+	+	+ +
15	Schizophyllum sp.	−	−	−	(+)	+
22	Trametes sp.	+	+	−	+	+ +
26	Coriolus sp.	−	+	(+)	+	+ +
28	Tinctoporia borbonica	+	+	+	+	+ +
31	Coriolus sp.	+	+	+	+	+ +
43	Trametes sp.	+	+	+	+	+ +
Pa64g	Unknown	+	+	+	ng.	+ +
207	Roseofomes sp.	+	+	+	+	+
211	Phellinus sp.	+	−	+	+	+ +
214	Phellinus sp.	+	−	+	+	+ +
236	Nigroporia sp.	−	+	−	(+)	+
350	Coriolus sp.	+	+	(+)	+	+ +
356	Ganoderma sp.	+	+	I	+	+ +
360	Unknown	+	+	+	+	+
364	Unknown	+	−	+	+	−
365	Amauroderma sp.	+	−	+	+	−
373	Lentinus sp.	+	+	+	ng.	−
374	Coriolus sp.	(+)	+	−	+	+ +
387	Tinctoporia sp.	+	+	+	+	+ +
390	Phellinus sp.	+	+	(+)	+	−
395	Phellinus sp.	+	+	+	ng.	−
396	Phellinus sp.	+	+	+	+	I
398	Ganoderma sp.	+	+	+	+	+ +
498	Coriolus sp.	+	+	(+)	+	+
C-71	Poria subacida	+	+	+	+	−
Cb-V-4	Collybia velutipes	−	−	+	+	+ +
F-la	Perenniporia livide	+	+	+	+	+ +
F-3b	Phellinus igniarius	+	+	+	+	−
F-50c	Rigidoporus zonalis	+	+	+	+	+ +
G-L-1	Ganoderma lucidum	+	+	+	+	+
P-36c	Trametes sp.	+	−	+	+	+ +
Ps-S-1	Pycnoporus (Polystic- tus) sanguineus	I	−	+	+	−
Typ	Tyromyces sp.	+	+	−	+	+ +
Uh	Unknown	+	+	+	+	+ +

[a] T.L. = thiolignin; O.L. = Cl₂-oxylignin; L.S.A. = Lignin Sulfonic Acid.

[b] + = positive result; − = negative result; (+) = questionable; ng. = no growth; + + = very good result.

From Fukuzumi, T., Nishida, A., Aoshima, K., and Minami, K., *Mokuzai Gakkaishi,* 23, 290, 1977. With permission.

mℓ of Koji juice (saccharide concentration, 14.9%), 2) with 10 g of glucose and 5 g of xylose, and 3) with no addition, in 1ℓ (final volume) of solution containing 750 mℓ of AE waste liquor. Composition of the basal medium is given in reference 5. Cultures were maintained on a shaker at 26.5°C for 2 weeks. Changes in color, pH, COD value, and saccharide concentration of the culture filtrates were followed.

The best results were obtained from the culture of *Tinctoporia borbonica* containing glucose and xylose. The fungus decolorized the waste liquor to a clear light yellow, decreasing the color by 99% after 4 days cultivation (Table 2, Figure 1). UV measure-

TABLE 2

Degrees of Decolorization and Decrements of Chemical Oxygen Demand and Saccharide Concentration During Cultivation of Selected Fungi on Kraft Waste Liquor Media

Number of days cultivated		4				7				14				Harvest cell weight (mg/10cc)
		pH	Decolor.[b] (%)	C.O.D. (ppm)	[Saccha.][c] mg/cc	pH	Decolor. (%)	C.O.D. (ppm)	[Saccha.] mg/cc	pH	Decolor. (%)	C.O.D. (ppm)	[Saccha.] mg/cc	
Schizophyllum sp. (No. 15)	A[a]	5.0	77	3,300	1.8	4.2	76	1,600	0.2	5.4	65	1,400	0.5	90
	B	4.4	82	3,900	4.6	4.4	76	2,200	1.4	5.0	83	950	0.1	83
	C	5.6	75	740	0.2	6.0	48	600	0.2	6.0	48	550	0.1	46
Trametes sp. (No. 22)	A	2.8	75	1,300	1.0	4.4	80	1,500	0.2	5.2	52	1,600	0.5	66
	B	2.2	88	1,000	0.2	2.2	61	1,300	0.2	2.8	47	1,300	0.1	54
	C	5.6	58	160	0.2	5.6	35	520	0.2	5.8	41	430	0.3	43
Tinctoporia borbonica (No. 28)	A	4.2	92	5,800	7.8	4.0	82	2,400	0.2	5.4	81	1,500	0.5	80
	B	4.0	99	6,100	11.0	2.4	85	4,000	0.1	3.6	91	1,100	0.1	70
	C	6.0	89	940	1.0	5.8	48	590	0.2	6.0	70	590	0.1	42
Ganoderma sp. (No. 398)	A	4.4	76	9,700	15.2	3.8	76	6,700	7.6	5.2	81	1,300	0.8	81
	B	5.2	74	8,900	15.4	4.8	70	8,500	13.8	3.4	89	3,900	7.7	72
	C	5.8	77	730	1.0	5.8	50	510	0.2	6.0	60	670	0.5	29
Tyromyces sp.	A	3.6	88	2,800	1.0	2.6	81	1,000	0.2	5.8	71	1,400	0.5	56
	B	3.8	91	3,600	4.6	2.4	86	2,300	1.4	3.8	88	870	0.3	65
	C	5.7	88	420	0.6	5.6	58	260	0.2	5.8	68	580	0.4	36

[a]   Medium A: with 75 cc of Koji juice (saccharide concentration, 14.9%), Medium B: with 10 g of glucose and 5 g of xylose, Medium C: with no addition to a 1 l solution containing 750 cc of waste liquor.

[b]   Decolorization — The color was measured by the absorbance of the medium at 457 nm. The original media had absorbances of 6.95, 5.50, and 3.25 for media A, B, and C, respectively. The figures in the table were obtained from the following equation:

$$1 - \frac{\text{absorbance of medium}}{\text{original absorbance}} \times 100\%$$

[c]   Saccharides — The numbers indicate the saccharide concentrations as measured by Sumner's method.

From Fukuzumi, T., Nishida, A., Aoshima, K., and Minami, K., Mokuzai Gakkaishi, 23, 290, 1977. With permission.

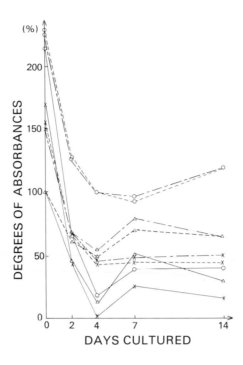

FIGURE 1.   Changes of absorbances of culture filtrates by cultivation of the fungus, *Tinctoporia* sp. (No. 28). o, Medium (A) (with 75 mℓ of Koji juice); x, Medium (B) (with 10 g of glucose and 5 g of xylose); Δ, Medium (C) (with no addition), all in a total volume of 1 ℓ of solution containing 750 mℓ of the waste liquor. Degrees of absorbances are expressed as percents in which the absorbance at 457 nm of medium (C) before cultivation was set at 100%. ——— ,457 nm; ---, 280 nm; —·—, 260 m. (From Fukuzumi, T., Nichida, A., Aoshima, K., and Miuami, K., *Mokozal Gakkaishi*, 23, 290, 1977. With permission.)

ment of the culture filtrate showed that the chlorine-oxylignin content also decreased with time, and measurement of the culture filtrate plus mycelial extract after 14 days cultivation showed that the total chlorine-oxylignin content decreased (Figure 2). All subsequent experiments were carried out using *Tinctoporia borbonica*.

The saccharides in the culture medium were almost totally consumed after 7 days of cultivation. Although the addition of saccharides to the medium initially caused a marked increase of COD value — 8620 ppm for medium B as opposed to 760 ppm for medium C — this COD value decreased to 1060 ppm after 2 weeks' cultivation of the fungus (see Figure 3). Furthermore, after 4 days, the culture solution with no added saccharides (C) decreased in color to a lesser extent than the culture solution to which saccharides were added (B) (Table 2).

C. Effects of Various Carbon Additions on Decolorization[5]

The effects of several carbon additions on decolorization were investigated because the waste liquor medium which contained glucose and xylose was best decolorized. The results observed are shown in Table 3. The addition of glucose or ethanol was

TABLE 3

Decolorization of kraft Waste Liquor in Media[a] Containing
Various Carbon Additions (Fungus: *Tinctoporia borbonica*)

Carbon source added	% W/V	Growth	Decolorization
Glucose	2	+ + + +	+ +
Xylose	2	+ + +	+ +
Ethanol	2	+ + + +	+ +
Acetaldehyde	2	−	−
Acetic acid	2	−	−
Methanol	2	+ +	±
Formaldehyde	2	−	−
Formic acid	2	−	−
Cellulose	1	+ +	±
None		+ +	±

[a] The basal medium contained 700 mℓ of the waste liquor
per liter.

From Fukuzumi, T., Nishida, A., Aoshima, K., and Minami,
K., *Mokuzai Gakkaishi*, 23, 290, 1977. With permission.

most effective for decolorization and growth of the fungus. For example, the UV absorption at 260 nm of the culture medium initially containing 2% ehtanol decreased
by approximately 35% during 4 days, incubation. Cultivation of the fungus in media
in which isolated chlorine-oxylignin was used instead of AE waste liquor and into
which glucose, xylose, and ethanol were added showed that glucose and ethanol were
effective carbon additions whereas xylose was not.

The above were carried out under aerobic conditions. If reduction of quinones to
phenols was the major reason for decolorization, it should take place under anaerobic
conditions. However, decolorization was shown not to occur under anaerobic conditions even in the best medium for aerobic decolorization.

D. Decolorization and Change of Chlorine-Oxylignin[6]

Next, changes of chlorine-oxylignin, a major component of AE waste liquor, were
followed through cultivation since the above experiments suggested that degradation
or some other change of structure in the oxylignin must have occurred. Chlorine-oxylignin was prepared by precipitation of acidified AE waste liquor.[7] The yields of precipitated and soluble oxylignin were 625 mg/ℓ and 344 mg/ℓ of waste liquor, respectively. *Tinctoporia borbonica* decolorized a medium containing oxylignin, precipitable
or soluble, containing 1% ethanol, but little decolorization was observed without the
addition of ethanol (Figure 4). Decolorization must have been due not only to the
consumption of chlorine-oxylignin, but also to structural changes in the lignin because
the decrease in weight of oxylignin in the culture (Figure 5) could not totally account
for the decrease in color measured at 457 nm (Figure 4).

Elemental analyses of chlorine-oxylignins isolated from cultures (Table 4) revealed
that the contents of carbon and phenolic hydroxyl groups were less and that the chlorine content was slightly more in the residual than that of the original lignin. In terms
of distribution by molecular weight, the residual chlorine-oxylignin obtained from a
120-hr culture with ethanol eluted from a Sephadex G-25 column at a lower molecular
weight than the original and low molecular weight aromatics in the sample disappeared. When a sample obtained from a culture without ethanol for the same length

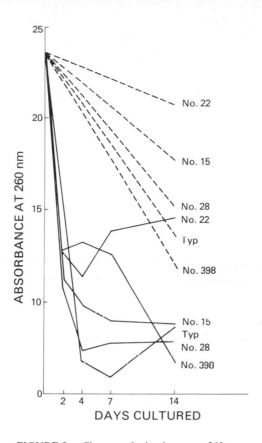

FIGURE 2. Changes of adsorbances at 260 nm of culture filtrates during cultivation of the fungi, and total absorbance including the mycelial extract. —— Culture filtrate, —— Dotted lines are interpolated between the initial value and combined value of the culture filtrate plus the extract of mycelium after 14 days of cultivation. Names of fungi, No. 15, 22, 28, 398 and Type are shown in Table 2. Extracts were prepared from the mycelial pellets recovered by centrifugation. 10 ml of methyl cellosolve was added. After ½ hr, the suspension was homogenized using a Polytron homogenizer, and the resultant mixture centrifuged to remova all debris. The debris was washed with ethanol. Combined supernatant and washings comprised the mycelial extract. (From Fukuzumi, T., Nishida, A., Aoshima, K., and Minami, K., *Mokuzai Gakkaishi,* 23, 290, 1977. With permission.)

of time was applied to the above gel-filtration, most of the oxylignin eluted as noticeably higher molecular weight polymers as shown in Figure 6. These results suggest that decolorization and degradation of chlorine-oxylignin by the fungus require a co-substrate such as ethanol or glucose.

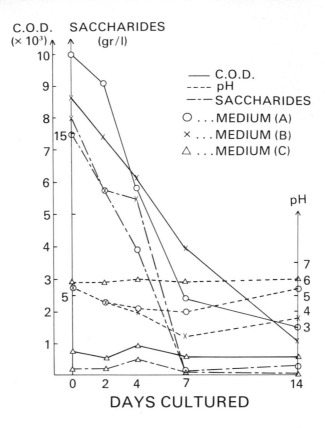

FIGURE 3.   Changes of COD, pH, and total saccharides in culture filtrates during cultivation of the fungus *Tinctoporia* sp. (No. 28). Cultivation and media are the same as in Figure 1. (From Fukuzumi, T., Nishida, A., Aoshima K., and Minami, F., *Mokuzai Gakkaishi*, 23, 290, 1977. With permission.)

E. Decolorization of Quinone Polymers by the Fungus[6]

The chromophoric structure of chlorine-oxylignin is believed to be mainly of a quinonoid nature. Therefore, quinone polymers were prepared and tested for decolorization by the fungus. *Tinctoporia borbonica* (No. 28) was cultivated in media containing polymers of methoxy-*p*-benzoquinone or of 2,6-dimethoxy-*p*-benzoquinone with 1% glucose or ethanol. As shown in Figure 7, the color of the media with these polymers decreased during cultivation of the fungus. However, even with the addition of glucose or ethanol, the main reason for decolorization of such media was strong adsorption by the mycelia.

Polymers derived from *o*-diphenols were not decolorized by the fungus even in the presence of 1% glucose.

F. Discussion

The first alkaline extract after chlorination in the kraft pulp bleaching process contains predominantly chlorine-oxylignin which has conjugated double bonds like muconic acid, quinone structures, and partially chlorinated structures resulting from oxidative chlorination of the lignin in wood.[8] The conjugated double-bond structures so closely resembled intermediates from the degradation of aromatics by microorganisms that chlorine-oxylignin seemed to be a favorable carbon source for white-rot fungi. Cultivation of *Tinctoporia borbonica* in media containing waste liquor suggested that

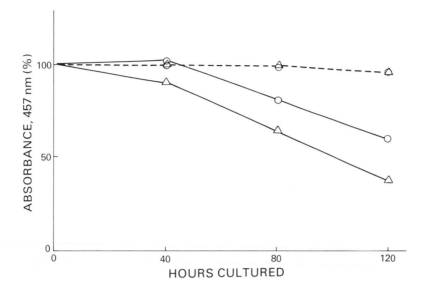

FIGURE 4.   Decolorization of chlorine-oxylignin during incubation with *Tincto-poria borbonica*. ———: medium containing 1% ethanol. ——: medium containing no carbon addition. O : acid-insoluble fraction. Δ : acid-soluble fraction.

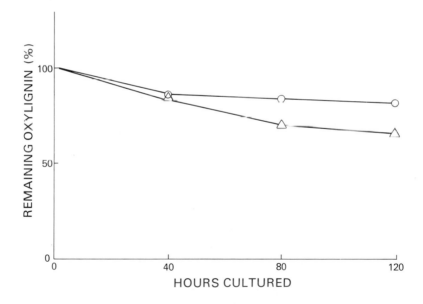

FIGURE 5.   Changes of the weights of chlorine-oxylignins during incubation with *Tinctoporia borbonica*. Oid-insoluble fraction. Δid-soluble fraction. Weights of ox-ylignins isolated from culture medium was set at 100% before cultivation.

not only does chlorine-oxylignin not inhibit the growth of the fungus, but it accelerates growth when glucose or ethanol is added to the medium.

The requirement of an additional carbon source such as glucose or ethanol for complete decolorization of the waste liquor by *Tinctoporia borbonica* may be due to the necessity of energy sources such as ATP, CoA, NAD or NADP, and of reducing agents

TABLE 4

Elemental Analysis of Chlorine-Oxylignins which were Isolated[a] from Culture Filtrates of *Tinctoporia borbonica*

	Hours cultivated	C (%)	H (%)	O (%)	Cl (%)	N (%)	Phenolic-OH (me9)
OLP[b]	0	45.7	4.8	42.1	7.1	0.3	0.26
	40	38.2	3.9	50.2	6.9	0.8	0.29
	80	39.1	3.8	48.0	8.4	0.7	0.24
	80	39.1	3.8	48.0	8.4	0.7	0.24
	120	37.9	3.7	50.0	7.7	0.7	0.22
OLS[c]	0	39.5	3.7	52.7	3.8	0.3	0.30
	40	35.3	3.4	55.4	5.2	0.7	0.22
	80	35.2	3.4	55.1	5.6	0.7	0.13
	120	33.1	3.4	55.6	7.3	0.6	0.09

[a]   Culture filtrates combined with washings of mycelial pellets were condensed to 25% volume. Proteins were precipitated by addition of an equal volume of a solution of trichloroacetic acid (0.1 *M*) + sodium acetate (0.2 *M*), pH 4.0. The filtrate was dialysed against water and evaporated to dryness.
[b]   OLP: acid-insoluble fraction.
[c]   OLS: acid-soluble fraction.

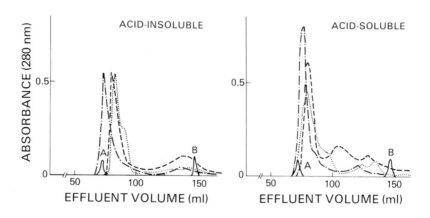

FIGURE 6.   Gel-filtration curves of acid-insoluble and -soluble chlorine oxylignins (Sephadex® G-25, H$_2$O). A = Blue Dextron (MW = 2 × 10^6); B = Vitamin B$_{12}$ (MW = 1355). ——-: before cultivation; — · — · —: after 120 hr of cultivation (no ethanol); ——: after 120 hr of cultivation (with 1% ethanol).

such as NADH, NADPH, and other electron-transport compounds produced from the metabolism of glucose or ethanol. Degradation or reduction of chromophoric structures of chlorine-oxylignin probably occurs on the mycelial surface since decolorization took place too quickly to be accounted for by consumption of the lignin.

III. DEFOAMING OF PULPING WASTE LIQUORS BY MICROORGANISMS

A. Screening of the Microorganisms

About 100 samples of microorganisms were gathered from wood chips, soils, acti-

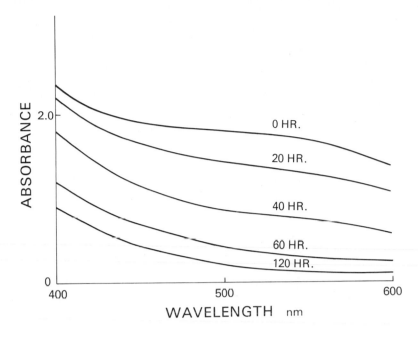

FIGURE 7. Change of visible absorption spectrum of the culture filtrate during cultivation of the fungus *Tinctoporia borbonica* on a medium containing 1% ethanol and 0.025% 2-methoxy-*p*-benzoquinone polymer.

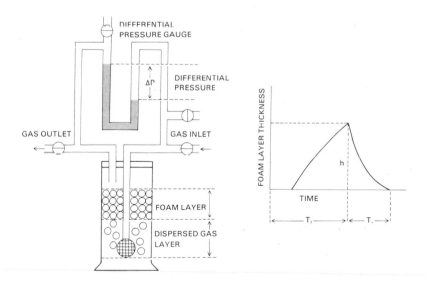

FIGURE 8. Procedure for measuring foam characteristics. Foam was generated by inletting air under constant differential pressure (ΔP) into a tube equipped with a glass filter of pore size G2 for T_f = 60 sec. Foaming power was measured in terms of foam height (h) and foam stability was represented by the lifetime (T_s) of the foam.

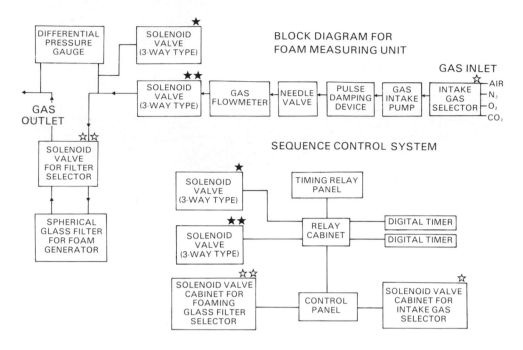

FIGURE 9.   A semiautomatic apparatus for measuring foam characteristics.

vated sludges, spent liquors, and effluents from several pulp and paper mills in Japan. Each of the samples gathered were cultivated two times successively on three kinds of agar media, one containing the first alkaline extract after chlorination in the kraft pulping process; another, kraft lignin; and the third, lignin sulfonate. Colonies showing good growth were grown up on yeast extract agar medium. The microorganisms were then cultivated individually with shaking in liquid media containing a spent liquor or a chemical found in spent liquor. Suitable microorganisms were selected by observing the ease of foam generation and the persistence of foam in the culture solution when the culture was shaken by hand.

B. Apparatus for Measurement of Foam Characteristics[9]

In order to evaluate the defoaming efficiencies of microorganism treatments of pulp mill waste liquors, it was essential to establish a reliable method for measuring foam characteristics. The foam was characterized by two factors: foaming power and foam stability. Foaming power was an indication of the ease with which foams were generated, as indicated by the foam height, and foam stability was a measure of the persistence of the foam, as indicated by its lifetime. The variation of foam height with elapsed time was recorded. The apparatus used is illustrated in Figure 8.

The length of time for foam generation was preset and controlled by solenoid valves connected with relays and time relays. The overall unit was designed to operate with a sequential control circuit of our own design. A block diagram for the foam-measuring unit and sequential-control system is illustrated in Figure 9. This sequential control system permitted semiautomatic operation of the unit and provided us with accurate and reproducible data under given conditions.

C. Defoaming of the Waste Liquors

Selected samples of microorganisms, which were expected on the basis of prelimi-

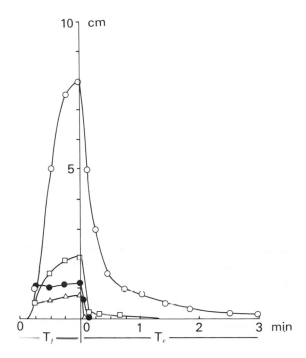

FIGURE 10.   Defoaming of spent sulfite liquor (pH 7.0) by microorganisms at 30°C. O = spent sulfite liquor filtrate components retained by an ultrafilter membrane of pore size 10^4 mol wt. ● = C-1, □ = C-3, Δ = C-4. Culture filtrates after incubation with the incubation with indicated microorganisms (Table 5) for 3 days.

nary examination, to have the ability to decompose the foaming materials in the culture without producing other foaming materials, were cultured individually in media containing a waste liquor or a chemical found in the waste liquor for 3 days at 30°C in test tubes with shaking. Each culture filtrate was measured for foaming power and foam stability on the semiautomatic apparatus. The results are shown in Table 5. About 10 samples of microorganisms were cultured in each waste liquor medium; two microbial samples showed especially strong defoaming ability (C-1 and C-4).

Of the bacterial samples in the table, C-1 was obtained from the effluent of a pulp mill, C-2 (Figure 11) from a muddy water puddle (naturally concentrated) of a pulp mill, C-3 and C-4 from activated sludges of kraft pulp mills, A-5 from the waste in the aeration stage of treatment with activated sludge, and B-6 from the effluents of a pulp and paper mill. Fungal sample C-7 was taken from the soil of a pulp mill, and C-8 and A-15 (Figure 12) was a bacterial sample obtained from a precipitation pond after treatment with activated sludge.

Representative measurements of foaming power and foam stability of the culture filtrates are illustrated in Figures 10, 11, and 12.

Bacterial sample C-4 best defoamed a wide variety of waste liquors. Cultivation of this microbial sample for 3 days reduced the foaming power and foam stability of the digesting liquor of the sulfite pulping process (mol wt > 10^4) by 90% and 99.6%, respectively, and of the first alkaline extract of the kraft pulping process (hardwood) by 61% and 66%, respectively. The sample was found to include two species of bacteria; one was *Pseudomonas putida* (FK-1), and the other has yet to be identified.

TABLE 5

Defoaming of Pulp and Paper Mill Waste Liquors with Selected Microorganisms (3 Day's, Incubation)

Medium[a]	pH	Inoculum	Cell weight (mg)[b]	ΔOD (%)[c]	Foaming power		Foam stability	
					Height (mm)	-Δh (%)	Lifetime (T_s)	$-\Delta T_s$ (%)
Spent sulfite liquor	MW>10⁴	7.0	Control[d]	0	(273nm)	80		230	
			C-4	5.4	-0.2	8	90	1	99.6
		5.5	Control	0		100		200	
			C-7	5.5	-10	15	85	7	96
	MW<10⁴	7.0	Control	0	(270nm)	40		130	
			C-1	3.1	+7	5	87	2	99
		5.5	Control	0		40		180	
			C-7	5.9	+6	10	75	0	100
	Hot alkaline extraction (softwood)	7.0	Control	0	"	55		180	
			C-1	5.6	"	5	91	1	99
		5.4	Control	0	"	30		180	
			C-7	2.9	"	7	76	3	98
Hemp digest liquor		7.0	Control	0	(266nm)	>100		>200	
			C-4	4.7	-26	15	>85	120	>40
		5.4	Control	0		100		200	
			C-8	4.4	+7	12	88	20	90
Spent kraft liquor	Alkaline extraction (softwood)	7.0	Control	"	(280nm)	12.0		40.0	
			B-6	"	-10.2	8.0	33	5.0	88
			A-5	"	+8.5	5.0	58	2.0	95
	Alkaline extraction (hardwood)	7.0	Control	"	(280nm)	14.0		6.0	
			C-3	"	-37.2	5.0	64	3.0	50
			C-4	"	-27.9	5.5	61	2.0	66

[a] Medium contained per liter: $(NH_4)_2HPO_4$, 3g; KH_2PO_4, 2g; $MgSO_4 \cdot 7H_2O$, 0.3 g; vitamin solution,[7] 5 mℓ; Hunters metal solution, 1 mℓ; and asparagine, 0.6 g. To this medium was added 3 g (solids) of the indicated liquors.

[b] Cell weight = weight of harvest

[c] ΔOD = change in optical density = $\left(\dfrac{OD_{test}}{OD_{control}} - 1 \right) \times 100\%$

Δh and ΔT_s were calculated analogously to ΔOD.

[d] Control = no inoculum.

[e] Measurements were not taken.

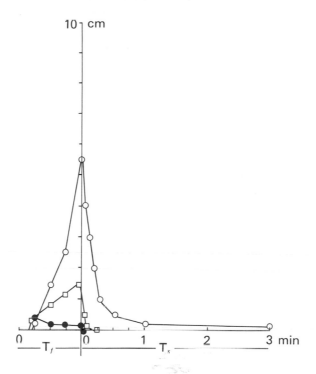

FIGURE 11.   Defoaming of the hot alkaline extract of spent sulfite liquior by microorganisms at 30°C. O = hot alkaline extract of spent sulfite liquor; ● = C-1, □ = C-2 (culture filtrates after growth with the indicated microorganisms (Table 5) for 3 days); T_1 = time for foam generation; T_2 = time representing foam stability.

Bacterial sample C-3 reduced the foaming power and foam stability of the first alkaline extract (hardwood) by 64% and 50%, respectively, and aromatics in the waste liquor by 37% as assessed by measurement of absorption at 280 nm, after 3 days. The bacterium in this sample was isolated and identified also as *Pseudomonas putida* (FK-2).

The spent liquors of the sulfite and kraft-pulping processes were easily defoamed by cultivation of microorganisms, as shown in Table 5. Microbial treatment of these spent liquors brought about a decrease in foaming characteristics of 60 to 99% of the original solution. Defoaming of white water was comparatively difficult, especially the extract of coated paper.

D. Discussion

Foam characteristics of waste liquors are dependent upon surface-active materials in the liquor. Microorganisms effective in defoaming these liquors were not only capable of removing them, but also did not produce foams, in contrast to the yeast in the making of beer.

The bacterial sample C-4 obtained from the activated sludge of a kraft pulp mill, has a strong and wide-ranging ability to decompose surface-active materials, substances of a certain, fairly high molecular weight (e.g., derivatives of lignin found in the spent liquors of pulping processes, polyacrylamide or glycylrrhizine). This ability is probably due to a synergism of the two bacteria, *Pseudomonas putida* FK-1 and the unidentified one,[10] since sample C-3 which included *Pseudomonas putida* FK-2 did not

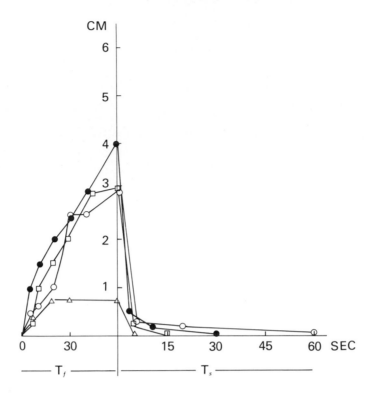

FIGURE 12.   Defoaming by microorganisms of a polyacrylamide solution used in coating papers (3 days' incubation at 30°C). ○ = control polyacrylamide solution; ● = A-15, □ = A-5, Δ = C-4 (culture filtrates after growth with the indicated microorganisms); T_f = time for foam generation; T_2 = time representing foam stability.

show as a wide range of activity, though it decomposed aromatics in the kraft waste liquor well. The laboratory strain of *Pseudomonas putida,* which has the ability to oxidize and assimilate low molecular weight aromatics, did not show the strong defoaming ability for waste liquors of the wild-type species.

IV. SUMMARY

Present methods for clarification of pulping waste liquors hardly exclude their color or foam. Both are caused by the presence of even a small amount of material in the effluent. We attempted to remove these materials from the waste liquor by cultivation of microorganisms.

Decolorization of kraft waste liquor (the first alkaline extract after chlorination) by selected white-rot fungi was studied. After 4 days of cultivation, *Tinctoporia borbonica* decreased the waste pigments by 70% or, when saccharides were added to the culture, by 99%. The fungus was then cultivated on media of waste liquor and of purified chlorine-oxylignin and various additional carbon sources were tested. Glucose and ethanol gave the best results. The ultraviolet (UV) and visible light (VL) absorptions of the oxylignin culture filtrate plus mycelial extracts decreased during the 4 days cultivation when ethanol was added to the medium (glucose not tested). Gel filtration of this residual oxylignin showed a decrease in the molecular weight of the oxylignin. This result suggests that decolorization was not due to surface adsorption of the waste lignin by the fungal cells, but rather to a partial degradation or reduction of the oxylignin.

With regard to microbial defoaming of pulping waste liquors, about 100 samples of microorganisms in waste and effluent liquors were gathered from several pulp mills in Japan. A special semiautomatic apparatus for measurement of foaming power and foam stability of the liquor was designed and constructed. After successive cultivation on agar media containing industrial lignins, suitable microorganisms were selected by aerobic cultivation in solution media containing waste liquors. The selected microorganisms were cultivated then in media containing a pulping waste liquor and the culture solution after 3 days incubation on a shaker, was filtered and measured for foaming power and foam stability.

Several microorganisms, five samples of bacteria and one sample of fungi, were effective in defoaming the waste liquor by cultivation. Of these, a sample of bacteria from activated sludge of a kraft pulp mill had the greatest ability to defoam waste liquor. Cultivation of this sample for 3 days reduced the foaming power and foam stability of the digesting liquor of the sulfite pulping process by 90% and 99.6%, respectively, and of the first alkaline extract of the kraft pulping process by 61% and 66%, respectively. One of the two bacteria in this sample was identified as *Pseudomonas putida* and the other has yet to be identified.

REFERENCES

1. Shibamoto, T., Fukuzumi, T., and Nakagawa, S., Degradation of thiolignin by microorganisms, *Mokuzai Gakkaishi*, J. Jpn. Wood Res. Soc., 7, 212, 1962.

2. Sundman, V. and Näse, L., A simple plate test for direct visualization of biological lignin degradation, *Pap. Puu*, 53, 67, 1971.

3. Tono, T., Tani, Y., and Ono, K., Microbial treatment of agricultural industrial waste, part I: adsorption of lignins and clarification of lignin-containing liquor by mould, *J. Ferment. Technol.*, 46, 569, 1968.

4. Marton, J., Stern, A. M., and Marton, T., Decolorization of kraft black liquor with *Polyporus versicolor*, a white rot fungus, *Tappi*, 52, 1975, 1969.

5. Fukuzumi, T., Nishida, A., Aoshima, K., and Minami, K., Decolorization of kraft waste liquor with white rot fungi, part I: screening of the fungi and culturing conditions for decolorization of kraft waste liquor, *Mokuzai Gakkaishi*, 23, 290, 1977.

6. Nishida, A., Fukuzumi, T., Shimazaki, T., and Minami, K., Decolorization of kraft waste liquor by a wood-rotting fungus, presented at 20th Symp. Lignin Chemistry, Nagoya, Japan, October 20 to 21, 1975, 21.

7. Samejima, K. and Kondo, T., Study on the color of waste liquor of pulp industry, part I: the relationship between the color of waste liquor in the kraft multistage bleaching and isolated Cl_2-oxylignin, *Mokuzai Gakkaishi*, 16, 347, 1970.

8. Ota, M., Durst, W. B., and Dence, C. W., Low molecular weight compounds in spent chlorination liquor, *Tappi*, 56(6), 139, 1973.

9. Onabe, F., Usuda, M., and Fukuzumi, T., presented at 25th Japan Wood Res. Soc., Fukuoka, Japan, April 6, 1975.

10. Yamazaki, M., Tamura, G., and Fukuzumi, T., unpublished data, 1978.

Chapter 12

REGULATION AND GENETICS OF THE BIODEGRADATION OF LIGNIN DERIVATIVES IN PULP MILL EFFLUENTS

Mirja Salkinoja-Salonen and Veronica Sundman

TABLE OF CONTENTS

I. INTRODUCTION

The increasing requirements of environmental control have raised the question of the fate of the lignin derivatives in pulp mill effluents and of their role in the natural environment. With effluents from the sulfite process, large amounts of lignin sulfonate are still being released into natural waters: 0.16 million tons were released into the Gulf of Bothnia from Swedish sulfite mills in 1975 — 13% (w/w) of the pulp produced.[1] The release of lignin derivatives in kraft mill effluents amounted to only 3% of the pulp produced, but a considerable part of it consisted of bleaching effluents and contained such poisonous compounds as chlorophenols.

For many years, the biodegradability of these lignin-related compounds has been under discussion. Lately, interest has focused on the chlorophenols in bleaching waste waters. These toxic pollutants, whose persistence in water systems is not well known, might be enriched in the food chain and, hence, might cause environmental damage in much lower concentrations than those which have visible toxic effects in bioassays.

For over a century, large amounts of chemically modified lignins have entered the environment either as lignin sulfonate or as kraft lignin, but their fate has never been satisfactorily determined. In the literature, proof of nonbiodegradability follow those of degradability. Recent results have favored the biodegradability of modified industrial waste lignins.[2-5] Yet, the question is still open, since these investigations were performed with wood-degrading fungi, which are not very likely to persist and function in aquatic environments. More work is therefore needed on bacterial transformations and degradation of waste lignins in aquatic systems.

This paper gives some conclusions from experiments on the biodegradation of lignin sulfonates and presents some characteristics of bacterial degradation of lignin-related phenolics. Results from an introductory survey of the fate of chlorophenols in kraft mill bleaching effluents are also presented.

II. BIODEGRADATION OF LIGNIN SULFONATES

A. Methodological Questions

During the last four decades a number of pitfalls have hampered studies of biodegradation of lignin sulfonates. The reasons for the mistakes made are easily seen in view of present-day knowledge. First of all, for the determination of lignin sulfonates, no definitive method has existed, on which assays of biodegradation could be based. The chlorine number method,[3] the nitroso method,[6] the method evaluating remaining lignin sulfonate with the aid of UV-absorbtion,[2,7] and other methods used — all suffer from the same weakness: they are influenced by structural changes in the lignin during biodegradation. Results obtained with such methods do not necessarily reflect changes in the lignin or lignin sulfonate content, since lignin biodegradation, as opposed to biodegradation of other biopolymers (proteins, carbohydrates, lipids, etc.), is not a hydrolytic reversal of the biosyntheses, but implies several changes in the molecular structure.[8] The obstacle was overcome only after ^{14}C-labeled synthetic and natural lignin and lignin sulfonates were introduced and their biodegradation analyzed according to the $^{14}CO_2$ released.[5,10-13]

A second pitfall in the study of the biodegradation of lignin sulfonates is the long-used approach of supplying them as the sole carbon source. Day et al.[14] showed, e.g., that even the white-rot fungus *Polyporus versicolor* (syn. *Trametes (Coriolus) versicolor)* was unable to metabolize sodium lignin sulfonate as the sole carbon source. With sulfonated lignin-related monomers as model compounds, later experiments have shown that sulfonation decreases the biodegradability of such monomers when they are given as the sole carbon source.[15] Fungal biodegradation of lignin sulfonate is enhanced by, or dependent on, other energy sources present, according to recent investigations.[2,4,7] To directly visualize the degradation of lignins with a large number of ligninolytic fungi, Sundman and Näse[16] used a simple plate test[17] and found that the biodegradation of kraft lignin and lignin sulfonates is more intense in a rich medium containing several additional carbon sources than in a meager medium with some additional yeast extract only. If one extrapolates to waste lignins the results available on lignin,[18] individual cosubstrates have markedly different effects on the biodegradation. The most profitable ones seem to be cellulose and cellobiose.[4,7,19] Fukuzumi, however, obtained with *Tinctoporia* sp. better degradation (decolorization) of kraft lignin in the presence of glucose than of cellulose[20] (see Volume II, Chapter 11).

A third possible pitfall could be restricting studies to pure cultures. Some utilization with mixed populations has been noted in several cases.[21-25] It might be profitable to concentrate on mixed cultures or to use natural microfloras, as Crawford et al.[13] recently did, in future work on the biodegradation of lignin sulfonates.

B. Microorganisms Involved

A large number of white-rot-fungi have been shown to degrade lignin sulfonate. In the direct plate test, Sundman and Näse[17] compared the ability of 21 typical white-rot species to degrade lignin and lignin sulfonate. Only *Armillaria mellea* was incapable of degrading lignin sulfonate. The medium contained glucose, tartrate, and malt extract. In several cases the ligninolytic capacity depended on the cosubstrates present. The results of Day et al.,[14] concerning the inability of *Trametes versicolor* to metabolize lignin sulfonate, were confirmed in a medium with no cosubstrate added and in one with yeast extract as the sole cosubstrate; degradation of lignin sulfonate was abundant in the medium containing glucose, tartrate, and malt extract.[16] In a similar screening of 35 species, Fukuzumi et al.[20] found only five of them to be incapable of metabolizing lignin sulfonate.

Gram-negative bacteria (*Pseudomonas* sp. in mixed[23] and in pure (Volume II, Chapter 8) cultures), and gram-positives in mixed cultures with protozoa of the class *Sarcodina*,[24] have also been shown to utilize lignin sulfonate. It seems justified to assume that the microfloras of soil and water, shown by Crawford et al.[13] to degrade maximally 25% of kraft lignin within a month, contain a wide variety of microbes which participate in the degradation of both kraft and sulfonated lignin.

C. Synergistic Action of Wood-Rotting Fungi

As visualized with the aid of the plate test of Sundman and Näse,[17] the biodegradation of lignin sulfonate is, in some cases, enhanced in areas where the mycelia of two different wood- or litter-degrading fungi interact.[16] Figure 1 shows the synergistic metabolization of Na-lignin sulfonate by two ligninolytic white-rot fungi. Of the two fungi used, *Trametes versicolor* was not capable per se of lignin sulfonate utilization on the medium employed; no clear zone is seen around its inoculation site. Nevertheless, a large area of degradation is seen in the center of the plate where the mycelia of *T. versicolor* and *Pleurotus ostreatus* interacted. The synergistic enhancement of lignin sulfonate (and kraft lignin) degradation of this kind is also found with pairs of fungi in which one is a brown-rot fungus and unable to degrade lignin. Due to synergistic action, these fungi, therefore, may play a role in decomposing lignin and lignin sulfonate in nature.

Compared with lignin, lignin sulfonate is more susceptible to synergistic action of pairs of fungi, which reflects the unfavorable effect of sulfonation on the biodegradability of lignin-like monomers, as reported by Watkins.[15] The lignin sulfonate was significantly ($P < 0.1\%$) more susceptible than lignin to synergistic metabolism when 225 pairs of fungi were screened for synergistic action on two different media.[16] On

FIGURE 1.   Synergistic metabolization of sodium lignin sulfonate by *Pleurotus ostreatus* (inoculated at the upper cross) and *Trametes versicolor* (inoculated at the lower cross). After the plate had been incubated for 7 days and the mycelial mats had been scraped off, the medium (0.025% lignin sulfonate) was flooded with a phenol reagent to visualize the remaining lignin sulfonate.

the other hand, the synergistic attack was also significantly more frequent on a meager yeast extract medium, unfavorable for lignin degradation, than on a rich glucose-maltose-tartrate medium, which favored ligninolysis.

D. Polymerization of Lignin Sulfonates During Biodegradation In Vitro — a Question of a Missing Co-Substrate

When microbes are grown on lignin sulfonate, a darkening of the medium is noted and concomitantly the molecular weight distribution of the substrate shifts towards higher molecular weights. This has been shown in cultures of white rot fungi[2,4,7,26] and with mixed microbial populations.[24,27] The polymerization and the darkening is assumed to result from the phenol-oxidizing enzyme activity in the cultures.

Selin and Sundman[2] showed lignin sulfonate to be strongly polymerized even in cultures of *Trametes versicolor,* in which a molecular weight fraction of lignin sulfonate served as the sole carbon source, and no measurable growth could be established. In the presence of glucose, growth occurred and the polymerization was repressed. Brunow et al. recently investigated the polymerization of lignin sulfonate with *T. versicolor.*[26] The action of the fungus was compared with that of horseradish peroxidase and hydrogen peroxide. Since the shift in the molecular weight distribution was much more pronounced with the fungus than with peroxidase/H_2O_2 dehydrogenation, it was concluded that the fungal polymerization cannot be caused by the phenol-oxidase-induced radical reaction only; some unknown polymerization reaction in the fungus could explain the difference.

It is uncertain whether the shift in the molecular weight distribution of lignin derivatives upon microbial growth is caused by preferential degradation of low molecular weight fractions or whether it is a result of real polymerization.[7,26] Strong evidence for real increase in the molecular weight is found in experiments in which the retention on an ultrafiltration membrane was compared before and after growth.[24] The filter had a molecular cutoff which allowed all lignin sulfonate to pass before growth. Some 10% of the lignin sulfonate in the culture supernatant was retained on the ultrafiltration membrane after 5-day growth of a mixed population of bacteria and protozoa.

The polymerization inherent in the microbial attack on lignin sulfonate is also seen in the metabolism of other lignin derivatives. To some extent, it has been studied in connection with bacterial degradation of the cyclolignan α-conidendrin.[28] In bacterial cultures with α-conidendrin as the sole carbon source, polymerization to a dark humus-like amorphous compound proceeds long after the growth of the bacteria has ceased (cf. the polymerization by *T. versicolor,* described above, which proceeds with no sign of fungal growth). The increase in molecular weight is evident in this case also.

Polymerization is repressed in the presence of carbohydrates. In centrally inoculated plates of *T. versicolor* on lignin sulfonate media, zonation appears if cellulose is included.[4] The inoculation site is surrounded by a bleached zone. Cellobiose: quinone oxidoreductase was isolated from such bleached areas.[29] Polymerization is limited where this enzyme and its carbohydrate substrate are present, since the phenoxy radicals produced by phenol-oxidizing enzymes are reduced back to the phenolic form.[4] Polymerization is prevented not only by cellulose and cellobiose but also by glucose, although with a weaker effect.[7,47] Fukuzumi found glucose and ethanol more effective than cellulose in fostering decolorization and depolymerization of oxylignin from kraft bleaching effluents by cultures of *Tinctoporia borbonica.*[20,30] In addition to the enzyme discovered by Westermark and Eriksson,[29] other oxidoreductase systems can probably link the oxidation of a nonaromatic substrate with the reduction of phenolic/quinoid radicals, thus preventing the polymerization of the phenolic substrate and even promoting the degradation of the polymerized material.

Polymerization seems to prevail whenever lignin derivatives are subjected to a microbial attack and easily oxidizable cosubstrates are lacking. Ecologically this could be of significance, since this type of polymerization leads to humus-like compounds and might contribute to the maintenance of a certain humus level in ecosystems.[28] Due to strong competition for available energy sources among microorganisms, easily oxidizable cosubstrates are not very likely to be present in natural surroundings and the polymerization of lignin derivatives will therefore be the prevailing function.

III. PLASMID INVOLVEMENT IN BACTERIAL DEGRADATION OF LIGNIN DERIVATIVES

A. Role of Plasmids in Biodegradation

Chakrabarty et al. first demonstrated the involvement of plasmids in catabolism. The CAM and SAL plasmids of camphor and salicylic acid catabolism, respectively, were described in 1972 and 1973.[31,32] Later *m*-toluic acid, naphthalene, octane, and xylene degradation were shown to be plasmid-coded in *Pseudomonas putida.* The catabolic plasmids in *Pseudomonas* have recently been reviewed.[33]

In 1975 Williams challenged soil with *meta*-toluate, and isolated thereafter *m*-toluate-degrading bacteria from nine different soil samples. He found that each of the 13 strains that were isolated was a *Pseudomonas* sp., and in each isolate *m*-toluate degradation was a plasmid-borne function (TOL-plasmid).[34] All strains degraded *m*-toluate, toluene, and xylenes (*meta* and *para*) to catechol, which was further metabolized by *meta*-cleavage of the benzene nucleus.[34,35] All strains could be cured of the TOL-plasmid, and in some but not all strains the TOL-plasmid was transmissible. Degradative plasmids, therefore, are ubiquitously involved in the catabolism of toluene and xylenes; in fact none of the strains isolated by Williams et al. appeared to possess a chromosomally-coded pathway for *m*-toluate utilization.[35] For naphthalene and salicylic acid both chromosomal and plasmid-coded metabolic routes are known; chromosomal genes code for the *ortho*-cleavage pathway, and plasmid genes usually code for the *meta*-cleavage pathway of catechol.[33]

Lignin is the richest source of aromatic carbon skeletons in nature. When lignin is removed from wood during the pulping process, a great diversity of aromatic compounds arises. In kraft pulping, most of the lignin is dissolved during cooking, and subsequently burned with sulphate to recover the sodium sulphide for continuation of the process. The remaining few percent of the lignin is removed and degraded by bleaching. Therefore, bleaching effluents are an important contributor to the lignin-originating pollution from pulp mills.

B. Genetic Instability of the Catabolism of Some Lignin-Related Phenols

We collected bacteria that thrive in pulp mill effluents, and asked the question whether their catabolism of lignin-originating phenolic compounds involves plasmids. Some 300 strains were isolated, using either undiluted spent bleaching liquor or simple hydroxy- and methoxy-substituted benzoic acid derivatives as the sole carbon source.

Table 1 shows some growth characteristics of the strains isolated on vanillic, veratric, ferulic, syringic, *para*- and *meta*-hydroxybenzoic acids as well as on spent liquor. The strains isolated on benzoic and salicylic acid (40 strains each) did not grow on any of the methoxylated aromatic acids. These strains, therefore, are not interesting when lignin degradation is considered — and were not included in the table.

It was found that undiluted kraft bleaching effluent selects for the utilization of vanillic, veratric, ferulic, syringic, protocatechuic, and *para*-hydroxybenzoic acid, but not for *ortho*- or *meta*-hydroxybenzoic acid, benzoic acid, or phenol.

When vanillic, veratric, ferulic, or syringic acid is used as sole carbon source, a similar collection of strains is obtained (Table 1). The majority of the strains will utilize at least three of the four acids, and most of them also grow on *para*-hydroxybenzoic and protocatechuic acids.

The choice of methods for the demonstration of plasmid involvement in a catabolic pathway is very limited. The frequency of spontaneous curing is often low — for instance, about 0.2% for the NAH and SAL plasmids and zero for many of the TOL plasmids. This makes it necessary to screen large numbers of strains for spontaneous curing. Mitomycin C (but not acridine orange or ethidium bromide) has been reported to greatly enhance loss of CAM, NAH, and SAL plasmids.[33] However, this method is laborious, and we applied it for a few strains only (see below). The benzoate curing employed by Williams[34,35] is based on the fact that the wild-type, plasmid-carrying strains will grow poorly on benzoate so that the cured derivatives will overgrow the culture in a few days in benzoate medium. The reason for this seems to be the dominance of the TOL plasmid-coded, ineffective pathway for benzoate catabolism over the more effective, chromosomally coded pathway. The chromosomal pathway is repressed when the plasmid is present.[35] Unfortunately, this method is so far limited to a single catabolic pathway.

Other agents that could be hoped to impair replication of the plasmid are those that harm the cell membrane or halt DNA replication. Aging, growth at elevated temperature, and chemicals that attack the membrane have been used. We tried both aging and growth on sodium dodecyl sulphate (SDS)-containing medium. Results of an SDS experiment are shown in Table 2.

The results show that many strains, after having survived the SDS-treatment, no longer utilize vanillic, ferulic, veratric, or syringic acid as the sole carbon source. About half of the strains no longer grew on *para*-hydroxybenzoic acid, the other half

TABLE 1

Growth of 243 Bacterial Strains Isolated From Kraft Mill Effluents on Lignin-Related Aromatics

Carbon source	Vanillic acid	Veratric acid	Ferulic acid	Syringic acid	p-Hydroxy-benzoic acid	m-Hydroxy-benzoic acid	Undiluted effluent
	Sole carbon source in the isolation medium[a]						
	Total number of strains isolated from each						
	40	37	37	34	17	20	58
	% of the isolates that showed good growth on carbon source						
Vanillic acid	100	92	38	79	41	90	100
Veratric acid	70	100	35	79	41	95	100
Ferulic acid	45	92	100	76	47	90	100
Syringic acid	15	46	49	100	35	90	100
p-OH-benzoic acid	75	91	57	94	100	55	85
m-OH-benzoic acid	13	51	41	32	35	100	15
o-OH-benzoic acid	5	5	7	11	17	5	0
Benzoic acid	23	46	59	20	100	5	0
Protocatechuic acid	70	100	89	89	100	100	100
Phenol	0	8	0	11	29	0	0
Number of isolates	40	37	37	34	17	20	58

[a] Carbon sources were added as neutralized, presterilized solutions to final concentrations of 100 mg/ℓ (50 mg/ℓ for benzoic acid). Other components of the medium were 0.4 g K_2HPO_4, 0.7 g KH_2PO_4, 0.15 g $MGSO_4$, 0.1 g NaCl, 2.0 g NH_4Cl, 30 mg $CaCl_2$, and 5 mg $FeSO_4 \cdot 6H_2O/\ell$ H_2O.

TABLE 2

Effect of Sodium Dodecyl Sulphate (SDS) in the Growth
Medium on Growth on Aromatic Compounds of Bacteria Isolated From Kraft Mill Effluents[a]

SDS added — primary isolation	None	None	1 %
— passage	None	1 %	None
Number of Strains	58	38	40
Substrate	Growth as % of all isolates		
Vanillic acid	100	13	8
Veratric acid	100	11	5
Ferulic acid	100	21	3
Syringic acid	100	8	12
p-OH-benzoic acid	85	3[b]	8
m-OH-benzoic acid	15	24	15
o-OH-benzoic acid	0	15	8
Benzoic acid	0	21	55
Protocatechuic acid	100	64	78
Phenol	0	0	3

[a]   Isolated for ability to grow on undiluted effluent as sole carbon source.
[b]   On p-hyroxybenzoic acid, 48% of the strains gave slight
growth.

grew poorly when compared with the parent strain, and only one of the 38 strains
(3%) showed normal growth. On the other hand, growth on *meta-* or *ortho*-hydroxy-
benzoic acid, benzoic, or protocatechuic acid was less impaired in the strains which
survived in the SDS medium.

In an analogous experiment with strains isolated on syringic acid, 7 out of 34 strains
survived in SDS medium, and, of these, 6 no longer grew on vanillic or ferulic acid
(results not included in Table 2).

In the reverse experiment (last column of Table 2), where strains were isolated from
SDS-containing medium, a population was obtained that resembles in many respects
the population treated with SDS upon isolation, except for growth on benzoate.

The loss of capacity to grow on methoxy-substituted aromatic acids was not regained
when the strains were grown through a few passages on yeast extract agar. Hence the
inability to grow on the aromatic compounds is not likely to be due to loss of some
vital cofactor, but rather, seems to be a permanent, genetic change. This result could
well be explained in terms of ubiquitous plasmid which codes for the catabolism of
the compounds studied.

C. A Degradative Plasmid for Cyclolignan and *para*-HydroxyBenzoic Acid?

Before we had the results indicated above, we had already noticed that some of the
strains that we had isolated on a cyclolignan, α-conidendrin, lost their ability to grow
on this substrate and on *para*-hydroxybenzoic acid spontaneously.[28] This happened
during storage of the strains as slants at·4°C. To see whether the spontaneous curing
could be enhanced, a mitomycin C experiment was carried out with one of the strains,
K17. The method was the same as that used by Chakrabarty to cure the salicylic acid
plasmid.[32]

It was found (Table 3) that 2 days' incubation in a medium which contained mito-

TABLE 3

Effect of Mitomycin-C on Ability of Bacterial Strain K17 to Metabolize p-*Hydroxybenzoic Acid*

C-source in Inoculum	Metomycin-C treatment[a]	Mitomycin-C concentration μg/ml	Colonies tested	Growth on Glucose	Growth on p-OH-benzoic acid	Growth on α-Conidendrin	Selection negative	Test negative
Yeast extract	p-OH-benz.	0	56	56	56	—	—	—
Yeast extract	Yeast extract	0	56	56	56	—	—	—
Yeast extract	p-OH-benzoic acid	10	120	98	76	—	22	22
Yeast extract	Yeast extract	10	72	72	2	—		
Glucose	p-OH-benzoic acid	0	84	84	—		—	—
Glucose	Glucose	0	123	123	—	123	—	—
Glucose	p-OH-benzoic acid	10	165	153	—	123	30	30
Glucose	Glucose	10	273	237	—	226	18	11

For the Yeast extract group: Selection = p-OH-benzoic acid negative; Test = α-Conidendrin negative. For the Glucose group: Selection = α-Conidendrin negative; Test = p-OH-benzoic acid negative.

[a] Two days' incubation in mitomycin-containing medium.

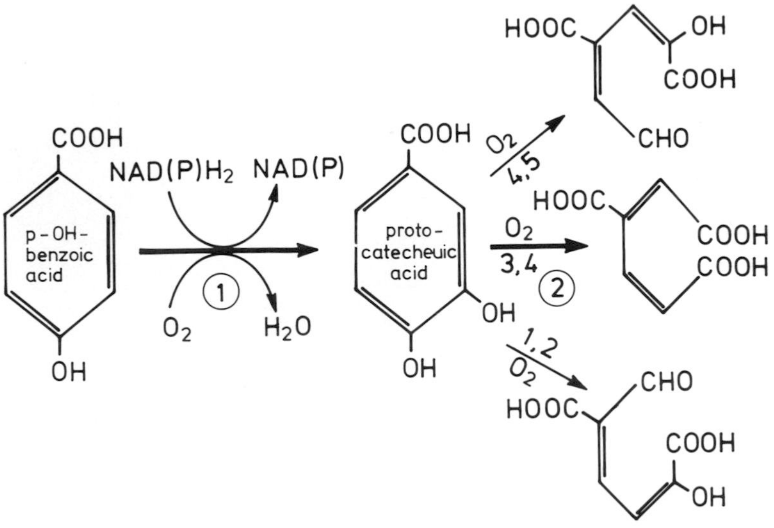

FIGURE 2.   The known metabolic pathways for bacterial oxidation
of *p*-hydroxybenzoic acid and protocatechuic acid. 1 = *p*-hydroxy-
benzoic acid hydroxylase; 2 = protocatechuic acid oxygenase.

mycin C (10 μg/ml) led to appearance of colonies that no longer grew with *para*-hy-
droxybenzoic acid as sole carbon source. Mitomycin C seems more effective if added
to *p*-hydroxybenzoic acid medium than to yeast extract or glucose-salts medium. Most
of the mutant colonies still do grow on the glucose-salts medium and therefore are
not, for example, amino acid auxotrophs. Several aromatic compounds, including benzoic acid, also supported growth (these results are not included in the table).

When 22 of the colonies that no longer grew on *p*-hydroxybenzoic acid were tested
for growth on the original substrate of isolation, α-conidendrin, none grew (Table 3).
In the reverse experiment where colonies not growing on α-conidendrin were selected
after mitomycin C treatment, the same correlation was found; of the 38 colonies tested
that did not grow on α-conidendrin, none showed growth on *p*-hydroxybenzoic acid
either. The loss of growth on *p*-hydroxybenzoic acid and on α-conidendrin was found
to be irreversible ($<10^{-10}$).

While the pathway for α-conidendrin degradation still is not fully understood, that
for *p*-hydroxybenzoic acid is well known in various microbial species and is depicted
in Figure 2. Of the enzymes involved, one is a mixed function oxidase (hydroxylase),
and another a dioxygenase. In order to see whether the production of these enzymes
was affected in the mutants "cured" for growth on *p*-hydroxybenzoic acid, and α-
conidendrin, the activity of these enzymes was measured in one mutant. The results
(Figure 3) show that no activity of either enzyme was induced in treated strain 201 in
conditions where the parent strain was fully induced. Thus *p*-hydroxybenzoic acid induces the corresponding hydroxylase in the parent only. The absence of protocatechuic
acid oxygenase activity in strain 201 was caused by absence of inducer, which is known
to be protocatechuic acid in the parent strain K17; full activity of protocatechuic acid
oxygenase was induced in strain 201 if protocatechuic acid was added to the medium
(Table 4). The activity was shown both by measurements of O_2 consumption with the
oxygen electrode and by spectrophotometric determinations of protocatechuic acid consumption, using extracts of protocatechuate-grown cells.[36] Protocatechuic acid and O_2
are used in a molar ratio of roughly 1:1 in both wild-type and mutant.

The results summarized above are compatible with the hypothesis that *p*-hydroxy-
benzoic acid hydroxylase is a plasmid-coded enzyme in the strain K17, and possibly in

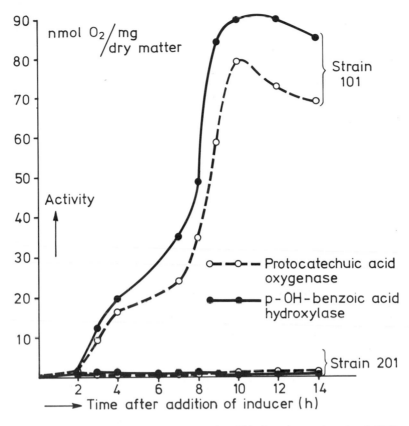

FIGURE 3.   Induction of p-hydroxybenzoic acid hydroxylase and protocatechuic acid oxygenase by p-hydroxybenzoic acid in the strain K17. 101 = wild type; 201 = cured mutant.

many other strains also. The data of Table 2 proved that it is growth on p-hydroxybenzoic acid only (but not on its metabolic product, protocatechuic acid) which is impaired by SDS in most strains.

In some organisms the aromatic ring-hydroxylating enzymes are Fe-enzymes; in others $FADH_2$ provides reduction equivalents to NAD(P)H. In K17, the p-hydroxybenzoic acid hydroxylase was shown to be NAD(P)H-dependent, but no dependence on $FADH_2$ could be demonstrated in cell-free extract (100,000 × g supernatant). Neither Tiron nor α,α-dipyridyl, known to inhibit Fe^{+++} and Fe^{++}-containing enzymes, respectively, inhibited the p-hydroxybenzoic acid hydroxylase of the strain K17 (Table 5). Next we tried SKF 525A, known to inhibit many mixed function oxygenases in animal systems.[37] It is an inhibitor of cytochrome P-450. SKF 525A was an active inhibitor (Table 5). It had no effect on the protocatechuic acid oxygenase of the same strain (Figure 4). It therefore seems that p-hydroxybenzoic acid hydroxylase employes cytochrome P-450 as a coenzyme, just as does the camphor hydroxylase that is plasmid-coded in *P. putida*.[31,37]

IV. BIODEGRADABILITY OF CHLOROPHENOLS IN KRAFT BLEACHING EFFLUENTS

A. Toxicity of Effluents

Currently, spent bleach liquor is considered to be the most harmful constituent in the effluents of pulp mills.[38-45] Chlorination stage effluent killed water fleas *(Daphnia*

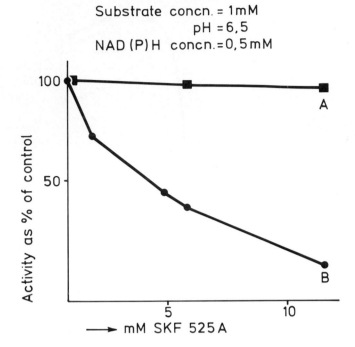

FIGURE 4.   Inhibition of *p*-hydroxybenzoic acid hydroxylase and protocatechuic acid oxygenase by the cytochrome P-450 inhibitor SKF 525A. A = protocatechuic acid oxygenase; B = *p*-hydroxybenzoic acid hydroxylase.

TABLE 4

Induction of Protocatechuic Acid Oxygenase by Protocatechuic Acid in the Strains K17101 (Wild Type) and K17201 (Cured Strain)[a]

		Protocatechuic acid oxidation	
		Activity of cell-free extract $(mg^{-1}$ protein $min^{-1})$	
Strain	Growth substrate	μMole O_2[b]	μMole protocate-chuic acid[c]
K17101	Glucose	0.04	0.03
	Protocatechuic acid	2.26	1.65
K17201	Glucose	0.00	0.00
	Protocatechuic acid	2.63	2.42

[a]   Phenylacetic acid-grown cells were transferred into a synthetic medium with protocatechuic acid and glucose as C-source (1 mg/mℓ). Activity of protocatechuic acid was measured in cell-free extracts of 14-h cultures.
[b] O_2-Electrode measurement.
[c]   Spectrophotometric measurement at 290 and 270 nm.

TABLE 5

Effects of Various Inhibitors on the Activity of *Para*-Hydroxybenzoic Acid Hydroxylase in Bacterial Strain K17101 (Wild Type)

Inhibitor	Concentration (mM)	Activity[a] (%)
None	—	100
α-α'-Dipyridyl	15.0	100
	37.5	96
	7.5	100
Tiron	7.5	100
	15.0	84
SKF 525A	4.8	44
	11.8	18

[a] Activity was measured as the oxygen uptake depending on $NAD(P)H_2$ (0.5 mM) and *para*-hydroxybenzoic acid (1 mM) with an O_2 electrode at pH 6.5 and 23°C.

pulex) within 5 min in 150-fold dilution.[41] The acute toxicity of the caustic extraction stage effluents is lower; the LD_{50} over a 96-hr period to fish was a 10 to 25-fold dilution of the effluent. Chronic exposure, however, affected fish at much lower dilutions.[41] By using [36]Cl-labeled effluent from the chlorination and caustic extraction stages of a pilot size bleach plant, Seppovaara and Hattula[44] were able to demonstrate the uptake of chlorinated organic compounds by a food chain consisting of bacteria, water fleas *(Daphnia pulex)*, molluscs *(Planorbarius)*, and guppies *(Lebister reticulatus)* in an aquarium.

The chemical nature of the toxic compounds in spent bleach liquor is not yet completely clear. This is due to the great spectrum of chlorinated organic compounds present in bleaching effluent, and analytical problems in separating or identifying them. Leach, Takhore, and Mueller[38,39,43] have identified 3,4,5-trichloroguaiacol, 3,4,5,6-tetrachloroguaiacol, monochlorodehydroabietic acid (two isomers) and dichlorodehydroabietic acid in spent bleach liquor. They found the median survival time (MST) of rainbow trout exposed to these compounds to be less than 96 hr in concentrations of 0.5 to 1 mg/ℓ. The concentrations of these compounds were reported by the same authors[38] in caustic bleachery effluents from a western Canadian mill to be between 0.3 and 0.75 mg/ℓ.

In Sweden Lindström and Nordin[46] reported concentrations of 25 to 170 µg/ℓ of 2,4,6-trichlorophenol, trichloroguaiacol (isomer not known), tetrachloroguaiacol and di- and tetrachlorocatechols each in the chlorination stage and caustic extraction stage effluents. Landner et al.[45] showed that 2,4,6-trichlorophenol and both tri- and tetrachloroguaiacol accumulated in the bodz of rainbow trout during exposure to sulfate pulp bleachery effluents, diluted 40 times with brackish Baltic Sea water. They found after 2 weeks' exposure that liver fat of the fish contained about 40 µg of 2,4,6-trichlorophenol and tetrachloroguaiacol per gram of liver fat each, and 130 µg/g of trichloroguaiacol.

The toxicity of spent bleach liquor constituents to fish and other organisms seems well documented, whereas nothing is known of the fate of these compounds in water ecosystems in situ. We were unable to find a single report on this matter in the literature. The experiments described below are an attempt to gain information on this question.

TABLE 6

Identified Chlorophenols From Kraft Bleaching Effluents of Three Kraft Mills

	Percentages of total amounts found			
Chlorophenols with identified structure	Mill A[a] birch	Mill A[a] pine	Mill B[a] birch	Mill C[a] pine
2,4-Dichlorophenol	0	3.9	0	0.5
2,4,6-Trichlorophenol	10.9	5.6	13	4
2,4,5-Trichlorophenol	0.3	<0.1	0	<0.1
4,5-Dichloroguiacol	73.1	87.0	78	93
4,5,6-Trichloroguiacol	13.0	2.0	4	1.8
2,3,4,6-Tetrachlorophenol	1	0.7	0.7	<0.1
3,4,5-Trichloro-2,6-dimethylphenol	1	0.6	4	<0.1
Pentachlorophenol	<0.1	<0.1	0.4	<0.1
Total amounts found (mg/m³)[b]	1021	5034	469	12,355

[a]   Bleaching sequence in Mill A, birch: C/D-E-H-D-E-D; in Mill A, pine: C-E-H-D-E-D; in Mill B: C-E-H-H-D-E-D; in Mill C: C/D-E-D-E-D. The letters indicate the sequence of the addition of chemicals: C = chlorine gas; D = chlorine dioxide; E = alkali (extraction stage); H = hypochlorite.

[b]   Includes identified chlorophenols only.

B. Analysis of the Chlorophenols

We extracted the chlorophenols from the water samples, removed the acidic substances (resin acids, etc.), and ethylated the phenols for gas liquid chromatography by the methods described by Lindström and Nordin.[46] We used a Carlo Erba gas liquid chromatograph (Model 2300 AC) fitted with a capillary column 80 to 100 m with OV-101 as the liquid phase. The analyses were performed isothermally at 170°C. Both electron capture (Ni-63) and flame ionization detectors were used.

The chlorophenols were identified with the aid of authentic model compounds synthesized for us by Dr. J. Knuutinen at the Department of Chemistry of the University of Jyväskylä, Finland. The ratio of the signals given by the flame ionization detector and the nickel detector is helpful for identification: the more chlorine the compound contains, the greater the ratio. Finally, one preparation of the chlorination and caustic extraction stage chlorophenols was analyzed by mass spectrometry by E. Simpura and K. Lindstrom at the Swedish Forest Research Institute with a Finnigan-9500 gas chromatograph — mass spectrograph-3200F system with an on-line computer with a data bank. This confirmed the identity of the chlorophenols presented in Tables 6 and 7. In addition to the compounds tabulated, we found peaks that coincided with authentic tetrachlorocatechol, 3,5-dichlorocatechol, 2,6-dichlorophenol, and tetrachloroveratrole. The quantitation of the chlorophenols was obtained by gas chromatophy, using response factors calculated for the authentic compounds. The response factors were as follows: 2,4-dichlorophenol, 2.8; 2,4,6-trichlorphenol, 38; 2,4,5-trichlorophenol, 74; 2,3,4,6-tetrachlorophenol, 123; 4,5-dichloroguaiacol, 3.2; 4,5,6-trichloroguaiacol, 3,4,5-trichloro, 2,6-dimethoxyphenol 70, and pentachlorophenol 127, each as millimeter peak height per nanogram. The response factors were rechecked weekly and found to vary 10 to 25%. Losses during extraction were less than 20%; no correction was calculated for this. A typical chromatogram of the chlorophenols from the mixed chlorination and caustic extraction stage effluents of pine is shown in Figure 5.

TABLE 7

Chlorophenols Identified in Recipient Waters of Kraft Bleaching Effluents

	Samples			Analysis of samples				
								Chlorophenols
Site	Layer	Month	Temperature, °C	DO^a (mg O$_2$/l)	BOD$_7$^{b} (mg O$_2$/l)	(μg/l)	Dominating compound(s)
1	Mixed	August	18	0.0		257	4,5-Dichloroguaiacol
2	Mixed	May	11	0.0	45	180	2,4-Dichlorophenol
	Top	August	19	0.0	91	92	2,4-Dichlorophenol
	Bottom^c	August	16	0.0	59	49	2,4-Dichlorophenol
3	Mixed	February	2.9	0.1	368	34	4,5-Dichloroguaiacol
	Top	August	19	0.2	8	9.2	4,5-Dichloroguaiacol, 2,4,6-trichlorophenol
	Bottom^c	August	16	0.8	9	2.6	4,5-Dichloroguaiacol, 2,4,6-trichlorophenol
	Top	February	0.4	14.2	0.1	1.2	2,4-Dichlorophenol, 4,5-dichloroguaiacol
	Bottom^c	February	1.4	12.6	3	4.4	2,4-Dichlorophenol, 2,4,6-trichlorophenol
4	Top	August	19	0.4	21	5.1	4,5-Dichloroguaiacol, 2,4,6-trichlorophenol
	Bottom^d	August	16	6.1	1	3.2	4,5-Dichloroguaiacol
	Top	March	0.5	14.3	0.8	2.2	Pentachlorophenol, 2,3,4,6-tetrachloro-phenol
	Bottom^d	March	1.7	6.7	7	10.5	4,5-Dichloroguaiacol, 2,4,6-trichlorophenol
5	Top	August	19	6.0	4	2.0	2,4-Dichlorophenol
	Bottom^e	August	16	8.3	1	4.3	2,4-Dichlorophenol, 4,5-dichloroguaiacol
6	Mixed	February	0.2	13.5	0.6	0.9	2,4-Dichlorophenol
	Mixed	February	0.2	13.3	1.3	4.7	4,5-Dichloroguaiacol, pentachlorophenol

^{a} Dissolved oxygen.
^{b} Biologial oxygen demand (7 days).
^{c} Depth = 9 m.
^{d} Depth = 13 m.
^{e} Depth = 15 m.

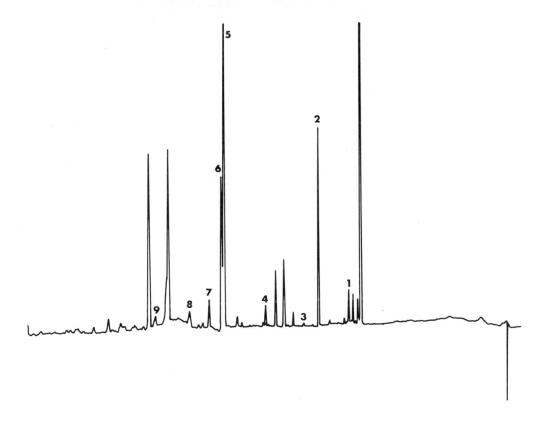

FIGURE 5.   Gas liquid chromatograph of chlorophenols from spent kraft bleach liquor. 1 = 2,4-dichlorophenol; 2 = 2,4,6-trichlorphenol; 3 = 2,4,5-trichlorophenol; 4 = 2,3,4,6-tetrachlorophenol; 5 = trichloroguaiacol (isomer unknown); 6 = 4,5,6-trichloroguaiacol; 7 = tetrachloroveratrole; 8 = 3,4,5-trichloro-2,6-dimethoxyphenol; 9 = pentachlorophenol. [Capillary column (80 to 100 m), OV-101; 175°C.][46]

Table 6 shows the analyses of the eight identified chlorophenols in the mixed effluent from bleacheries of three different pulp mills in Finland. The total amount varied between 0.5 and 12 ppm and corresponds well to what has been reported in other countries.[36,46-48] The distribution of the different chlorophenols appears very similar in all mills, so that the major difference would be the volume of water used by the bleach plant (figures not available).

C. Fate of the Chlorophenols in Recipient Waters

Figure 6 shows the location at the lake side of Mills A and B and that of the effluent pipes. Mill B has two sewers, a lakeside one for pulping (cooking) effluents, a riverside one for bleaching effluents. The numbers in the map show where the samples were collected.

Table 7 shows the amounts of chlorophenols, dissolved oxygen, and biological oxygen demand (7 days) of the samples from the lake and the river in summer and winter 1977-1978. It is seen that the amount and type of chlorophenols found showed seasonal variation. In the sampling point 2, close to the sewer of Mill A, 180 μg/ℓ of the eight chlorophenols were found in May 1977, but only 49 to 92 μg/ℓ in August when the water was warm.

In spent bleach liquor it is the chlorinated guaiacols and 2,4,6-trichlorophenol which dominate (Table 6), whereas in the effluent recipient it is 2,4-dichlorophenol. This could mean that dichlorophenol is less biodegradable than the major chlorophenols,

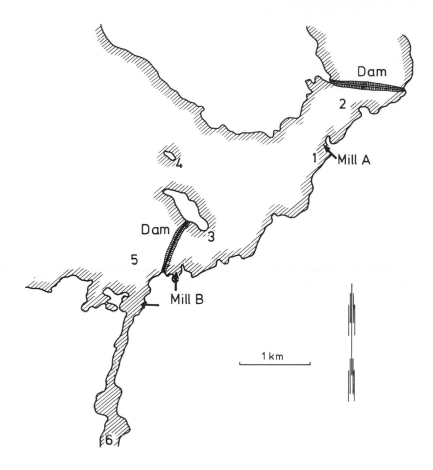

FIGURE 6.   Location of the sample collecting sites downstream from two pulp mills.

or that it is being formed by biological or (photo) chemical degradation of the poly-chlorinated phenols and guaiacols. The involvement of biological degradation is indicated by the observation that in winter, when the water temperature is low (0.2 to 1.5°C), both BOD_7 and dichloroguaiacol tend to accumulate; this is seen at the sampling points 2, 4, 5, and 6. Interpretation of the results was somewhat complicated by a 3-week period in February 1978 when Mill A was out of operation and did not contribute to the accumulation of chlorophenols in the recipient.

Total volume of bleaching effluents from the two mills comprise about 0.3% of the total flow of the river depicted in Figure 6. Thus the concentrations of the chlorophenols found in the river in February 1978 are about equal to what could be calculated from the dilution factor alone, indicating little biodegradation in wintertime of the chlorophenols in the natural ecosystem.

In many samples — most of which were not included in Table 7 — the amount of pentachlorophenol surprised us. Concentrations which were 2 to 20 times higher in the natural waters than in the undiluted total pulp mill effluents (sampling point 1 in the map) were not uncommon. We therefore conclude that although some (usually 1 to 5 $\mu g/\ell$) pentachlorophenol is present in kraft bleaching effluent, this is not its main source in the water ecosystem. In rain water (mixed samples from a 30-day period in winter 1978) we found concentrations as high as 20 $\mu g/\ell$, which indicates that the source of pentachlorophenol should also be sought away from the actual polluted area.

In the present analysis we investigated the monomeric chlorinated phenols only.

However, they constitute only a minor fraction of the total pollution of organochlorine compounds from the kraft bleaching plant. Chlorophenols analyzed in the present work amounted to 250 g/ton of pulp produced. According to Hardell and de Sousa, the total organic carbon of the effluents is much higher, 9 kg in the chlorination stage and 20 kg in the caustic extraction stage per ton of pulp produced.[49] From the total input of chlorine in the bleaching plant, less what is found as inorganic chlorine in the effluents, it can be calculated that, with high-molecular-weight material included, 3 to 4 chlorine atoms on an average are bound per phenylpropane unit present.

The main part of the chlorolignin probably is of high molecular weight. Hardell and de Sousa[49] analyzed by ultrafiltration the dry-matter molecular weight of spent bleach liquor and found that the molecular weight of about 40% of the chlorination extract and of about 75% of the caustic extract organic carbon is higher than 10,000. No information is available about the behavior of this high molecular weight material in the water ecosystem.

V. SUMMARY

The results obtained in studies of the biodegradation of lignin sulfonates are surveyed and the reasons for previous failures in demonstrating its degradability analyzed. The microorganisms capable of lignin sulfonate degradation include wood-rotting fungi, bacteria (particularly pseudomonads), and mixed populations of bacteria and protozoa. In several wood-degrading fungi, the degradation of lignin sulfonates is enhanced by pairs of fungi in a synergistic way.

The nature of lignin polymerization, important in lignin sulfonate degradation in vitro, and evidence suggesting that the increased molecular weight does not reflect preferential consumption of small molecules, is discussed.

Bacteria, isolated from soil as capable of utilizing α-conidendrin (a natural lignan occurring in sulfite waste liquor) as sole carbon source, frequently lost their ability to grow on this substrate. The ability to grow on p-OH-benzoic acid was concomitantly lost, but many other aromatics were still used as carbon source. Enzymological and genetic evidence for extra-chromosomal coding for the catabolism of p-OH-benzoic acid is presented.

Lignin is transformed into chlorolignin and chlorinated phenolic compounds during bleaching in the pulping process. The bleachery effluents of three Finnish pulp mills were analyzed for chlorophenols by capillary gas chromatography. The metabolic fate of the effluent chlorophenols in the receiving water was traced during summer and winter 1977—1978. Some of the chlorophenols were found more resistant than others to microbial degradation.

ACKNOWLEDGMENTS

Bacteriological analyses were performed by Raili Paasivuo and Kirsi Nikkinen, enzyme analyses by Mirja Laukkanen, and gas chromatography by Riitta Boeck.

REFERENCES

1. **Laake, M.,** Avfallsligniner fra celluloseindustrien, effekter, spredning og analyse. Preliminary study report, Nordforsk's project group on investigation of waste lignins of the pulp industry, Nordforsk, Helsinki, 1976, 34.

2. **Selin, J.-F., Sundman, V., and Räihä, M.**, Utilization and polymerization of lignosulfonates by wood-rotting fungi, *Arch. Microbiol.*, 103, 63, 1975.
3. **Hiroi, T. and Eriksson, K.-E.**, Microbial degradation of lignin. I. Influence of cellulose on the degradation of lignins by the white-rot fungus *Pleurotus ostreatus, Sven. Papperstidn.*, 79, 157, 1976.
4. **Hiroi, T., Eriksson, K.-E., and Stenlund, B.**, Microbial degradation of lignin. II. Influence of cellulose upon the degradation of calcium lignosulfonate of various molecular sizes by the white-rot fungus *Pleurotus ostreatus, Sven. Papperstidn.*, 79, 162, 1976.
5. **Lundquist, K., Kirk, T. K., and Connors, W. J.**, Fungal degradation of kraft lignin and lignin sulfonates prepared from synthetic ^{14}C-lignins, *Arch. Microbiol.*, 112, 291, 1977.
6. **Selin, J.-F. and Sundman, V.**, Analysis of fungal degradation of lignosulphonates in solid media, *Arch. Microbiol.*, 81, 383, 1972.
7. **Huttermann, A., Gebauer, M., Volger, C., and Rösger, C.**, Polymerization und Abbau von Natriumlignosulfonate durch *Fomes annosus* (Fr.) Cooke, *Holzforschung*, 31, 83, 1977.
8. **Kirk, T. K.**, Chemistry of lignin degradation by wood-destroying fungi, in *Biological Transformation of Wood by Microorganisms*, Liese, W., Ed., Springer-Verlag, New York, 1975, 153.
9. **Kirk, T. K., Connors, W. J., Bleam, R. D., Hackett, W. F., and Zeikus, J. G.**, Preparation and microbial decomposition of synthetic [^{14}C] lignins, *Proc. Natl. Acad. Sci. U.S.A.*, 72, 2215, 1975.
10. **Haider, K. and Trojanowski, J.**, Decomposition of specifically ^{14}C-labelled phenols and dehydropolymers of coniferyl alcohol as models for lignin degradation by soft and white-rot fungi, *Arch. Microbiol.*, 105, 33, 1975.
11. **Crawford, D. L. and Crawford, R. L.**, Microbial degradation of lignocellulose: the lignin compound, *Appl. Environ. Microbiol.*, 31, 714, 1976.
12. **Trojanowski, J., Haider, K., and Sundman, V.**, Decomposition of ^{14}C-labelled lignin and phenols by a *Nocardia* sp., *Arch. Microbiol.*, 114, 149, 1977.
13. **Crawford, D. L., Floyd, S., Pometto, A. L., III, and Crawford, R. L.**, Degradation of natural and kraft lignins by the microflora of soil and water, *Can. J. Microbiol.*, 23, 434, 177.
14. **Day, W. C., Gottlieb, S., and Pelczar, M. J., Jr.**, The biological degradation of lignin. IV. The inability of *Polyporus versicolor* to metabolize sodium lignosulfonate, *Appl. Microbiol.*, 1, 78, 1953.
15. **Watkins, S. H.**, Bacterial degreadation of lignosulfonates and related model compounds, *J. Water Pollut. Control Fed.*, 42, 1247, 1970.
16. **Sundman, V. and Näse, L.**, The synergistic ability of some wood-degrading fungi to transform lignins and lignosulfonates on various media, *Arch. Microbiol.*, 86, 339, 1972.
17. **Sundman, V. and Näse, L.**, A simple plate test for direct visualization of biological lignin degradation, *Paper and Timber*, 53, 67, 1971.
18. **Kirk, T. K., Connors, W. J., and Zeikus, J. G.**, Requirement for a growth substrate during lignin decomposition by two wood-rotting fungi, *Appl. Environ. Microbiol.*, 32, 192, 1976.
19. **Ander, P. and Eriksson, K.-E.**, Influence of carbohydrates on lignin degradation by the white-rot fungus *Sporotrichum pulverulentum, Sven. Papperstidn.*, 78, 643, 1975.
20. **Fukuzumi, T., Nishida, A., Aoshima, K., and Minami, K.**, Decolourization of kraft waste liquor with white rot fungi. I. Screening of the fungi and culturing condition for decolourization of kraft waste liquor, *Mokuzai Gakkaishi*, 23, 290, 1977.
21. **Kleinert, T. N. and Joyce, C. S.**, Lignin sulfonic acids as mould nutrients, *Sven. Papperstidn.*, 62, 37, 1959.
22. **Abernathy, A.**, Microbial Degradation of Lignin Sulfonate, Ph.D. thesis, University of North Carolina at Chapel Hill, University Microfilms No. 64-1826, Ann Arbor, Mich., 1963.
23. **Ferm, R. and Nilsson, A.-C.**, Microbiological degradation of a commercial lignosulfonate, *Sven. Papperstidn.*, 72, 531, 1969.
24. **Sundman, V. and Selin, J.-F.**, Microbial utilization of lignosulfonates of various molecular sizes, *Paper and Timber*, 52, 473, 1970.
25. **Selin, J.-F. and Sundman, V.**, Microbial action on lignosulfonates, *Finska kemists. Medd.*, 80, 11, 1971.
26. **Brunow, G., Wallin, H., and Sundman, V.**, A comparison of the effects of a white-rot fungus and H_2O_2-horseradish peroxidase on a lignosulfonate, *Holzforschung*, in press, 1978.
27. **Ferm, R. and Nilsson, R.**, Analysis of microbially degraded lignosulfonates by thin-layer chromatography and ultrafiltration, *Sven. Papperstidn.*, 73, 283, 1970.
28. **Sundman, V.**, A study of lignanolytic soil bacteria with special reference to α-conidendrin decomposition, *Acta Polyt. Scand.*, Ch. 40, 75, 1965.
29. **Westermark, U. and Eriksson, K.-E.**, Cellobiose-quinone oxidoreductase, a new wood-degrading enzyme from white-rot fungi, *Acta Chem. Scand.*, Ser. B, 28, 209, 1974.
30. **Fukuzumi, T.**, Microbial decolorization and defoaming of pulping waste liquors, in *Lignin Biodegradation: Microbiology, Chemistry, and Applications*, Kirk, T. K., Ed., CRC Press, Cleveland, 1979.
31. **Rheinwald, J. G., Chakrabarty, A. M., and Gunsalus, I. C.**, A transmissible plasmid controlling camphor oxidation in *Pseudomonas putida, Proc. Natl. Acad. Sci. U.S.A.*, 70, 885, 1973.

32. **Chakrabarty, A. M.,** Genetic basis of the degradation of salicylate in *Pseudomonas, J. Bacteriol.,* 112, 815, 1972.
33. **Chakrabarty, A. M.,** Plasmids in *Pseudomonas, Annu. Rev. Genet.,* 10, 7, 1976.
34. **Williams, P. A. and Worsey, M. J.,** Ubiquity of plasmids in soil bacteria: evidence for the existence of new TOL plasmids, *J. Bacteriol.,* 125, 818, 1976.
35. **Duggleby, C. J., Bayley, S. A., Worsey, M. J., and Williams, P. A.,** Molecular sizes and relationships of TOL plasmids in *Pseudomonas, J. Bacteriol.,* 130, 1274, 197.
36. **Gibson, D. T.,** Assay of enzymes of armoatic metabolism, in *Methods in Microbiology,* Vol. 6A, Norris, J. R. and Ribbons, D. W., Eds., Academic Press, New York, 1971, 462.
37. **Gunsalus, I. C., Pederson, T. C., and Sligar, S. G.,** Oxygenase catalyzed biological hydroxylations, *Annu. Rev. Biochem.,* 44, 377, 1975.
38. **Leach, J. M. and Takhore, A. M.,** Isolation and identification of constituents toxic to juvenile rainbow trout *(Salmo gairdneri)* in caustic extraction effluents from pulp mill bleach plants, *J. Fish. Res. Board Can.,* 32, 1249, 1975.
39. **Leach, J. M., and Takhore, A. N.,** Identification of the toxic constituents in kraft mill bleach plant effluents, CPAR Rep. No. 245-2, Canadian Forestry Service, Ottawa, Ontario, 1975.
40. **Walden, C. C.,** The toxicity of pulp and paper mill effluents and corresponding measurement procedures, Review Paper, *Water Res.,* 10, 639, 1976.
41. **Seppovaara, O.,** Toxiciteten hos Blekeriavloppsvatten i Finska Förhållanden, organiska miljogifter i Vatten. *XII Nordiska Symposiet om Vattenforskning,* Nordforsk, 1976, 113.
42. **Katz, B. M. and Cohen, G. M.** Toxicities of excessively chlorinated organic compounds, *Bull. Environ. Contam. Toxicol.,* 15, 644, 1977.
43. **Mueller, J. C., Leach, J. M., and Walden, C. C.,** Detoxification of bleached kraft mill effluents — a manageable problem, TAPPI Environ. Conf., Chicago, Illinois, 77, 1977.
44. **Seppovaara, O. and Hattula, T.,** The accumulation of chlorinated constituents from pre-bleaching effluents in a food chain in water, *Pap. Puu,* 8, 489, 1977.
45. **Landner, L., Lindström, K., Karlsson, M., Nordin, J., and Sorensen, L.,** Bioaccumulation in fish of chlorinated phenols from kraft pulp mill bleachery effluents, STFI-meddelande ser. B, Nr 437, SCANforsk Rapport Nr 126, Stockholm, 1977.
46. **Lindström, K., and Nordin, J.,** Gas chromatography-mass spectrometry of chlorophenols in spent bleach liquors, *J. Chromatogr.,* 128, 13, 1976.
47. **Gellerstedt, G.,** Blekeriseminarium — Dagens Teknik och Framtida Lösningar, *Sven. Papperstidn.,* 4, 95, 1978.
48. **Erickson, M. and Dence, C. W.,** Phenolic and chlorophenolic oligomers in chlorinated pine kraft pulp and bleach plant effluents, *Sven. Papperstidn.,* 10, 316, 1976.
49. **Hardell, H.-L. and de Sousa, P.,** Karakterisering av blekeriavlutar, Del 1, Klor-och alkalistegsavlut från förblekning av tallsulfatmassa, STFI-meddelande ser. B, Nr 332, SCANforsk Rapport Nr 81, Stockholm, 1975.

Chapter 13

LIGNIN BIODEGRADATION AND THE PRODUCTION OF ETHYL ALCOHOL FROM CELLULOSE

S. L. Rosenberg and C. R. Wilke

TABLE OF CONTENTS

I. INTRODUCTION

A. Present Process

During the last several years, our group has been engaged in developing a biochemical process for the conversion of lignocellulosic materials to ethyl alcohol.[1-3] A diagram of the process as presently conceived is shown in Figure 1.[4] Improvements are still being made, however, so this diagram does not represent the final configuration.

Briefly, the process operates as follows. Lignocellulosic material (in this case, corn stover) is milled and then extracted with a hot solution of dilute sulfuric acid. This treatment solubilizes most of the hemicellulose sugars, the primary component of which is xylose. We are currently working on a process for the conversion of xylose to ethanol, but this is not shown in the flow sheet and is not considered in the economic analysis of the process.

The treated solids, composed mainly of lignin and cellulose, are contacted with a cellulase-containing supernatant from a culture of the fungus *Trichoderma reesii (viride)*, strain QM 9414, in the hydrolysis step, and soluble sugars are produced. Cellulase enzymes are initially adsorbed on the cellulosic substrate but are released in the course of cellulose digestion. These enzymes are recovered and reused by contacting fresh acid-extracted substrate. The enzyme-free, glucose-containing hydrolyzate is then concentrated and fermented to ethanol and CO_2 by a continuously growing culture of the yeast *Saccharomyces cerevisiae*, ATCC 4126. The effluent from this fermentation is centrifuged, and part of the cells are recyled to the fermenter to increase the cell density and, thus, the productivity of the fermenter. The rest of the cells are dried for use as an animal feed supplement. Ethanol is recovered from the culture fluid by distillation. Residual unfermented sugars are removed from the waste stream by anaerobic digestion and converted to methane and CO_2. The methane is used to generate steam and electricity for the process. The unhydrolyzed residue from the hydrolysis step which consists mainly of lignin and some cellulose is also recovered and used to generate steam and power for the process.

In the enzyme production step, a continuous culture of *T. reesii* is grown with pure cellulose as the substrate, and cellulase and β-glucosidase enzymes are induced. In the separation step, the mycelium is removed from the enzyme-containing culture fluid as a slurry. Part of the mycelium is recycled to the fermenter. The rest of the mycelium plus residual cellulose is dewatered and recovered for use as an animal feed supplement. It has been calculated[4] that 23 gal of 95% ethanol can be obtained per ton of corn stover at a processing cost of about $1.80 per gallon exclusive of by-product credits.

B. Importance of Delignification

Lignin is recognized as a barrier to cellulose degradation by cellulases.[5] In the present scheme, chemically delignified cellulose is used only as a substrate for mold growth and enzyme induction. The substrate used for saccharification is not delignified because of the cost of such a chemical treatment. A cheap delignification step would benefit the process in two ways. It would increase the amount of cellulose accessible to cellulase hydrolysis and thus increase the overall yield of the process, and it would reduce the amount of inert material carried through the process, reducing the size of the plant and thus the capital cost.

C. Biodelignification

Two processes have been demonstrated for the partial delignification of wood using living cultures of white-rot fungi. The first involves the use of cellulase-less mutants which metabolize lignin and hemicellulose but spare cellulose.[6,7] The second involves

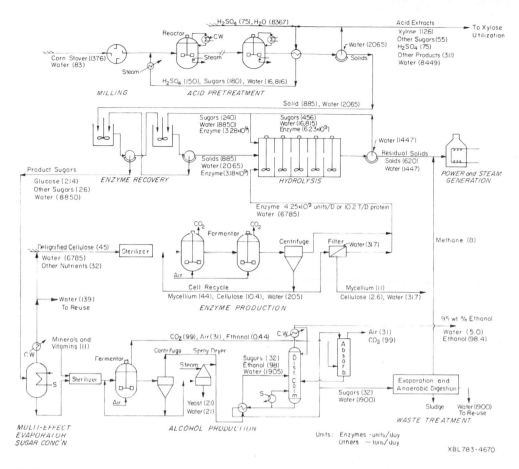

FIGURE 1.   Process for the biochemical conversion of corn stover to ethyl alcohol. (From Cysewski, G. R. and Wilke, C. R., *Biotechnol, Bioeng.*, 18, 1297, 1976. With permission.)

the use of natural strains which preferentially ("selectively") degrade lignin in the early stages of their attack on wood.[5,8] In the first approach, significant amounts of hemicellulose sugars are removed during lignin degradation. In the second approach, only small amounts of lignin are degraded before carbohydrate loss begins. In both cases, the degradation products are consumed by the culture and are unavailable for other uses.

C. Potential of Enzymatic Delignification

While at the present time only chemical processes are used to delignify lignocellulosic materials, an enzymatic process might be attractive if a cell-free enzyme system could be developed. Use of an enzyme system should significantly reduce energy requirements for delignification. Such a process would also allow recovery and reuse of the enzymes and possible production of useful chemicals from the dissolved lignin.

The enzymology of lignin degradation has recently been reviewed.[9] No cell-free system for the solubilization of lignin has been demonstrated, although both fungal laccase[10] and peroxidase[11] preparations have been shown to carry out a simultaneous depolymerization and repolymerization of lignin (see Volume II, Chapter 2).

II. WORK WITH THERMOPHILIC AND THERMOTOLERANT FUNGI

A. Basic Attractiveness

In our effort to demonstrate a cell-free enzyme system capable of rapid lignin degradation, we focused our attention on thermophilic organisms because previous work[12,13] suggested that they would be able to degrade lignin and cellulose faster than their mesophilic counterparts. We further restricted our studies to the fungi, because convincing demonstrations of lignin degradation only occurred within this group.

While the thermophilic and thermotolerant[14] fungi had been fairly well characterized with respect to their ability to degrade pure cellulose,[15-18] little was known about their ability to degrade lignocellulose except for two strains, *Chrysosporium pruinosum* and *Sporotrichum pulverulentum,* which had been reported to degrade both lignin and cellulose by Nilsson[19] and Ander and Eriksson.[7] These strains are now characterized as imperfect forms of the mold *Phanerochaete chrysosporium.*[20]

B. Cellulose and Lignocellulose Degradation

Table 1 shows the results of experiments in which growth on and degradation of hanging filter paper (cellulose) or newsprint (lignocellulose) strips were noted for a large assemblage of taxonomically characterized thermophilic and thermotolerant fungi.[21] The paper strips were suspended from hooks in Erlemeyer flasks with the lower 5 to 10 mm of each strip submerged in a mineral medium supplemented with 0.01% yeast extract. Each culture was adjusted to the optimum pH and incubated at the optimum growth temperature as determined previously.[22]

The results obtained allowed the organisms tested to be divided into four groups based on their ability to grow with and degrade cellulose and lignocellulose. Group I fungi neither degraded nor grew on cellulose or lignocellulose. Group II fungi grew on and degraded cellulose but not lignocellulose. An apparent exception is *Torula thermophila* which showed growth on cellulose but no obvious degradation. The degradation which must have occurred was probably too diffuse to be noted. Group III fungi showed good growth on both cellulose and lignocellulose, but obvious degradation was restricted to the cellulose. The growth on lignocellulose may have actually been supported by the supplementary cellulose fibers incorporated in the newsprint (see below). Group IV fungi were able to grow on and degrade both cellulose and lignocellulose.

Several interesting findings emerged from this work. With respect to cellulose, three organisms in Group II appeared to show regional preferences for growth on and degradation of the paper strip. *Humicola insolens, Malbranchea pulchella* var. *sulfurea* and *Myriococcum albomyces* grew best on nonsubmerged regions of the cellulose, while the other organisms which grew with cellulose showed growth or degradation above and below the liquid surface.

With respect to newsprint, which is a mixture of 85% ground wood fibers (lignocellulose) and 15% cellulose fibers, a number of organisms appeared to grow on all regions of the strip (Group III). But only *Chrysosporium pruinosum* and *Sporotrichum pulverulentum* showed the ability to cause significant degradation (thinning) of the substrate, and this degradation occurred characteristically only in regions of the paper strip above the liquid meniscus.

For those organisms that showed it, restriction of growth or degradation to nonsubmerged regions of the substrate is probably related to the local concentration of one or more factors. Important factors may include degradative enzymes, cofactors, metabolic intermediates, minerals, and oxygen.

TABLE 1

Growth on and Degradation of Cellulose (Filter Paper) and Lignocellulose (Newsprint) by Thermophilic and Thermotolerant Fungi

Organism	Cellulose		Lignocellulose	
	Growth[a]	Degradation[b]	Growth[a]	Degradation[b]
		I		
Humicola lanuginosa	− (5)[c]	− (5)	− (5)	− (5)
Mucor miehei	− (5)	− (5)	− (5)	− (5)
M. pusillus	− (5)	− (5)	− (5)	− (5)
Stilbella thermophila	− (5)	− (5)	− (5)	− (5)
Talaromyces thermophilus	− (5)	− (5)	− (5)	− (5)
Thermoascus aurantiacus	− (5)	− (5)	− (5)	− (5)
Thermomyces stellatus	− (5)	− (5)	− (5)	− (5)
		II		
Chaetomium thermophile v. coprophile	+ a[d] (32)	+ b (32)	− (5)	− (5)
C. thermophile v. dissitum	+ a (19)	+ b (19)	− (5)	− (5)
Humicola grisea v. thermoidea	+ ab (32)	+ b (32)	− (5)	− (5)
H. insolens	+ b (32)	+ b (32)	− (5)	− (5)
Malbranchea pulchella v. sulfurea	+ + bc (32)	+ c (32)	− (5)	− (5)
Myricoccum albomyces	+ b (32)	+ b (32)	− (5)	− (5)
Torula thermophila	+ ab (19)	− (5)	− (5)	− (5)
		III		
Allescheria terrestris	+ + abc (32)	+ + abc (32)	+ abc (8)	− (5)
Aspergillus fumigatus	+ + abc (12)	+ abc (12)	+ abc (5)	− (5)
Sporotrichum thermophile	+ + abc (32)	+ + abc (32)	+ abc (5)	− (5)
Talaromyces emersonii	+ + abc (12)	+ + abc (12)	+ abc (5)	− (5)
Thielavia thermophila	+ + abc (12)	+ + abc (12)	+ abc (5)	− (5)
		IV		
Chrysosporium pruinosum	+ + abc (32)	+ + abc (32)	+ + abc (32)	+ + bc (32)
Sporotrichum pulverulentum	+ + ab (12)	+ + ab (12)	+ + ab (12)	+ + b (12)

[a] (−) Indicates no growth in excess of that seen in control flask. (+) and (+ +) indicate relative amounts of growth in excess of that seen in control flask.

[b] (−) Indicates no obvious degradation compared to uninoculated control. (+) and (+ +) indicate relative degree of thinning or rotting of paper strip.

[c] Number in parentheses indicates day after which no further growth or degradation was observed. Observations were made on days 5, 8, 12, 19, and 32.

[d] a, b, and c indicate regions of the paper strip displaying most growth or degradation. a = submerged, b = 5 mm band above liquid meniscus, c = all area above b.

From Rosenberg, S. L., *Mycologia*, 70, 1, 1978. With permission.

III. PREFERRED CONDITIONS FOR RAPID LIGNOCELLULOSE DEGRADATION

A. Culture Design

Since our qualitative experiments indicated that rapid and extensive lignocellulose degradation could only be expected in nonsubmerged substrates, we set about developing a culturing system which would allow us to produce large amounts of mycelium and degraded substrate both for quantitative measurement of lignin and carbohydrate degradation and for the isolation of lignin-degrading enzymes.

The system which we developed utilizes a 25 mm high × 150 mm diameter Petri dish half-filled with mineral-agar medium supplemented with thiamin. Up to four sterile 47 mm diameter Nuclepore membrane filters (5 μm pore size) can be laid on the agar surface; 200 mg of dry, sterile lignocellulose fiber are spread on each of the filters, a drop of inoculum is added, and the cultures are incubated.

The lignocellulose fiber used was prepared by washing cattle manure obtained from a feed lot. The procedure was designed to remove both sand and water-soluble materials and yield a particulate residue which was rich in lignin. Much of the carbohydrate present in the original feed grain and straw was solubilized during fermentation in the rumen. The dry, washed manure fiber contained 37 ± 1.5% lignin, 37 ± 2% reducing sugar (cellulose plus hemicellulose), 10 ± 0.5% crude protein and 14 ± 1% ash. Lignin and carbohydrate were analyzed by modifications of the 72% sulfuric acid method of Moore and Johnson[24] and the anthrone method of Fairbairn.[26]. Details of the techniques and analytical methods used will be reported in a forthcoming publication.[23]

Figure 2 shows how various fractions of the lignocellulosic substrate are lost with time in a series of replicate agar plate cultures inoculated with *C. pruinosum*, ATCC 24782.[23] Organics are defined as that fraction of the dry residue which volatilize at 550°C.

Substrate degradation appears to cease after about 12 days with 75% of the organics, 80% of the carbohydrate, and 50% of the lignin degraded. During the period of active substrate degradation (3 to 12 days), the moisture content of the culture varies between 77 and 82% w/w.[23] Over the first 24 days of the experiment, the protein content (Kjeldahl nitrogen × 6.25) of the original residue doubles from 11.6% to 23.3%.[23]

Based on the amount of time required for equal percentages of degradation, the rate of lignin degradation displayed by *C. pruinosum* growing under these conditions is about four times greater than the rate reported by Kirk and Moore[5] for the mesophilic white rot *Polyporus giganteus* growing on wafers of aspen wood. The rate of lignin degradation displayed by *P. giganteus* is the highest previously reported.

B. Submerged Cultivation-Inhibition of Lignin Degradation

Figure 3 shows how degradation proceeds under conditions of submerged cultivation in a mineral medium using the same organism and substrate in shaking flask cultures.[23] Initial substrate losses are due to solubilization caused by autoclaving the fiber in the mineral medium. Here, little or no lignin was degraded, and losses of carbohydrate and organics were limited to 50 and 40%, respectively. The rate of carbohydrate degradation was also lower under these conditions than in the agar plate cultures. It is possible that shaking the submerged culture prohibits close contact between the mycelium and the substrate, and this may be responsible for the absence of significant lignin degradation (See Volume II, Chapter 4).

Figure 4 shows the results of an experiment in which stationary, submerged fiber cultures were analyzed. A tight mycelial mat formed on the bottom of each flask which encompassed and immobilized almost all of the fiber particles, indicating that close contact between mycelium and substrate was established. The results are similar to those shown in Figure 3 except that somewhat less carbohydrate was degraded. This may be due to the fact that in these cultures the fiber and medium were autoclaved separately and then combined.

These quantitative data confirmed our initial observations (Table 1) that significant lignocellulose degradation by *C. pruinosum* does not occur in a submerged substrate.

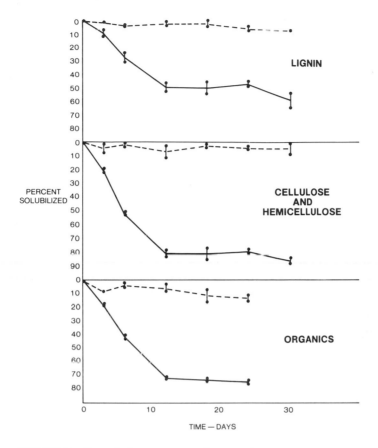

FIGURE 2. Degradation of manure fiber constituents by *Chrysosporium pruinosum* growing in damp fiber on the surface of mineral agar plates. Each data point represents the contents of one filter. All experiments were run in duplicate. Dashed lines indicate uninoculated controls.

C. Effect of Oxygen

It has been suggested that lignin degradation may be inhibited under submerged conditions by a low oxygen tension caused by fungal metabolism. To examine this possibility, a series of submerged standing cultures was prepared and incubated in an atmosphere of pure oxygen. Data in Figure 5 indicate that there was a small but significant increase in lignin degraded under these conditions and a much larger increase in carbohydrate loss (see Figure 4). Degradation appears to be essentially complete after 30 days of incubation (see also Volume II, Chapter 4).

Experiments were also carried out in which shaking submerged cultures were incubated in a pure oxygen atmosphere. Results similar to those shown in Figure 4 were obtained suggesting that both close hyphal-substrate contact and high oxygen tensions are required for significant lignin degradation to occur under conditions of submerged cultivation. We are presently investigating the effects of greater than atmospheric pressures of oxygen on submerged lignin and cellulose degradation.

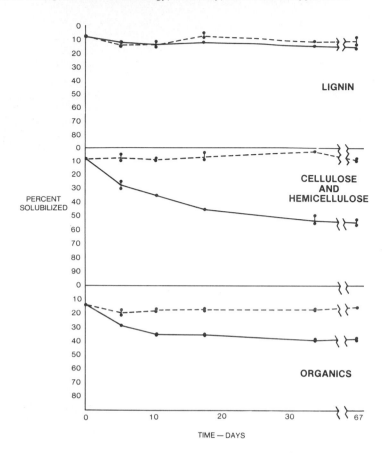

FIGURE 3.   Degradation of manure fiber constituents by *Chrysosporium pruinosum* growing in submerged (shake flask) cultures. Each data point represents the combined contents of two shake flasks. Single points indicate that replicates give identical values. Dashed lines indicate uninoculated controls.

IV. ATTEMPTS TO DEMONSTRATE CELL-FREE LIGNIN DEGRADATION

A. Extraction of Cultures

Having developed the plate-culturing technique for *C. pruinosum* which allowed rapid lignin degradation to proceed, we next attempted to derive a cell-free lignin-degrading enzyme system from these cultures.

In our first approach a number of 4-day-old plate cultures were harvested and mixed with half-strength mineral medium[21] at 0°C for 30 min to extract enzymes. The suspension was then centrifuged. The recovered supernate represented a 1:10 dilution of the liquid present in the original fiber cultures. The supernate was concentrated at 4°C for 24 hr in an Amicon pressure cell using a UM-2 membrane (approximate molecular weight cutoff = 1000). A 26-fold concentration of the supernate was achieved. The concentrate was centrifuged, and the supernate was sterilized by vacuum filtration through a 0.4 μm pore size Nucleopore filter. Filtration was very slow, requiring 4 hr at 20°C. Aliquots of the sterile filtrate were added to 200 mg samples of sterile fiber in test tubes, mixed to coat the wall of the tube, and incubated in a humidified incubator at 40°C for 6 days. The aliquot size was chosen to reproduce in the fiber the

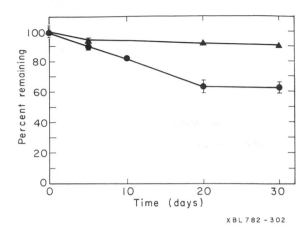

FIGURE 4. Lignin ▲—▲ and carbohydrate •—• loss in stationary submerged cultures of *Chrysosporium pruinosum* grown with manure fiber.

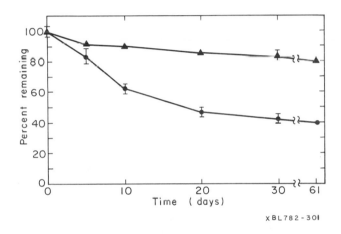

FIGURE 5. Lignin ▲—▲ and carbohydrate •—• loss in stationary submerged cultures of *Chrysosporium pruinosum* grown with manure fiber in an oxygen atmosphere.

TABLE 2

Residual Lignin and Carbohydrate (Expressed as Glucose Equivalent) in Manure Fiber Treated with Concentrated Culture Supernate

Addition to fiber	Average % lignin remaining ± average deviation[a]	Average % carbohydrate remaining ± average deviation[a]
Concentrated supernate	98.0 ± 0.7	83.8 ± 3.3
Boiled concentrated supernate	98.7 ± 1.3	97.8 ± 3.5
Mineral medium	97.4 ± 0	100.3 ± 1.5
None	100.0 ± 1.3	100.0 ± 3.4

[a] "No addition" control = 100%; four replicates analyzed.

TABLE 3

Residual Lignin and Carbohydrate (Glucose Equivalent) in Manure Fiber Treated with Expressed Culture Fluid

Addition to fiber	Average % lignin remaining ± average deviation[a]	Average % carbohydrate remaining ± average deviation[a]
Culture fluid	100.7 ± 0	94.4 ± 0.7
Boiled culture fluid	100.0 ± 0.7	100.0 ± 2.1

[a] Boiled culture fluid control = 100%; two replicates analyzed.

moisture content of an agar plate fiber culture (∼80% w/w). Table 2 shows the results obtained. While carbohydrate was degraded, no lignin-solubilizing activity was detected.

There are a number of problems with the above approach. Enzymes may be inactivated during dilution, low molecular weight cofactors may be lost in the concentration step, and significant losses in activity may occur due to the time required for concentration. In order to minimize or eliminate these problems, a different procedure was devised.

B. Use of Expressed Culture Fluid

Agar plate cultures were grown for 5 days and then scraped off the filters directly into the barrel of a plastic syringe. The plunger was inserted and pushed in tightly using a press, and culture fluid was collected. A small disc of polyurethane foam was placed in the bottom of the syringe barrel before loading to prevent clogging of the orifice. In this way, 45% of the liquid content of the cultures could be recovered. The expressed fluid was centrifuged, filter-sterilized, and then added to fiber tubes as before. Table 3 shows the results of this experiment. Again, no lignin loss is seen, and only a small amount of carbohydrate was solubilized.

It is obvious from the results of these experiments that the lignin degradation system produced by this organism does not consist solely of an easily separable set of soluble extracellular enzymes, as is the case for the cellulase system.[25]

C. Use of Diffusion Cultures

Since it was not clear whether the lignin-degrading system produced by this organism was diffusible or required the presence of living mycelium for activity, an apparatus was constructed to try to answer these questions. The apparatus (Figure 6), termed a diffusion chamber, consists of two glass tubes glued to and separated by a 47 mm diameter Nuclepore membrane filter of 0.2 μm pore size. The tubes were 20 to 30 mm long with an O.D. of 46 mm and an I.D. of 44 mm. The ends of the tubes were plugged with well-washed polyurethane sponge plugs. The bottom chamber contained 200 mg of sterile manure fiber moistened with 0.8 mℓ of mineral medium plus thiamin. The top chamber was inoculated with 0.2 mℓ of a culture suspension (inoculated filter). The cultures were incubated in a humidified incubator, and 0.6 mℓ aliquots of mineral medium were added to the top surface weekly to prevent drying and provide minerals for culture growth. These chambers allowed growth of the mold on the top surface of the filter without hyphal invasion of the substrate below the filter. As controls, one series of chambers was prepared, but not inoculated, while another series was prepared with the fiber inoculated. The results of these experiments are shown in Table 4.

FIGURE 6.   Diffusion chamber. Top plug raised to show details of construction.

It is apparent that organisms growing in direct contact with the lignocellulose fiber (Lines 6 and 7) are able to degrade significant amounts of both lignin and carbohydrate within the first 2 weeks of incubation, although neither lignin nor carbohydrate losses are as complete as in the agar plate cultures (Figure 2).

Cultures where the inoculum is separated from the substrate (Lines 4 and 5) show a mat of mycelial growth and conidia on the top surface of the filter. The fiber below

TABLE 4

Residual Lignin and Carbohydrate (Glucose Equivalent) in Manure Fiber From Diffusion Cultures

Sample	Incubation time-days	Average % lignin remaining ± average deviation[a]	Average % carbohydrate remaining ± average deviation[a]
Uninoculated control	0	100.0 ± 0.8	100.0 ± 0.1
	14	96.7 ± 0.2	94.3 ± 2.8
	31	96.5 ± 0	89.8 ± 0.4
Inoculated filter	14	94.5 ± 1.0	78.1 ± 0.8
	31	92.0 ± 2.5	54.9 ± 4.2
Inoculated fiber	14	77.5 ± 1.0	26.9 ± 7.1
	31	76.0 ± 2.8	27.4 ± 2.5

[a]   Relative to average value for zero time uninoculated control. Two replicates analyzed.

remains sterile. Under these conditions an appreciable amount of carbohydrate is solubilized. Growth of the organism above coupled with solubilization of carbohydrate below indicates that both enzymes and soluble products are diffusing through the separating membrane. Lignin loss, while somewhat higher than in the uninoculated controls, is not believed to be significant, since losses of approximately 10% are seen in the uninoculated controls from other experiments (Figures 2, 3).

Taken together, the data from the enzyme-isolation and diffusion experiments suggest that the lignin-degradation system or one or more of its components produced by this organism is either not induced, unstable, nondiffusible (e.g., bound to the cell wall), or inactive, at small distances (about 1 mm) from growing hyphae.

V. PRESENT DIFFUSION CULTURE STUDIES

A. Selection of Mutants of *Chrysosporium pruinosum*

Although we have not yet been able to demonstrate a cell-free lignin-degrading system, the diffusion culture technique may be useful in further attacking this problem. We are presently using two approaches in our attempt to demonstrate a cell-free system. Cultures of *C. pruinosum* are being mutagenized in order to obtain a strain which produces a diffusible system. The selection technique involves the use of a fiber substrate which was previously exhaustively degraded by the wild-type in a diffusion culture. Any organisms able to grow on this partially degraded material in a diffusion culture would presumably possess either a diffusible lignin-degradation system or an altered cellulase activity. Preliminary experiments have shown that the wild type displays only very slight growth in diffusion cultures using previously degraded fiber as a substrate. This suggests that the heavy growth seen using undegraded substrate occurs primarily at the expense of the "diffusate" and not to any significant degree at the expense of nutrients carried in the inoculum.

B. Testing of Mesophilic Lignin-Degrading Fungi

We are also using diffusion chambers to test a number of mesophilic lignin-degrading molds which differ from one another both taxonomically and with respect to the way in which they have been shown to degrade lignin. By analogy with common metabolic pathways described in other microorganisms, it is reasonable to expect that there may be alternative biochemical approaches to the degradation of lignin and that some of these organisms may produce a diffusible system.

VI. SUMMARY

During the last few years our group has been engaged in developing a biochemical process for the conversion of lignocellulosic materials to ethyl alcohol. The present process involves pretreatment of the substrate with hot, dilute acid to remove hemicellulose and increase cellulose reactivity, treatment of the extracted lignocellulose with cellulases from *Trichoderma reesii* to produce glucose, and fermentation of the glucose to ethanol and CO_2 by *Saccharomyces cerevisiae*.

Lignin is a barrier to complete cellulose saccharification in this process, but chemical and physical delignification steps are too expensive to be used at the present time. An enzymatic delignification process might be attractive for several reasons: little energy would be expected to be needed, enzymes could be recovered and reused, and useful chemicals might be produced from dissolved lignin.

We examined a number of thermophilic and thermotolerant fungi for the ability to rapidly degrade lignocellulose in order to find an organism which produced an active lignin-degrading enzyme system. *Chrysosporium pruinosum* and *Sporotrichum pulverulentum* (now recognized as imperfect forms of the mold *Phanerochaete chrysosporium*) were found to be active lignocellulose degraders, and *C. pruinosum* was chosen for further study.

It was found that in shake flask cultures using a lignocellulosic substrate, carbohydrate (cellulose and hemicellulose) was degraded, but lignin was not. Lignin and carbohydrate were degraded when the substrate remained moistened by, but not submerged in, the liquid medium. Carbohydrate degradation was much more extensive in moist cultures than in submerged cultures. In static submerged cultures both lignin and carbohydrate degradation could be stimulated by an atmosphere of pure oxygen.

Attempts were made to demonstrate a cell-free lignin degrading system by both extraction and pressing of cultures grown on moist lignocellulose. Carbohydrate-degrading activity was found but not lignin-degrading activity. This led us to ask whether diffusible lignin-degrading activity could be demonstrated in this organism. Using an apparatus in which the culture was separated from the damp substrate by a bacteriological membrane filter, we found that mycelial growth could occur on the inoculated side of the membrane at the expense of carbohydrate degraded on the substrate side, but little or no lignin was lost.

The data indicate that the lignin degradation system, or one or more of its components, produced by this organism is either unstable, non-diffusible, or inactive at small distances (about 1 mm) from growing hyphae.

At present, studies are being conducted using diffusion cultures to select mutants of *C. pruinosum* that do produce a diffusible lignin degradation system. We are also examining a number of mesophilic lignin-degrading molds for this ability.

ACKNOWLEDGMENTS

This work was supported in part by the General Electric Company, the Department of Energy, and the National Science Foundation.

REFERENCES

1. **Cysewski, G. R. and Wilke, C. R.,** Utilization of cellulosic materials through enzymatic hydrolysis. I. Fermentation of hydrolysate to ethanol and single cell protein, *Biotechnol. Bioeng.,* 18, 1297, 1976.
2. **Wilke, C. R., Cysewski, G. R., Yang, R. D., and von Stockar, U.,** Utilization of cellulosic materials through enzymatic hydrolysis. II. Preliminary assessment of an integrated processing scheme, *Biotechnol. Bioeng.,* 18, 1315, 1976.
3. **Wilke, C. R., Yang, R. D., and von Stockar, U.,** Preliminary cost analysis for enzymatic hydrolysis of newsprint, *Biotechnol. Bioeng.,* Symp. No. 6, 155, 1976.
4. **Wilke, C. R., Yang, R. D., Sciamanna, A. S., and Freitas, R.,** Raw materials evaluation and process development studies for conversion of biomass to sugars and ethanol, paper presented at the 2nd Ann. Symp. on Fuels from Biomass, Rensselaer Polytechnic Institute, Troy, N.Y., June 20 to 22, 1978.
5. **Kirk, T. K. and Moore, W. E.,** Removing lignin from wood with white-rot fungi and digestibility of resulting wood, *Wood Fiber,* 4, 72, 1972.
6. **Eriksson, K.-E. and Goodell, E. W.,** Pleiotropic mutants of the wood-rotting fungus *Polyporus adjustus* lacking cellulase, mannanase and xylanase, *Can. J. Microbiol.,* 20, 371, 1974.
7. **Ander, P. and Eriksson, K.-E.,** Influence of carbohydrates on lignin degradation by the white-rot fungus *Sporotrichum pulverulentum, Sven. Papperstidn.,* 78, 643, 1975.
8. **Ander, P. and Eriksson, K.-E.,** Selective degradation of wood components by white-rot fungi, *Physiol. Plant,* 41, 239, 1977.
9. **Ander, P. and Eriksson, K.-E.,** Lignin degradation and utilization by microorganisms, *Prog. Ind. Microbiol.,* 14, 2, 1977.
10. **Ishihara, T. and Miyazaki, M.,** Oxidation of milled wood lignin by fungal laccase, *Mokuzai Gakkaishi,* 18, 415, 1972.
11. **Sopko, R. S.,** Degradation of spruce and sweetgum lignins by *Polyporus versicolor* and *Pleurotus ostreatus,* Master's thesis, North Carolina State University at Raleigh, 1972.
12. **Brock, T. D.,** Life at high temperatures, *Science,* 158, 1012, 1967.
13. **Waksman, S. A. and Gerretsen, F. C.,** Influence of temperature and moisture upon the nature and extent of decomposition of plant residues by micro-organisms, *Ecology,* 12, 33, 1931.
14. **Cooney, D. G. and Emerson, R.,** *Thermophilic Fungi,* W. H. Freeman, San Francisco, 1964.
15. **Chang, Y.,** The fungi of wheat straw compost. II. Biochemical and physiological studies, *Trans. Br. Mycol. Soc.,* 50, 667, 1967.
16. **Eriksson, K.-E. and Rzedowski, W.,** Extracellular enzyme system utilized by the fungus *Chrysosporium lignorum* for the breakdown of cellulose. I. Studies on the enzyme production, *Arch. Biochem. Biophys.,* 129, 683, 1969.
17. **Fergus, C. L.,** The cellulolytic activity of thermophilic fungi and actinomycetes, *Mycologia,* 61, 120, 1969.
18. **Tansey, M. R.,** Agar-Diffusion assay of cellulolytic ability of thermophilic fungi, *Arch Mikrobiol.,* 77, 1, 1971.
19. **Nilsson, T.,** Mikroorganismer i Flisstackar, *Sven. Papperstidn.,* 68, 495, 1965.
20. **Burdsall, H. H., Jr. and Eslyn, W. E.,** A new *Phanerochaete* with a *Chrysosporium* imperfect state, *Mycotaxon,* 1, 123, 1974.
21. **Rosenberg, S. L.,** Cellulose and lignocellulose degradation by thermophilic and thermotolerant fungi, *Mycologia,* 70, 1, 1978.
22. **Rosenberg, S. L.,** Temperature and pH optima for 21 species of thermophilic and thermotolerant fungi, *Can. J. Microbiol.,* 21, 1535, 1975.
23. **Rosenberg, S. L.,** Physiological studies of lignocellulose degradation by the thermotolerant mold *Chrysosporium pruinosum, Developments in Industrial Microbiology,* 20, 133, 1979.
24. **Moore, W. E. and Johnson, D. P.,** Procedures for the chemical analysis of wood and wood products, Forest Products Laboratory, U.S. Department of Agriculture, 1967.
25. **Streamer, M., Eriksson, K.-E., and Pettersson, B.,** Extracellular enzyme system utilized by the fungus *Sporotrichum pulverulentum (Chrysosporium lignorum)* for the breakdown of cellulose, *Eur. J. Biochem.,* 59, 607, 1975.
26. **Fairbarn, N. J.,** A modified anthrone reagent, *Chem. Ind. (London),* No. 4, 86.

Chapter 14

BIOMECHANICAL PULPING

Karl-Erik Eriksson and Lars Vallander

TABLE OF CONTENTS

I. INTRODUCTION

It has long been known that the degradation of wood in nature is caused mainly by fungi. The strong wood-degrading effect that fungi have depends, in part, upon the organization of their hyphae, which gives the organisms a penetration capacity. Different types of fungi give rise to different types of wood rot and are normally classified as either soft-rot, brown-rot, or white-rot fungi. The blue-staining fungi are also associated with wood damage. They do not, however, cause wood decomposition.

The term soft-rot emanates from the fact that there is a softening of the surface layer when wood is attacked by this group of fungi. In the secondary wall of the attacked wood, there appear cylindrical cavities with conical ends. The term soft-rot is now used whenever this characteristic cavity pattern appears, even if no softening of the surface layer has taken place. Soft-rot is caused by fungi belonging to Ascomycetes and Fungi Imperfecti. They mainly cause a degradation of the polysaccharides in wood. Like all other wood-degrading fungi, soft-rot fungi demethylate lignin. This seems to be the main mechanism of attack on lignin by both soft- and brown-rot fungi.

Brown-rot and white-rot attack on wood is mainly caused by fungi belonging to Basidiomycetes. The hyphae of the fungi are normally localized in the cell lumen and these hyphae penetrate from one cell to another through openings or by producing boreholes themselves. Brown-rot fungi degrade cellulose and the hemicelluloses in wood. In an early stage of degradation, they depolymerize cellulose faster than the degradation products are utilized. Brown-rotters begin the degradation process from the hyphae within the cell lumen and first degrade the S_2 and then, later, the S_1 layer of the cell walls.

The white-rot fungi are a rather heterogeneous group of organisms. These organisms have in common that they can degrade lignin as well as all the other wood components. They also have in common the ability to produce extracellular enzymes which oxidize phenolic compounds related to lignin. The relative amounts of lignin and cellulose degraded and utilized by these fungi vary, and so does the order of preferential attack. The normal method of wood degradation by white-rot fungi is that the cellulose and the lignin are attacked simultaneously. Although examples of lignin degradation prior to cellulose degradation do exist,[1] the results from our work and the work of other researchers[2-5] indicate that one of the polysaccharides in wood must be degraded simultaneously with lignin. It appears that the energy requirement for lignin degradation is so large that white-rot fungi require an additional, easily accessible energy source; thus, the wood polysaccharides serve as cosubstrates.

The bioconversion of wood by microorganisms is an important development to which more and more attention is being paid. From what is said above, it is obvious that white-rot fungi, which have the ability to degrade all the wood components including lignin, are particularly useful in this context. Even if an absolutely specific attack on the lignin is not obtained, it has been demonstrated[6] that enough lignin can be degraded to cause a decrease in the energy demand for production of thermomechanical pulp if the wood chips are pretreated with cellulase-less mutants of white-rot fungi. Low molecular weight sugars and some of the xylan in wood are used as cosubstrates.[2]

The first genetic experiments to produce cellulase-less mutants were carried out by Eriksson and Goodell.[7] The principle is to irradiate a spore suspension with UV-light until approximately a 2% survival is obtained. The spores are then spread on cellulose agar plates supplied with small amounts of glucose to allow cellulase-less mutants to grow (Figure 1). Since colonial growth is essential, a chemical causing this type of growth is normally also added. After 2 weeks a clear zone develops around the colonies indicating cellulose degradation. Colonies without clear zones are further tested for

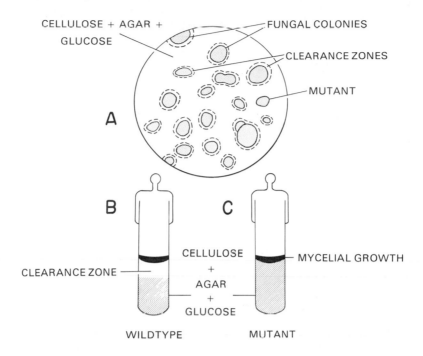

FIGURE 1.   Selection of cellulase-less mutants.

cellulase production by transferring these organisms to test tubes (Figure 1). By using the hemicelluloses xylan and glucomannan in test tubes, it was also found that cellulase-less mutants did not form the enzymes mannanase or xylanese either, or if so, only in very small amounts.[7]

II. CHARACTERISTICS OF DIFFERENT WILD-TYPE AND CELLULASE-LESS MUTANTS OF WHITE-ROT FUNGI

The fungi and their cellulase-less mutants selected for treatment of wood chips prior to mechanical pulping were wild-type and mutant strains: *Sporotrichum pulverulentum*, (= *Phanerochaete chrysosporium*) and Cel 44, *Phlebia radiata* and Cel 26, and *Phlebia gigantea* and Cel 50. These fungi and mutants were all investigated for pH- and temperature optima, as well as for growth rate in wood. For the mutants, the optimal nitrogen source in different kinds of wood was also determined.

A. pH-Optima
The pH-curves for all the strains are very similar, as can be seen in Figure 2. *S. pulverulentum* and Cel 44 have optimum growth rates at pH 4.7; *P. radiata* and Cel 26, at pH 4.9-5.0; and *P. gigantea* and Cel 50, at pH 4.8. The investigated strains grow at pH-values up to 6.5, and on the acidic side (using other media) they grow down to pH-values slightly above three.

B. Temperature Optima
Temperature optima for the strains studied differ more, as can be seen in Figure 3. *S. pulverulentum* and Cel 44 have optimal growth rates between 38 to 39°C; *P. radiata* and Cel 26, at 26°C; and *P. gigantea* and Cel 50, at 28°C. The fact that *S. pulverulentum* and its mutant have such high temperature optimum compared to the other strains studied makes them of course particularly interesting.

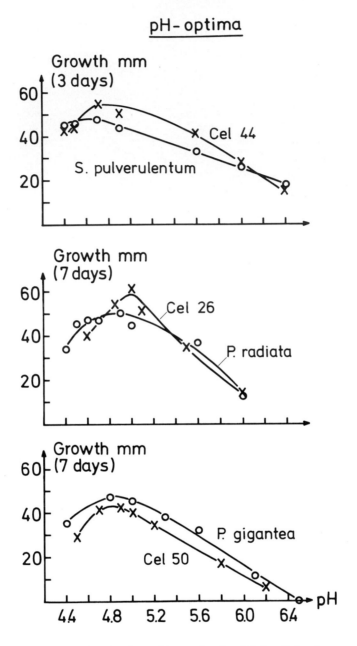

FIGURE 2.   pH-Optima for three white-rot fungi and their cellulase-less mutants grown on malt agar plates.

C. Growth Rate in Wood

All the strains, wild types as well as mutants, were examined for their growth rates in birch, pine, and spruce wood. Wood specimens, 20 and 60 mm in length, were inoculated on one of the cross-sectional faces and stored at 20°C. At different time intervals the opposite cross sectional face was pressed against the surface of a malt agar plate. The plates were incubated at 25°C, and by observing when growth first appeared on the plates, the time required for the different fungi to grow through the wood could be estimated.

By growing the fungi on malt agar plates, the growth rate on this medium was deter-

Temperature optima

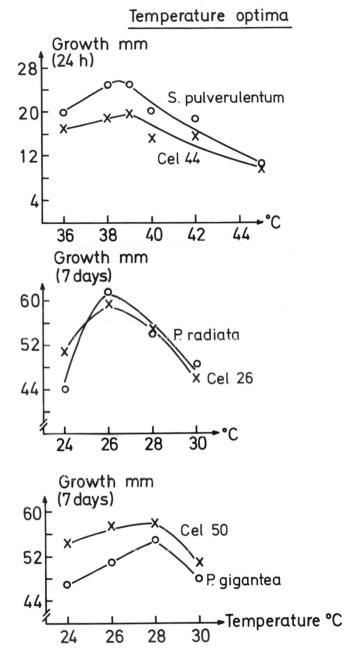

FIGURE 3.   Temperature optima for three white-rot fungi and their cellulase-less
mutants grown on malt agar plates.

mined. In 1 day, *S. pulverulentum* and Cel 44 grew 20 mm, whereas the other strains
required 3.5 to 4 days for the same amount of growth. Lag time is included in these
results. Maximal growth speed is obtained after 2 to 3 days. These results are approximately the same as those obtained for fungi growing in wood. In wood, however, the
mutants grow somewhat slower than the wild types. Rypáček and Navrátilová[8] earlier
found that *Polyporus (Coriolus) versicolor* grew in beech wood with the same speed
as on a malt agar plate.

C:N	urea/NH₄H₂PO₄ 0/100	10/90	20/80		Mutant	Pine urea/NH₄H₂PO₄	C:N	Spruce urea/NH₄H₂PO₄	C:N
160:1		x							
200:1	x	x	x		Cel 44 (S.p.)	10/90	160:1	10/90	200:1
250:1	x	x							
400:1		x							

C:N	10/90	20/80	25/75	30/70	Mutant	Pine urea/NH₄H₂PO₄	C:N	Spruce urea/NH₄H₂PO₄	C:N
160:1		x			Cel 26 (P.rad)	20/80	250:1	20/80	400:1
250:1		x							
400:1	x	x	x	x					
650:1		x			Cel 50 (P.gig)	20/80	250:1	25/75	400:1

FIGURE 4.   Optimal nitrogen sources for the three cellulase-less mutants grown on wood flour. The two small tables to the left in the figure show the C:N ratio studied. These ratios were obtained by the addition of different amounts of nitrogen to plates containing 1.5 g of wood flour.

The maximum growth rates on agar plates are the following:

- *S. pulverulentum* at 39°C, 45 mm/day; at 25°C, 25 mm/day.
- *P. gigantea* at 28°C, 10 mm/day.
- *P. radiata* at 26°C, 13 mm/day.

If these figures are valid also for wood it would mean that the growth rate hardly is the rate-limiting factor in the rotting process. To establish more reliable results, further growths studies of both wild types and mutants are being carried out in wood blocks.

D. Optimal Nitrogen Source

Optimal carbon/nitrogen (C/N) ratio as well as optimal source of nitrogen were evaluated for the three cellulase-less mutants. It was thereby determined under which nitrogen conditions the highest weight loss was obtained after three weeks' cultivation of the mutant strains at 25°C on Petri plates with pine and spruce wood flour as carbon source. The results are given in Figure 4. It can be seen that the ratio urea/NH₄H₂PO₄ varies from mutant to mutant and so does the C/N ratio. It seems to be a general rule that the lower the C/N ratio, the faster the growth rate.

III. MICROSCOPIC INVESTIGATION OF THE ATTACK ON WOOD*

A. Growth of *S. pulverulentum* and Cel 44 in Wood Cells

Scanning electron microscopy has been used to visualize fungal growth in wood. The pictures given in Figures 5 and 6 show that Cel 44 grows in the lumen of the vessels and tracheids. Branching of the fungal hyphae within the cells is common for both wild type and mutant fungi. In birch (Figure 5) the fungus grows from one vessel to another through the perforated plate. In pine (Figure 7), the fungus spreads into the ray cells not only via pores from the tracheids, but also from ray cell to ray cell. Penetration of fungal hyphae through cell walls, at least from wild types of white-rot fungi, has also been observed to take place (Figure 8).

---

* See also Volume II, Chapter 15.

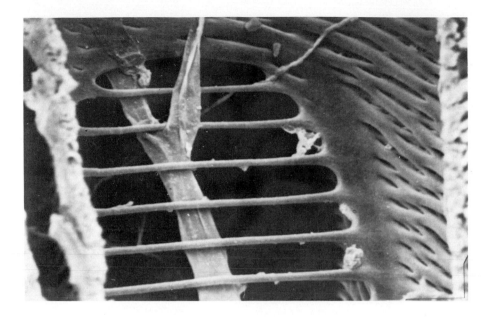

FIGURE 5. Cel 44 growing through the perforation plate between two birch vessels.

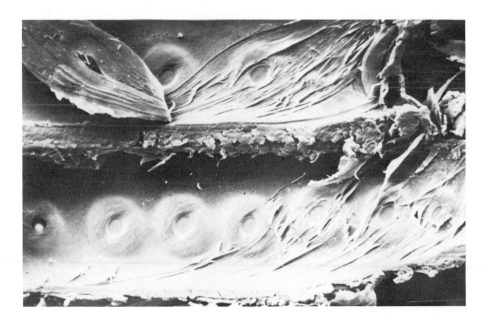

FIGURE 6. Cel 44 causing degradation in tracheids of pine.

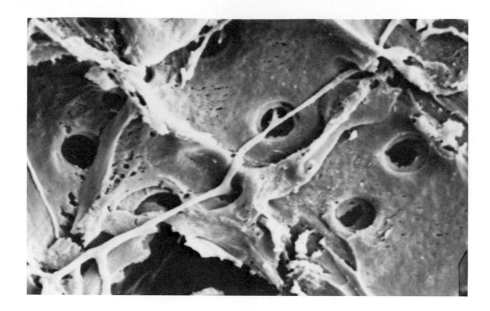

FIGURE 7.   Cel 44 growing through pores between ray cells and between ray cell and tracheid in pine.

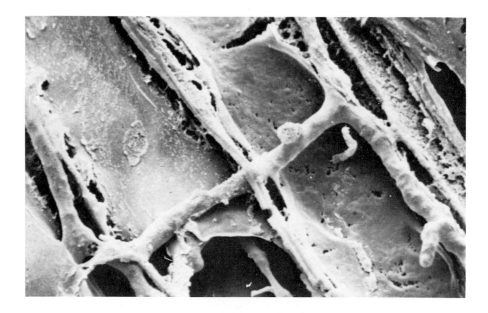

FIGURE 8.   Ray cell wall in spruce penetrated by *S. pulverulentum*.

B. Changes of the Lignin Content in Wood Cell Walls Attacked By *S. pulverulentum* and Cel 44

Sections of pine and spruce were taken from wood rotted for 7 weeks by either *S. pulverulentum* or Cel 44. The lignin content in the cell walls of the rotted wood was then determined by UV-microscopy. The extinction was measured at 280 nm at points 1 μm apart from the lumen of one cell, through the middle lamella, to the lumen of the neighboring cell.

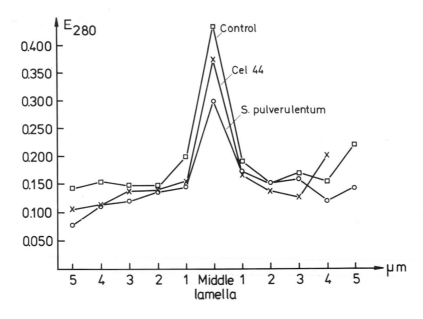

FIGURE 9.   Lignin content in pine cell wall measured by UV-microscopy at points 1 μm apart, from lumen of one cell, through the middle lamella, to the lumen of the neighboring cell. The points are the mean values of ten measurements. The pine wood has been rotted by *S. pulverulentum* and Cel 44 for 7 weeks.

The results given in Figure 9 show that the pretreatment of the wood with both *S. pulverulentum* and its cellulase-less mutant Cel 44 decreases the lignin content in the cell walls. The wild type has greater delignifying ability than the mutant. This observation is in full accordance with earlier observations.[2]

IV. BENCH COMPOSTING APPARATUS

A. Construction and Operation

The rotting of the wood chips takes place in four cylinders each with a volume of 20 ℓ. The flow rate of air and the temperature can be automatically regulated separately for every cylinder. At the beginning of an experiment the cylinders are filled to 75% capacity with wood chips, autoclaved, and inoculated with the fungus. Air of high humidity is continuously passed through the cylinders and analyzed after passage for its CO_2 content, which is automatically recorded. At the start of each experiment the cylinders are intermittently rotated to ensure proper mixing of chips and fungus. Experiments are usually run for 2 weeks.

B. Rotting Results

The first experiments with the bench-comlosting apparatus were devoted to the development of an improved inoculation technique. This has resulted in a much faster rotting. The experiments were first carried out with birch, which is the most rapidly degraded wood tested. The first experiments gave a CO_2 evolution of 10 ℓ in 12 days. This was later increased to 44 ℓ in the same time. Later, studies have been carried out mainly with pine wood. A CO_2 evolution of 66 ℓ in 2 weeks with pine chips have been recorded. These figures correspond to production of 0.2, 0.9, and 1.4 moles of CO_2/ kg of wood, respectively. In terms of weight losses, the figures are 0.4, 2.2, and 3.5%. With a continuously increasing knowledge of optimum conditions for rotting by the

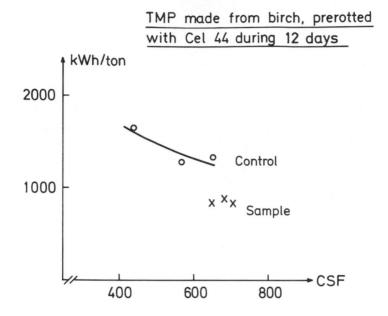

FIGURE 10.   Energy consumption for thermomechanical pulp production from birch, rotted for 12 days with Cel 44.

mutants, better knowledge of enzyme mechanisms involved in lignin degradation, and certainly genetic improvements of the mutant strains, faster degradation can safely be predicted.

C. Energy-Saving Results By Rotting in the Bench Composting Apparatus Prior to Refining

Birch chips were rotted for 12 days with Cel 44 to a weight loss of 2%. From these chips, thermomechanical pulp was made in two steps, pretreatment at 127°C for 3 minutes followed by refining in a 12-in. refiner. With this refiner, it was difficult to obtain a high enough energy transfer. Consequently, the pulp that was obtained was very coarse, 650 to 700 CSF (Canadian Standard Freeness). In Figure 10, the results from the pretreated and control birch chips are given. At 650 CSF, the energy consumption for the rotted sample is 850 kWh/ton compared to 1300 kWh/ton for the control. This suggests an energy saving by the pretreatment process of approximately 30%.

In an experiment with pine chips, it was necessary to have a power input as high as 11 to 12 kW to observe energy saving for the pretreatment process. Difficulties in obtaining a uniform feed rate of chips into the refiner caused the power consumption to fluctuate. This gave rise to a very high energy consumption and made it difficult to accurately determine the energy saving from the pretreatment process.

D. Pulp and Paper Qualities

Birch wood was used in the preliminary experiments to determine the optimal rotting conditions, since it is biologically degraded faster than either pine or spruce wood. However, mechanical pulp produced from birch wood gives rise to a paper with very low strength values. No difference in strengh values were obtained between rotted and nonrotted chips. The only observed difference was that paper sheets from rotted chips had a 3 to 4% lower brightness value.

Subsequent studies have mainly been carried out with pine wood chips. The strength

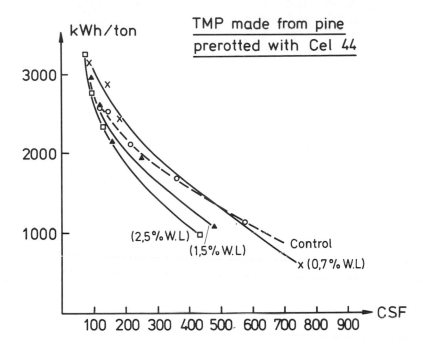

FIGURE 11. Energy consumption for thermomechanical pulp production from pine wood chips pretreated with Cel 44.

properties, tear and tensile indices were the same for pulp from rotted and nonrotted pine chips. The brightness (ISO) of pulp from rotted chips was, however, 7 to 8% lower. The light absorption coefficients were the same but the light scattering seemed to be somewhat higher.

The decreased brightness obtained with rotted chips may be a drawback. It is possible, however, that the biological degradation has introduced structures in the lignin which are more easily bleached. This possibility will be studied in the near future.

V. SCALING UP OF THE PRETREATMENT EQUIPMENT

In order to obtain more reliable data on the energy saving obtainable by a microbiological pretreatment of wood chips, the experimental equipment was scaled up and a pretreatment tank with a volume of 1.5 m³ was used.

In a preliminary experiment, 100 kg of pine chips (dry weight) were sterilized by steam in the pretreatment tank and then inoculated with a 10 -ℓ suspension of Cel 44. The pretreatment tank was aerated with 500 ℓ/h of sterile air. Samples (ca. 30 kg) were removed from the pretreatment tank at weight losses of 0.7, 1.5, and 2.5%. Thermomechanical pulp was produced in a two-stage refining process. The results are given in Figure 11. It can be seen that, as expected, the higher the weight loss, the lower the energy consumption. The energy saving is particularly obvious for the first-stage pulps (indicated by the point with the highest freeness value on each curve). At a freeness of 435 CSF the energy consumption is around 1000 kWh/ton, which is slightly more than 30% lower than for the control (approximately 1500 kWh/on). The energy saving determined from the second-stage pulp is, however, smaller. If the chips are refined to a freeness of 125 CSF, only a 10% energy saving is obtained with the biologically pretreated chips. It should, however, be pointed out that these results are from preliminary experiments using the large-scale pretreatment equipment. Experiments to optimize the reaction conditions, the inoculation technique, and the reactor design are under way.

VI. SUMMARY

About 10 days' pretreatment of wood chips by cellulase-less mutants of white-rot fungi significantly decreases the energy consumption in the production of mechanical pulp. Most of the work described here has been carried out with Cel 44, a cellulase-less mutant of the thermotolerant fungus *Sporotrichum pulverulentum*. This mutant operates optimally at 39°C and grows in wood at a speed of 22 mm/day. Pretreatment of wood chips has been carried out with this and other organisms in a bench-composting apparatus where about 10 kg of wood chips (dry weight) can be treated simultaneously. The pretreated chips have been processed to pulp in a laboratory refiner system and the energy input measured. These investigations indicate a saving of 30% in the energy demand as a result of the biological pretreatment. A scale-up of the pretreatment process to cubic meter scale is at present being undertaken to allow more accurate determination of different parameters of interest.

REFERENCES

1. **Ander, P. and Eriksson, K.-E.,** Lignin degradation and utilization by micro-oranisms, in *Progress in Industrial Microbiology*, Vol. 14, Bull, M. J., Ed., Elsevier, Amsterdam, 1978, 1.
2. **Ander, P. and Eriksson, K.-E.,** Influence of carbohydrates on lignin degradation by the white-rot fungus *Sporotrichum pulverulentum, Sven. Papperstidn.*, 78, 643, 1975.
3. **Ander, P. and Eriksson, K.-E.,** The importance of phenol oxidase activity in lignin degradation by the white-rot fungus *Sporotrichum pulverulentum, Arch. Microbiol.*, 109, 1, 1976.
4. **Hiroi, T. and Eriksson, K.-E.,** Microbiological degradation of lignin. I. Influence of cellulose on the degradation of lignins by the white-rot fungus *Pleurotus ostreatus, Sven. Papperstidn.*, 79, 157, 1976.
5. **Kirk, T. K. and Moore, W. E.,** Removing lignin in wood with white-rot fungi and digestability of resulting wood, *Wood Fiber*, 4(2), 72, 1972.
6. **Ander, P. and Eriksson, K.-E.,** Mekanisk massa från förrötad flis — en inledande undersökning (English summary), *Sven. Papperstidn.*, 78, 641, 1975.
7. **Eriksson, K.-E. and Goodell, E. W.,** Pleiotropic mutants of the wood-rotting fungus *Polyporus adustus* lacking cellulase, mannanase, and xylanase, *Can. J. Microbiol.*, 20, 371, 1974.
8. **Rypáček, V. and Navrátilová, Z.,** Das Wachstum der Pilze in Holz, *Drev. Vysk.*, 16, 115, 1971.

Chapter 15

ULTRASTRUCTURAL CHANGES IN THE WALL OF SPRUCE TRACHEIDS DEGRADED BY *SPOROTRICHUM PULVERULENTUM*

Katia Ruel and Fernand Barnoud

TABLE OF CONTENTS

I. INTRODUCTION

The subject of wood degradation by fungi of the Basidiomycete group (white-rot type or brown-rot type) has been explored during the last three decades by the different techniques of electron microscopy.[1-5] This research has revealed several interesting aspects on the way the hyphae progress into tracheids, fibers, vessels, and ray parenchyma and also on the images of lysis of wood substance. Important general articles covering this field were published in 1970[6] and 1973.[7] Unfortunately, in most of the cases the photomicrographs were presented with an enlargement inadequate to elucidate the changes at the macromolecular level in the wood cell wall.

Until recently, the conventional method of contrasting positively the different components of lignified walls have not permitted the exploration of the interrelations of the three main types of macromolecules: cellulose, hemicelluloses, and lignin. Using the negative contrast obtained with uranyl acetate, Heyn[8] showed, 15 years ago, that cellulose in the wall of pine tracheids is present in the form of separated individual elementary fibrils 3.5 nm in diameter, but nothing was known at this period about the relation of these fibrils to the lignin and hemicelluloses which together represent 50% of the wall. In 1975, Kerr and Goring[9] presented electron photomicrographs showing lignin distribution in the walls of spruce and fir which, in addition to other physical-chemical data, have led to a new concept of the entire structure of the wall of a conifer tracheid. The presence of interrupted lamellae of lignin with a tangential orientation is the most characteristic feature of the proposed structure.

More recently, Ruel et al.[10] have reexamined this concept and established with more evidence from data of scanning transmission electron microcopy that lignin in spruce and fir tracheids is actually distributed as interrupted lamellae between layers of cellulose fibrils and hemicelluloses. The average interval between the axis of two consecutive layers of lignin observed on transverse sections was found to be 7.1 nm in spruce and 8.4 nm in fir. The arrangement of cellulose proto-fibrils and the interrelations between the lignin and the hemicelluloses are at the level of the "unit structure." These subjects require further investigations.

Applying the same method of observation to hardwood fibers and graminae fibers, Ruel et al.[11] have also shown, using conventional transmission electron microscopy (E.M.), that the same basic features are present in these lignocellulosic fibers, which are quite different in many aspects from conifer tracheids. The authors thought that a series of observations on lignified walls attacked by a white-rot fungi could provide further information concerning the way the lignin is morphologically degraded in the parts of the wall digested by the hyphae.

For this study, samples of spruce infected by *Sporotrichum pulverulentum* (= *Phanerochaete chrysosporium*) were kindly provided by Dr. K. E. Eriksson. Both the wild type and the mutant cellulase-less Cel 44 were studied, the latter being particularly interesting in the fact that cellulose is not destroyed, whereas the lignin and the hemicelluloses, on the contrary are degraded by both strains[12] (see Volume II, Chapter 1, 14).

An examination at the highest enlargement possible of the only parts of the walls degraded by the action of the fungus was then begun. Most observations were made after fixing the samples by the permanganate method according to Kerr and Goring[9] and so only the lignin was contrasted by deposition of MnO_2 grains. To examine the polysaccharide moiety of the wall, the method of periodic acid-thiosemicarbazide-silver proteinate PATAg[13]) was used, and also uranyl acetate according to Cox and Juniper.[14] Tissues were embedded in Epon-MNA 1:1. The ultra-thin sections were obtained with an LKB-Ultratome III® equipped with a diamond knife. The electron microscope used was a Philips 300 working at a tension of 80 kV. Most of our observations were made on spruce samples incubated during 8 weeks in presence of *S. pulverulentum* wild-type or Cel 44. The data presented here have been established by the examination of more than 500 E.M. negatives. At the point of these investigations, no important differences have been noticed between the two strains of *S. pulverulentum*.

II. RESULTS AND DISCUSSION

The characteristic structure of the S2 and S3 layers of spruce sound wood is given by the Figure 1A*, which represents a transverse section. The wall appears as a compact system of thin and parallel lamellae visualized by MnO_2 (or Mn_2O_3) grains. The selective fixation of manganese oxides must be related to the reactive phenolic groups of the lignin molecule and also to carboxylic groups generated by permanganate oxidation of the α-carbon of the sidechain of the phenylpropane units. The small size of the grains (35Å) provides one of the best electron dense elements which can be used for contrasting lignin at the present time. The validity of this method has been discussed in several papers.[9,15-17] In the S2 and S3 layers, the stratae have a uniform thickness. It must be recalled that, in these conditions of observation, the fibrils of cellulose which are cut transversely are not visualized. If they were, it would be by an effect of negative contrast.

In Figure 1B the photomicrograph shows some aspects of heavily degraded zone. The S2 layer is strongly disorganized. It can be seen that the parallel rows of lignin lamellae are progressively disrupted up to a point at which they disappear completely. Nevertheless, in these areas of the most severe attack, the lignin is still visible as dark spots of 50 to 60 Å.

The swelling and disorganizing effect of the enzymes secreted by the fungi are clearly seen in Figure 1B. The disorganization of the compact system of lamellae can be interpreted as a result of the breakdown of the hemicelluloses. Several chemical data on the action of *S. pulverulentum*.[12,18] have shown that glucomannans, which are the most accessible part of the wall (noncrystalline), are degraded much faster than the fibrillar cellulose. This is also the meaning of several of our E.M. photomicrographs, which show that the cellulase-less mutant Cel 44 is very active in degrading the walls. Furthermore, when using a method of contrasting the carbohydrate moiety of the wall, such as the PAT-Ag system, we have observed in wood samples decayed by Cel 44 (Figures 2A and 2B) that the reactivity of the wall is much greater, an observation which shows that the hemicelluloses have been degraded. Other data obtained by Ruel and Barnoud[19,20] on reed *(Arundo donax)* fibers, have established that enzymatic solubilization of hemicelluloses is responsible for a deep physical modification of compact cellulosic walls.

Another set of data is presented in Figures 3A and 3B. In the S1 layer shown in Figure 3A, there is an image of the swelling effect of the diffused enzymes (probably polysaccharidases). The size of the zones of progressive disorganization by swelling and rupturing the compact system of lamellae as measured at low magnification, can be estimated at 2 to 3 μ. Figure 3B shows other aspects of lysis. In that situation there is an almost complete dissolution of the S1 layer at the point of contact of a hypha. In these areas, the large dark granules around the hypha and in contact with the S1 and S2 layers might correspond to aggregates of enzymes of a glycoproteic nature.

Longitudinal sections (Figure 4) in the zones of lysis are also of interest and show that the degradation follows the orientation of the cellulose microfibrils in the S2 layer. Reference and degraded wall are visible on the same figure, and the lamellar longitu-

---

* All figures appear at the end of the text.

dinal arrangement of lignin in the right side contrasts with the deep disorganization of the portions of the secondary wall which are still visible near the S1 layer, and where the lamellae of lignin are disrupted and finally reduced to dark spots of 70 to 80 Å in diameter.

III. CONCLUSION

The main ultrastructural changes caused by the enzymatic degradation of the wall components during the growth of *S. pulverulentum* (wild type and cellulase-less mutant Cel 44) on spruce wood were established by observations on ultra-thin sections of wood (25 nm) in which the lignin moiety was contrasted with potassium permanganate. As observed previously by many authors,[6,7,20] the diffusion of the enzymes is responsible for the disorganization of the secondary wall mainly in the S1 layer or in the near part of the S2 layer and also, but less frequently, in the middle lamella. We have observed that the compact and uniform system of tangential lamellae of the lignin hemicelluloses-cellulose complex, which is structured on a repetitive basis of 71Å, is progressively disorganized.

The most characteristic features of this disorganization that have been established by observations at high magnifications are an enlargement of the carbohydrate layers and a consecutive rupturing of the compact and well-ordered system of lamellae. The swelling observed in the spruce tracheids attacked by the cellulase-less mutant Cel 44 provides experimental evidence that the breakdown of the hemicellulose part of the wall (glucomannans) is the first step of the action of the hyphae invading the lignified walls.

In the swollen and disorganized region, the lignin is still stainable with manganese oxides, but the lamellae are resolved into small granules of 70—80Å in the part of the wall where the lysis is particularly advanced. These aspects of wall lysis are not necessarily observed in the immediate vicinity of hyphae or microhyphae. All our observations tend to establish that lignin breakdown is not truly a contact phenomenon, but this question needs further investigation.

These E. M. investigations have shown that the processes of lignin degradation can be followed at the ultrastructural level. Using fungi more active than *S. pulverulentum* on the lignin part of wood,[18] this type of *in situ* observation at the level of the "unit structure" of wood can be a complementary method for a comprehensive view of lignin degradation under biological conditions.

ACKNOWLEDGMENTS

The authors are indebted to Dr. K. E. Eriksson, Swedish Forest Products Research Laboratory (Stockholm), for providing samples of wood decayed by *S. pulverulentum* wild-type and mutant Cel 44. Thanks are also addressed to Dr. H. Chanzy, Centre de Recherches sur les Macromolécules Végétales, (Grenoble, France) and Dr. D. A. I. Goring, Pulp and Paper Research Institute of Canada, (Pointe Claire, Quebec), for helpful criticism in the interpretation of electron photomicrographs.

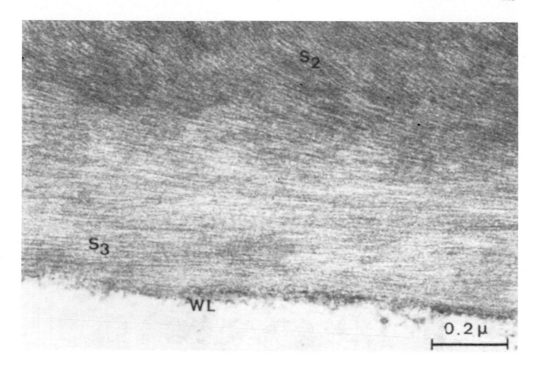

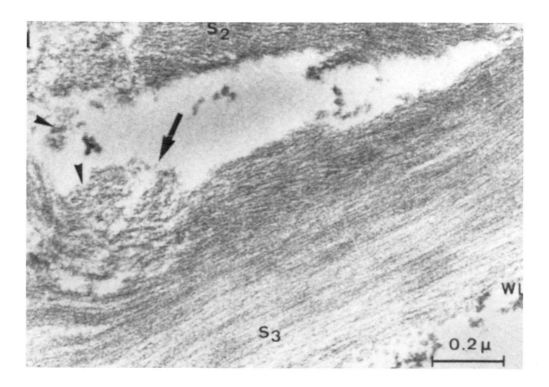

FIGURE 1. Symbols: S_2, S_3 = middle and inner layers, respectively, of secondary wall; WL = warty layer. (A) Transverse section of fiber wall of sound spruce wood treated with $KMnO_4$. The orientation of lamellae is different in the S_2 and S_3 layers, and is similar to that established for cellulose microfibrils. (Magnification × 92,000.) (B) Transverse section of fiber wall decayed by Cel 44. The interlamellar spaces are enlarged, and parallel rows of lignin lamellae are progressively disrupted (arrow) and finally resolved into small nodules of about 50 Å in diameter (arrowheads). (Magnification × 91,000.)

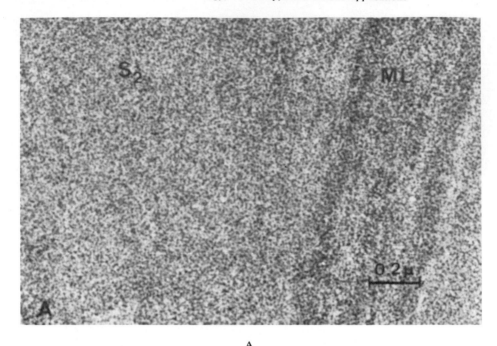

A

B

FIGURE 2.   Symbols: ML = middle lamella; H = hypha; S₂, S₃ = middle and inner layers, respectively, of secondary wall; WL = warty layer. (A) Transverse section of sound fiber wall after staining with silver proteinate (PATAg). After oxidation by periodic acid, the wall is uniformly contrasted by silver grains. The middle lamella region is more densely stained than the S₂ layer. (Magnification × 72,000.) (B) Transverse section of a fiber wall after Cel 44 degradation. The reactivity of the wall for silver proteinate is higher than that of the control. Arrows indicate regions of loss of substance through fungal degradation. (Magnification × 72,000.)

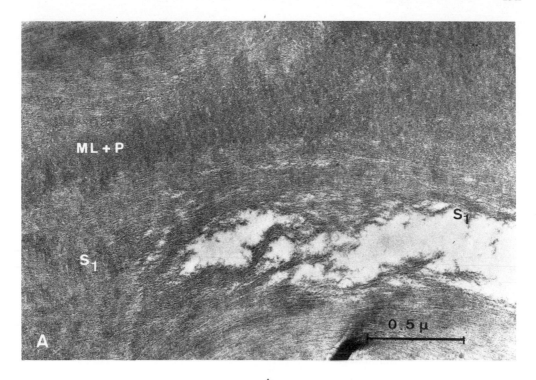

A

B

FIGURE 3.   Symbols: P = primary wall; S_1 = outer layer of secondary wall; S_2, S_3 = middle and inner layers, respectively, of secondary wall; WL = warty layer. (A) Transverse section of cell wall degraded by Cel 44 and stained with $KMnO_4$. Into the S_1 layer there is a progressive disruption of the lamellae due to an enlargement of the interlamellar space. (Magnification × 57,000.) (B) Transverse section of a portion of fiber wall invaded by hypha of Cel 44. Middle lamella and primary wall are completely digested by the fungus. The entire wall is broken (arrows). (Magnification × 57,000.)

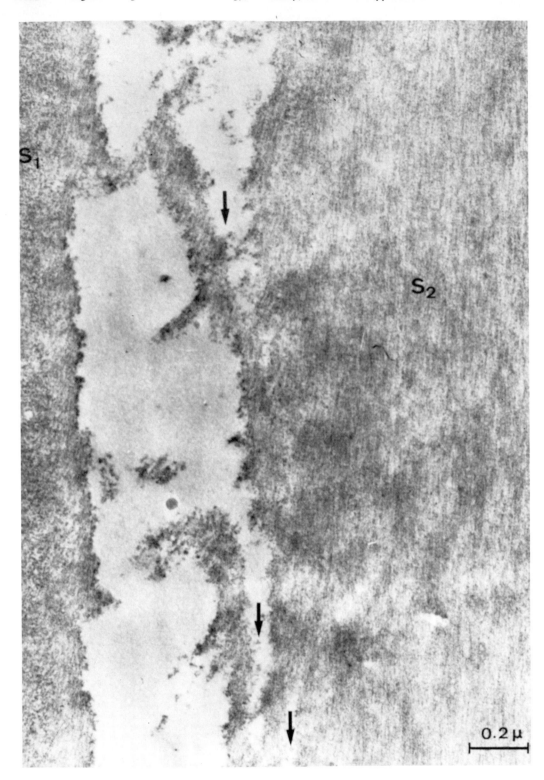

FIGURE 4. S_2, S_3 = middle and inner layers, respectively, of secondary wall; WL = warty layer. Longitudinal section of tracheid wall degraded by Cel 44 and stained with $KMnO_4$. The S_2 layer shows the longitudinal aspect of degradation. Lamellae are interrupted and resolved into clear areas which still contain electron-opaque granulations identical to those observed in transverse section. (Magnification × 69,000.)

REFERENCES

1. **Liese, W. and Schmid, R.** Submicroscopical changes of cell wall structures by wood-destroying fungi, in *Fifth International Congress for Electron Microscopy,* Academic Press, New York, 1962.

2. **Levi, M. P. and Preston, R. D.,** A chemical and microscopic examination of the action of the soft-rot fungus *Chaetomium globosum* on beechwood (Fagus Sylvatica), *Holzforshung,* 19, 183, 1965.

3. **Sachs, I. B., Nair, V. M. G., and Kuntz, J. E.,** Penetration and degradation of cell walls in oaks infected with *Ceratocystis Fagacearum, Phytopathology,* 60(9), 1399, 1970.

4. **Chou, C. K. and Levi, M. P.,** An electron microscopical study of the penetration and decomposition of tracheid walls of *Pinus Sylvestris* by *Poria Vaillantii, Holzforschung,* 25, 107, 1971.

5. **Jutte, S. M.,** Some examples of the interrelation between wood structure and the mode of attack of microorganisms and cellulase preparations, *Acta Bot. Neerl.,* 22(4), 360, 1973.

6. **Liese, W.,** Ultrastructural aspects of woody tissue disintegration, *Annu. Rev. Phytopathol.,* 8, 231, 1970.

7. **Wilcox, W. W.,** Degradation in relation to wood structure, in *Wood Deterioration and Its Prevention by Preservative Treatments,* Vol. I. Nicholas, D. D., Ed., Syracuse University Press, Syracuse, New York, 1973, 107.

8. **Heyn, A. N. J.,** The elementary fibril and supermolecular structure of cellulose in soft wood fiber, *J. Ultrastruct. Res.,* 26, 52, 1969.

9. **Kerr, A. J. and Goring, D. A. I.,** The ultrastructural arrangement of the wood cell wall, *Cellul. Chem. Technol.,* 9, 563, 1975.

10. **Ruel, K., Barnoud, F., and Goring, D. A. I.,** Lamellation in the S2 layer of softwood tracheids as demonstrated by scanning transmission electron microscopy, *Wood Sci. and Technol.,* 12, 287, 1978.

11. **Ruel, K., Goring, D. A. I., Barnoud, F.,** Ultrastructural aspect of hardwood and graminae fibers in relation with lignin. In preparation.

12. **Ander, P. and Eriksson, K. E.,** Influence of carbohydrates on lignin degradation by the white-rot fungus *Sporotrichum Pulverulentum, Sven. Papperstidn.,* 78, 643, 1975a.

13. **Seligman, A. M., Hanker, J. S., Wasserkrug, H., Dmochowsky, H., and Katzoff, L.,** Histochemical demonstration of some oxidized macromolecules with thiocarbohydrazide (T.C.H.) or thiosemicarbazide (T.S.H.) and osmium tetroxyde, *J. Histochem. Cytochem.,* 13, 629, 1965.

14. **Cox, G. and Juniper, B.,** Electron microscopy of cellulose in entire tissue, *J. Microsc. (Oxford),* 97, 343, 1973.

15. **Hepler, P. K., Foskett, D. E., and Newcomb, E. H.,** Lignification during cell wall formation in *Coleus:* an electron microscopic study, Am. J. Bot., 57, 85, 1970.

16. **Bland, D. E., Foster, R. C., and Logan, A. F.,** The mechanism of permanganate and osmium tetroxide fixation and the distribution of lignin in the cell wall of *Pinus radiata, Holzforschung,* 25, 137, 1971.

17. **Hoffmann, P. and Parameswaran, N.,** On the ultrastructural localization of hemicelluloses within dilignified tracheids of spruce, *Holzforschung,* 30, 62, 1976.

18. **Ander, P. and Eriksson, K. E.,** Selective degradation of wood components by white rot fungi, *Physiol. Plant.,* 41, 239, 1977.

19. **Ruel, K., Comtat, J., and Barnoud, F.,** Localisation histologique et ultrastructurale des xylanes dans les parois primaires des tissus d' *Arundo donax, C. R. Acad. Sci. Paris., Ser. D,* t.284, 1421, 1977.

20. **Ruel, K., Joseleau, J. P., Comtat, J., and Barnoud, F.,** Ultrastructural localization of xylans in the developing cell wall of graminae fibers by the use of an endoxylanase, *Journal Applied Polymer Symposium,* John Wiley & Sons, New York, 28, 1976, 971.

21. **Cowling, E. B.,** Microorganisms and microbial enzyme systems as selective tool in wood anatomy, in *Cellular Ultrastructure of Woody Plants,* Cote, W. A. Jr., Ed., Syracuse University Press, New York, 341, 1964.

Chapter 16

LIGNIN BIODEGRADATION: SUMMARY AND PERSPECTIVES

T. Kent Kirk, Takayoshi Higuchi, and Hou-min Chang

TABLE OF CONTENTS

I. INTRODUCTION

Studies of lignin biodegradation have been conducted for many years,[1-5] but in the past 4 years or so research has accelerated markedly. Indeed, approximately half of the citations in this book, of which there are about 1000, are to papers published in 1974 or later. As mentioned in the Foreword, the increased research effort can be traced to establishment of a firm understanding of the basic chemistry of lignin (Chapter 1, Volume I), and to increased research support spurred by hopes of eventually using ligninolytic systems for man's betterment. This book summarizes the progress that has resulted. Good theoretical, experimental, and methodological foundations have been or are being built that will provide the framework for further in-depth research. In this final chapter, an attempt is made to place past research into perspective in condensed form, based primarily on the preceding chapters, but also on the discussions that took place at the Madison seminar, and, inevitably, on the authors own experiences and prejudices.

II. MICROBIOLOGY

A. Methodology

The recent development of ^{14}C-lignin methodologies (Volume I, Chapter 3,) has made it possible to begin to assess the roles of various microbial taxa in lignin biodegradation. Many earlier studies, unfortunately, must be viewed with skepticism. ^{14}C-lignins have also been applied via the most probable number technique to estimate the numbers of ligninolytic propagules in various environments (Volume I, Chapter 3). The labeled lignins can also be of use in screening isolated organisms for ability to degrade lignin (Volume I, Chapter 3; Volume II, Chapter 7), although better methods are needed for primary screening (Volume I, Chapter 7).

Despite their usefulness, biodegradation assays based on ^{14}C-lignins must be used cautiously (Volume I, Chapter 3). The researcher must prepare his lignins carefully and establish their integrity. A few percent conversion to $^{14}CO_2$ by a microbe might represent degradation only of a low molecular weight fraction, or even of a nonlignin contaminant in the case of natural lignin prepared by feeding precursors to plants. Although relatively resistant to chemical and biological degradation, lignin is not an unusually stable polymer. Long-term incubations in cultures can lead to a certain amount of nonbiological fragmentation, producing low molecular weight materials which could be readily metabolized. Nonbiological polymerization can also occur. For these reasons, molecular weight distribution, used with radiolabeled as well as nonlabeled lignins to evidence biodegradation, certainly must be used with caution, and appropriate noninoculated controls included.

B. Bacteria

Substantial effort has been expended in the past few years to determine the role of bacteria in lignin biodegradation. That bacteria degrade a wide array of low molecular weight, lignin-related aromatic compounds is clear (Volume I, Chapters 2, 6, and 8; Volume II, Chapters 6, 8, 9, and 12); some studies have demonstrated facile bacterial metabolism of lignin-related "dimers" of molecular weight up to about 350 (Volume I, Chapter 8; Volume II, Chapters 6 and 8). The probable fates of the C_1 units resulting from demethylation or demethoxylation of aryl methoxyl groups has been discussed by Kuwahara (Volume II, Chapter 9). Sundman (Volume II, Chapter 12) has presented evidence that plasmid-encoded pathways (see also Volume I, Chapter 2) are important in the bacterial degradation of lignin-related phenols.

Bacterial degradation of the lignin macromolecule is more difficult to assess than is degradation of low molecular weight aromatics, as pointed out in Volume I, Chapter 6. Methodological difficulties cloud interpretation of most of the older studies purporting to show bacterial degradation. Nevertheless, in the past 2 to 3 years, convincing data have been obtained with radiolabeled lignins, and it is now apparent that bacteria, including actinomycetes, can degrade lignin to some extent (Volume I, Chapters 3 and 6; Volume II, Chapters 7 and 8). Rapid bacterial degradation (in comparison to white-rot fungi) has not yet been observed, but studies with adequate techniques are only beginning.

Recent work with ^{14}C methodology has failed to support earlier reports[5] of anaerobic metabolism of lignin, although metabolism of single-ring aromatics related to lignin in O_2-free environments has been established (Volume I, Chapter 5). It would be of interest to determine the molecular size limits of the anaerobic breakdown of lignin-related aromatics, and also whether anaerobes can modify the lignin polymer at all.

C. Yeasts and Imperfect Fungi

Degradation of lignins by yeasts has not been demonstrated, nor has it been seriously studied. It is interesting to note that Iwahara (Volume I, Chapter 8) isolated several yeasts capable of growing on dilignols. This should be pursued.

Imperfect fungi (so classified because their sexual stages have been lost in evolution or have not yet been described), are actually too heterogeneous a group to be treated together. At least one white-rot fungus, *Phanerochaete chrysosporium* (Basidiomycete), was classified until recently as an imperfect species (as *Sporotrichum pulverulentum, S. pruinosum,* and *Chrysosporium lignorum.*)[6] Nevertheless, various imperfect fungi such as in the genera *Fusarium, Penicillium,* and *Aspergillus* are so ubiquitous in soils and plant residues that their possible role in lignin biodegradation is an important question. At the present time, the roles of the three genera mentioned and the myriad of other imperfect fungi inhabiting lignin-rich environments are not known. Cain has described his extensive studies of aromatic catabolism by these fungi (Volume I, Chapter 2), and Iwahara (Volume I, Chapter 8) and Higuchi (Volume I, Chapter 9) have shown that several fusaria including *Fusarium solani* can grow on dilignols, and degrade at least the lower molecular weight portions of dehydrogenation polymers (DHPs) and a milled wood lignin (MWL). On the other hand, Rosenberg and Wilke (Volume II, Chapter 13) observed no lignin degradation in newsprint by a number of imperfect fungi. The fungi and bacteria inhabiting soils and plant residues are certain to be of importance as lignin modifiers even if they cannot efficiently convert the macromolecule to CO_2 in axenic culture. This needs further study.

D. Soft-Rot and Brown-Rot Wood Decay Fungi

Although materials balance studies have indicated that soft-rot fungi such as certain *Chaetomium* sp. can degrade lignin, the report by Haider (Volume I, Chapter 6) is the first demonstrating substantial degradation to CO_2. This group also needs further study, and perhaps offers strains suitable for detailed chemical and biochemical investigation and, perhaps, utilization.

It has been known for years that brown-rot fungi cause some alterations in lignin, primarily demethylation and limited oxidation,[5] as they decompose wood. Thus the degradation of ^{14}C-lignins to $^{14}CO_2$ reported by Haider and Trojanowski: (Volume I, Chapter 6) was unexpected. This report raises a number of questions: (1) why lignin degradation during brown-rot of wood is apparently limited to demethylation and some slight oxidations; (2) what the physiological basis is for the very marked differences between white- and brown-rot fungi in their effects on wood;[5] (3) whether brown- and white-rot fungi in fact have similar effects on lignin; and (4) whether the suppressed (?) oxidative effects of brown-rot fungi on lignin in wood are related to the very destructive oxidative[7] degradation of the cellulose in wood. Even partial answers to these questions can lead not only to a better understanding of lignin biodegradation, but also to a more rational approach to the control of these most economically important of all destroyers of wood products.

E. Mixed Microflora

Microbes rarely act alone in nature. In the case of lignin biodegradation, certain organisms in a mixed population no doubt cause limited degradation, leaving a modified residue that can then be attacked by other members of the population, and so on to CO_2 and/or humus. The formidable microbiology and methodology involved in studying such interactions presents perhaps the greatest experimental challenge in lignin biodegradation research. Martin and Haider (Volume I, Chapter 4) have pioneered in documenting the fate of lignin and lignin-related compounds in soils, as well as in assessing the role of lignin, polysaccharides, and microbially-synthesized material in the formation of humic materials.

F. White-Rot Fungi

In contradiction to the statement made above, white-rot fungi frequently can be found as the sole organisms decaying a section of wood in nature. Thus it is not surprising that these fungi are found in axenic culture to degrade lignin to CO_2 more efficiently than pure cultures of other organisms unaccustomed to solitary confinement. Nor is it surprising that the white-rot fungi have been the most studied lignin-degraders, for their ability to decompose lignin in wood has been apparent for half a century. They are clearly of great importance in cycling the carbon of lignin in nature. Despite the fact that they apparently can act singly, white-rot fungi may act synergistically with brown-rot or other white-rot fungi in degrading lignin (Volume II, Chapter 12)— another point needing further study.

All this notwithstanding, only a small percentage of known white-rot fungi has been studied, and it is unlikely that scientists have happened to choose for study the best that nature has to offer. Attempts are under way to isolate and identify more potently ligninolytic strains and species, and some success has been reported by Setliff and Eudy (Volume I, Chapter 7).

Until very recently, research on lignin metabolism by white-rot fungi had concentrated on the chemistry of the process. Two approaches have been used: (1) characterization of white-rotted lignins; and primarily (2) elucidation of the effects of the fungi on lignin-related aromatics. These approaches have provided most of what is known about the chemistry of lignin metabolism (see Section III), and continue to be very useful (Volume I, Chapters 9 and 11; Volume II, Chapters 1, 3, and 6).

With the recent development of ^{14}C-methodology (Volume I, Chapter 3), other aspects of lignin metabolism by white-rot fungi are now being studied. For the first time, the culture parameters favoring lignin metabolism -- which are rather exacting -- have been elucidated (for *Phanerocheate chrysosporium*) (Volume II, Chapter 4). It is now possible to achieve relatively rapid lignin metabolism in a simple defined medium, and to assess ligninolytic activity (^{14}C-lignin $\rightarrow$ $^{14}CO_2$) of cultures in a matter of hours (Volume II, Chapter 4). Physiological studies with optimized cultures have already provided some unexpected findings: ligninolytic activity appears in cultures irrespective of the presence of lignin, and is not induced by lignin; also, activity appears in response to nutrient nitrogen starvation, in what can be viewed broadly as a "secondary metabolic" event (Volume II, Chapter 4).

Failure to achieve primary growth and lignin catabolism simultaneously probably explains in part why at least some white-rot fungi do not use lignin as carbon source (Volume II, Chapter 4). This needs further investigation, however; the white-rot fungus *Poria subacida* reportedly has been cultured on Brauns' lignin (Volume II, Chapter 6), and several fusaria were isolated for ability to grow on DHP (Volume I, Chapter 8). That various fungi grow at the expense of *p*-hydroxybenzoic acid and related compounds has been shown by Cain (Volume I, Chapter 2), although known ligninolytic fungi have not been studied in this regard.

Genetic investigations (Volume II, Chapters 1 and 5) have been initiated (in *Phanerochaete chrysosporium),* with development of mutant selection techniques, isolation of some auxotrophic mutants, and description of techniques for obtaining the sexual fruiting stage. The genetic approaches promise to add a very powerful dimension to research on the chemistry, biochemistry, and physiology (and eventually to the application) of ligninolytic systems. Already genetic studies have indicated a need for phenol-oxidizing enzymes in lignin metabolism by white-rot fungi (Volume II, Chapters 1 and 5).

III. CHEMISTRY AND BIOCHEMISTRY

A. Chemistry

Studies of isolated and purified white-rotted lignin polymers have shown the degradative process to be oxidative (Volume I, Chapter 11; Volume II, Chapter 3). The degraded polymers contain α-carbonyl groups and carboxyl groups — both aromatic and aliphatic. Further details of structural changes, deduced indirectly from chemical, spectroscopic, and analytical data, disclosed a nearly complete absence of arylglycerol-β-aryl ether substructures, alteration of phenylcoumaran substructures, and oxidative cleavage of aromatic nuclei in the polymer. Research is under way to attempt to identify specific structures in the degraded lignins (Volume I, Chapter 11).

Shimada (Volume I, Chapter 10) has focused attention on the importance of reconciling the racemic nature of the asymmetric carbons in all of the various lignin substructures with the suspected and probable stereospecificity of the degradative enzymes. He has pointed out that α carbinol oxidation in sidechains (known to occur, Volume I, Chapter 11) destroys one asymmetric center (at C_α) and that enolization between the α- and β-carbons in such oxidized sidechains destroys a second center (at C_β). Treatment of $^3H/^{14}C$-double-labeled DHPs with the phenol-oxidizing enzyme laccase resulted in a substantial and equal loss of ^3H from both C_α and C_β (Volume I, Chapter 10), (Laccase is known to oxidize α-carbinol groups in units containing free phenolic hydroxyl groups).

Higuchi and Shimada (Volume I, Chapters 9 and 10) have presented an attractive hypothesis of oxygenative cleavage between the C_α- and C_β-carbons in internal sidechains. They suggested that this oxidation would take place in sidechains having α-carbonyl groups in arylgylcerol-β-aryl ether substructures, and a double bond between C_α and C_β in phenylcoumaran substructures (the latter the result of dehydration). Higuchi (Volume I, Chapter 9) has isolated and identified a compound, 5-acetylvanillyl alcohol, from among the *F. solani* degradation products of the phenylcoumaran dilignol dehydrodiconiferyl alcohol. The origin of this compound is suspected to be via oxygenolytic cleavage between C_α and C_β in the sidechain; this observation thus provides the first experimental support for such cleavage.

The new hypothesis is most attractive in its simplicity and provides a clear rationale for experimental testing, because the degradation products can be predicted, and would be of low molecular weight (Volume I, Chapter 9). The major predicted product would be vanillic acid, making the study of catabolism of this compound by white-rot fungi (Volume II, Chapter 1) of considerable relevance. Use of isotope dilution techniques with ^{14}C-lignins will be a powerful method for assessing the importance of various suspected intermediates of degradation.

It does not seem possible to explain the various characterizing data for white-rotted lignins (Volume I, Chapter 11) solely on the basis of the above hypothetical oxygenative sidechain cleavages. As pointed out in Volume I, Chapter 11), characterization of white-rotted ligin indicates the presence of ring cleavage fragments. Thus, one of the fundamental questions concerning the chemistry of lignin metabolism by white-rot fungi is whether (1) the lignin is significantly depolymerized to low molecular weight aromatics (Volume I, Chapters 9 and 10), (2) aromatic ring cleavage occurs in the polymer (Volume I, Chapter 11), or (3) both modes (1) and (2) are operative. Study of the metabolism of lignin-related aromatics bound retrievably to biologically inert soluble polymers[8] offers one approach to answering this question, as does the search for ring cleavage fragments in white-rotted lignin (mentioned in Volume I, Chapter 11).

Relatively much experimental and speculative attention has been directed at how the arylglycerol-β-aryl ether linkages are cleaved during lignin metabolism, because these are the single most abundant type of interunit linkage. Fukuzumi (Volume II, Chapter 11) has reported a β-ether-cleaving enzyme which requires reduced glutathione and NADH. The crude enzyme released guaiacol and guaiacylglycerol from guaiacylglycerol-β-(o-methoxyphenyl) ether (= "-β-guaiacyl" ether). Fukuzumi suggested that the guaiacylglycerol is formed by a reduction of the expected monooxygenase cleavage product, guaiacyldihydroxyacetone, by a reductase in the crude enzyme preparation. Further study of this system is certainly needed.

Fusarium solani converts guaiacylglycerol-β-coniferyl ether to guaiacylglycerol-β-vanillin ether during growth on the former, but the mechanism of subsequent β-ether cleavage has not yet been elucidated (Volume I, Chapter 9). Bacterial metabolism of β-aryl ether models has also been studied; β-hydroxypropiovanillone and coniferyl alcohol have been isolated as products from guaiacylglycerol-β-coniferyl ether (Volume II, Chapter 11) and guaiacol from guaiacylglycerol-β-guaiacyl ether.[9] Here, too, however, the chemistry of cleavage has not been determined.

Fukuzumi has also studied bacterial metabolism of several other lignin-related "dimers" and identified some of the intermediates. Such studies should be pursued, with ligninolytic fungi as well as bacteria.

However, investigations of the metabolism of low molecular weight compounds cannot automatically be extrapolated to the lignin polymer. The possibility must be entertained that different pathways for certain substructures exist inside and outside of cells, even though the underlying principles and chemical constraints governing microbial aromatic catabolism (Volume I, Chapter 2) will obtain in both cases.

B. Biochemistry

The enzymes participating directly in lignin metabolism have yet to be identified. Attempts to detect cell-free activity (Volume II, Chapters 3 and 13) have not been successful. Perhaps one reason for this is that contact between hyphae and lignin is required; electron microscopic studies suggest that this is the case (Volume II, Chapter 15). Enzymological studies per se will of course be more likely to succeed following identification of specific catabolic reactions, and development of assays for each.

The role of phenol-oxidizing enzymes in lignin biodegradation has always been a favorite research subject,[2-5] and will continue to be now that genetic studies (Volume II, Chapters 1 and 5) have apparently established their necessity. Their role is still unclear. Ander and Eriksson[10] have suggested that phenol-oxidizing enzymes play some kind of regulatory role (as they do in sexual fruiting of certain Basidiomycetes). A more direct role is also advocated. Effects on lignin include polymerization, limited demethoxylation, α-carbinol oxidation, and limited sidechain eliminations (Volume I,

Chapter 10; Volume II, Chapter 2). As far as is known, the enzymes act only indirectly through single electron oxidation of phenolic hydroxyl groups. As mentioned above, Shimada (Volume I, Chapter 10) has proposed an important central role for phenol-oxidizing enzymes in oxidizing α-carbinols in sidechains.

Cellobiose-quinone oxidoreductase is considered by Eriksson (Volume II, Chapter 1) to be important in lignin metabolism by white-rot fungi. The exact function of this intriguing activity, as that of phenol-oxidizing enzymes, however, remains a mystery.

Cain (Volume I, Chapter 2) has provided an excellent perspective and overview of current understanding of the biochemistry of the microbial catabolism of low molecular weight, lignin-related aromatics. Much of the reported work with fungi is from his laboratory and previously unpublished. Cain has emphasized the importance of the convergence of catabolic pathways on common intermediates. Metabolism of these intermediates, particularly protocatechuic acid (PCA) and catechol, has received the bulk of research attention. Many fascinating differences between microbial taxa, particularly between bacteria and fungi, have come to light. For example fungi have no meta cleavage pathway (which is apparently plasmid-encoded), and, without exception, fungi and bacteria differ in the cyclization product of 3-carboxy-*cis,cis*-muconate in PCA metabolism. Much has already been learned about the regulation of aromatic catabolic pathways in microorganisms, and again the differences between bacteria and fungi are substantial. The evolutionary significance of these differences is being debated. Many of the key enzymes of aromatic catabolism have been isolated and characterized, and those of PCA catabolism in bacteria are even being sequenced. Aromatic transport systems are now known in fungi as well as in bacteria.

Despite the enormous amount of research that has gone into the study of aromatic catabolism, however, relatively few investigations have been made with microorganisms known to be ligninolytic. Such studies are needed, and hopefully will be made in the very near future.

IV. APPLICATION OF LIGNINOLYTIC SYSTEMS

To write about applying a system (or systems) that we know relatively little about might seem premature. However, the possibilities are intriguing enough to justify the boldness, and of course microbiological processes were applied for centuries before they were even remotely understood -- long before microbes were discovered. Biological processes are inherently attractive because they often have one or more advantages over classical chemical processes: (1) greater efficiency, (2) greater specificity, (3) opportunity for novel transformations, (4) lower pollution, (5) lower energy consumption, and (6) absence of noxious or toxic reagents.

Some of the possibilities for applying ligninolytic systems include:

Microbial delignification — This may be employed for production of pulp, animal feeds, and substrates for other microbial or enzymatic processing; for improving mechanical and other high-yield pulps; for improving vegetable food products, etc. Eriksson and Vallander (Volume II, Chapter 14) have reported good progress in reducing the energy requirements for mechanical pulping by pretreating wood chips with a lignin-degrading fungus, mutated to eliminate its cellulolytic properties. Rosenberg and Wilke (Volume II, Chapter 13) have described their research aimed at biodelignifying fibers to improve cellulase digestibility.

Microbial bleaching of pulps — This could lead to energy and chemical savings, and perhaps to improvements in pulp properties in addition to brightness.

Microbial treatment of effluents containing lignin-derived wastes — Considerable interest has been shown the biodegradability of lignin sulfonates and lignin-derived pulp bleach plant effluents (Volume II, Chapters 11 and 12). Improved biological treatments including foam and color control can be the result of the use of microbes specifically chosen for ability to metabolize lignin and lignin-derived waste products. Sundman (Volume II, Chapter 12) Fukuzumi (Volume II, Chapter 11) and Haider (Volume II, Chapter 6) and Trojanowski have discussed these possibilities.

Conversion of lignin to useful chemicals — This is an attractive possibility (Volume I, Chapter 7; Volume II, Chapters 3 and 9) which does, however, depend on an improved understanding of what chemicals are produced during lignin biodegradation by various microbes.

Finally, a very interesting use of improved understanding is in the design of synthetic polymers with desirable properties but which are rendered biodegradable by inclusion of structures susceptible to ligninolytic systems. Haraguchi and Hatakeyama (Volume II, Chapter 10) have made the first step in this direction with their lignin-related polystyrenes. Biodegradability introduced by analogy to lignin -- which is oxidatively biodegraded in nature -- should allow polymer properties not allowed by biohydrolyzable linkages introduced by analogy to other biopolymers.[11]

Clearly, all of the above potential applications of ligninolytic systems can be pursued better with a more substantial knowledge base. Important from the practical standpoint are the chemical reactions involved, the number, stabilities, and cofactor requirements of the ligninolytic enzymes catalyzing those reactions, the rate-limiting steps, possibilities to isolate better organisms and genetically to improve the best, and techniques for producing optimally active ligninolytic inoculum or enzymes.

V. RESEARCH NEEDS

It is clear from the foregoing that lignin biodegradation is a complex process, that considerable progress has been made in understanding it, but that many unknowns remain. A list of present research needs places the current state of knowledge in perspective:

1. Development of standard ^{14}C-assays for ligninolytic activity, i.e., activity against the lignin polymer so that various microbes can be compared; ideally, a commercial source of ^{14}C-lignins should be available
2. Elucidation of the ligninolytic capacities of bacteria, yeasts, and imperfect fungi that inhabit plant residues and soils; emphasis must be placed on the lignin *polymer*
3. Elucidation of the pathway of lignin metabolism by white-rot fungi, and by other microbes, such as soft-rot fungi, demonstrated to be significantly ligninolytic; again, emphasis should be on the lignin polymer, and on the initial reactions
4. Elucidation of the pathway(s) of metabolism by ligninolytic microbes of low molecular weight aromatics related to lignin
5. Characterization of the enzymes catalyzing the specific degradative reactions in lignin, and in its degradation products such as vanillic acid, by ligninolytic microbes
6. Determination of the genetic and regulatory properties of the ligninolytic system(s)
7. Determination of the interrelationships between carbohydrate metabolism, lignin metabolism, and nitrogen metabolism in white-rot fungi
8. Further study of the chemistry of humic substances in various environments to firmly document the role of lignin in their formation

VI. REFERENCES

1. **Waksman, S. A.**, Decomposition of lignin, in *Wood Chemistry,* Wise, L. E., Ed., Reinhold, New York, 1944, 853.
2. **Gottlieb, S. and Pelczar, J. J., Jr.**, Microbiological aspects of lignin degradation, *Bacteriol. Rev.,* 15, 55, 1951.
3. **Higuchi, T.**, Decomposition of lignin by wood rotting fungi, *Bull. Nagoya Reg. For. Off.,* 6, 134, 1954.
4. **Higuchi, T.**, Formation and biological degradation of lignins, *Advanced Enzymology,* Vol. 34, Nord, F. F., Ed., Interscience, New York, 1971, 207.
5. **Kirk, T. K.**, Effects of microorganisms on lignin, *Ann. Rev. Phytopathol.,* 9, 185, 1971.
6. **Burdsall, H. H., Jr. and Esyln, W. E.**, A new *Phanerochaete* with a *chrysosporium* imperfect state. *Mycotaxon,* 1, 123, 1974.
7. **Highley, T.**, Requirements for cellulose degradation by a brown-rot fungus, *Material Org.,* 12, 25, 1977.
8. **Connors, W. J., Brunow, G., and Kirk, T. K.**, Fungal degradation of lignin-related aromatics attached to biologically inert polymers, Proc. TAPPI Conference, Forest Biology/Wood Chemistry, Madison, Wis., June 1977, 163.
9. **Crawford, R. L., Kirk, T. K., and McCoy, E.**, Dissimilation of the lignin model compound veratryl-glycerol-β-(*o*-methoxyphenyl) ether by *Pseudomonas acidovorans:* initial transformations, *Can. J. Microbiol.,* 21, 577, 1975.
10. **Ander, P., and Eriksson, K.-E.**, The importance of phenol oxidase activity in lignin degradation by the white-rot fungus *Sporotrichum pulverulentum, Arch. Microbiol.,* 109, 1, 1976.
11. **Huang, S. J., Bell, J. P., Knox, J. R., Atwood, H., Bansleben, D., Bitritto, M., Borghard, W., Chapin, T., Leong, K. W., Natarjau, K., Nepumuceno, J., Roby, M., Soboslai, J., and Shoemaker, N.**, Design, synthesis and degradation of polymers susceptible to hydrolysis by proteolytic enzymes, *Biodeterioration of Materials,* 3, 731, 1975.

SUBJECT INDEX

O

P

CHEMICAL INDEX

A

Acetaldehyde, II: 166

Acetic acid, I: 81; II: 56, 166

Acetoguaiacone, II: 108

Acetosyringone, II: 108

Acetoveratrone, II: 108

5-Acetylvanillic acid, I: 191

5-Acetylvanillyl alcohol, I: 177, 179, 181, 188
 NMR specturum, I: 178

Aconitic acid, II: 56

cis-Aconitic acid, II: 106

Alcohol dehydrogenase, II: 135

Alcohol oxidase, II: 135

Anisaldehyde, II: 104, 108

Anisic acid, I: 118, 121, 122, 123, 125; II: 104, 108

o-Anisic acid, II: 115

p-Anisic acid, I: 14; II: 113, 115—116, 129, 133—134

o-Anisidine, II: 69

Anisole, II: 105

l-Arabinose, I: 16

Arylglycerol-β-guaiacyl ether, I: 207

Aryglycerol-β-aryl-ether, I: 172

L-Asparagine, II: 56

Avicel®, II: 35

Azide, I: 45

2,2'-Azo-bis-isobutyronitrile, II: 148

B

Benzene, I: 52

Benzoate, I: 38

Benzoic acid, I: 113, 122—123; II: 106, 140, 184—186
 decomposition, I: 81

p-Benzoquinone, II: 25

Benzyl alcohol, I: 155; II: 105

Bikaverin, II: 61

α, α'-Bipyridyl, II: 9

1,2-Bisguaiacyl-1,3-propiodiol unit, I: 9

C

Caffeic acid, I: 80, 81, 87; II: 109, 129

Camphor, II: 183

Carboxydilactone, I: 54

3-Carboxy-5-hydroxy-cis,cis-muconic acid
 monomethyl ester, II: 157

2-Carboxymuconate, I: 54; II: 157
 uptake by Pseudomonas, I: 43—44

2-Carboxy-cis,cis-muconate, I: 54

3-Carboxymuconate, I: 50, 51

3-Carboxy-cis,cis-muconate, I: 27, 30, 48—50; II: 241

Carboxymuconate cyclase, I: 3, 41, 49, 51
 reaction mechanism, I: 37

Carboxymuconate cycloisomerase, I: 43

3-Carboxy-cis,cis-muconate cycloisomerase, I: 30

3-Carboxy-cis,cis-muconate lactonizing enzyme, I: 39—40

β-Carboxymuconic acid, II: 154, 156
 IR spectrum, II: 155

β-Carboxy-cis,cis-muconic acid, II: 11, 157

3-Carboxymuconolactone, I: 30—33, 43, 46, 50—51

4-Carboxymuconolactone, I: 27, 30—31, 48—51

4-Carboxymuconolactone decarboxylase, I: 30, 39—40, 43

3-Carboxymuconolactone hydrolase, I: 31, 41, 49
 mechanism, I: 34—37

5-Carboxyvanillic acid, I: 191

5-Carboxyvanillyl alcohol, I: 177, 181

Catalase, II: 135

Catechol, I: 26, 43, 53, 81—82, 87, 107; II: 29, 106—107, 140, 142

Catechol-1,2-oxygenase, I: 30, 32, 43; II: 141

Catechol 2,3-oxygenase, I: 53

Catechuic acid-5-oxygenase, II: 157

Cellobiono-δ-lactone, II: 2

Cellobiose, II: 2

Cellobiose:quinone oxidoreductase, I: 200, 216; II: 2—3, 13, 27—29, 39—40, 83, 183, 241

Cellulose, II: 6, 166, 183
 ethyl alcohol production, see Ethanol, production from cellulose

Chymotrypsin, I: 67, 69

Cinnamate, I: 26

Cinnamic acid, I: 3, 107; II: 76, 84

trans-Cinnamic acid, I: 3

α-Conidendrin, II: 88, 183, 186
 degradative plasmid, II: 186—189

Coniferaldehyde, I: 26, 173, 184; II: 108

Coniferyl alcohol, I: 2, 5, 7, 15, 25, 63, 79—84, 87, 102, 112, 118, 122—123, 172—174, 182; II: 46, 85—86, 88, 105, 108, 240
 ^{14}C-labeled, II: 68
 dehydrogenation, I: 6
 formation, II: 76
 phenolic derivatives, I: 80

β-Coniferyl ether, I: 9

p-Coumarate, I: 26

p-Coumaric acid, I: 2, 3, 80—81, 87, 112; II: 105, 108—109

4-Coumaroyl CoA ligase, I: 196

p-Coumaryl alcohol, I: 2, 7, 15, 63, 79, 83—84, 87, 102

[14]C-labeled, I: 64
 phenolic derivatives, I: 80

Cresol, I: 107

m-Cresol, I: 91

p-Cresol, I: 91

Cresorsellinic acid, I: 79, 87

Cumic acid, II: 140

Cycloheximide, I: 44; II: 58, 61

p-Cymene, II: 140